Placental Communications: Biochemical, Morphological and Cellular Aspects

Communications placentaires: aspects biochimique, morphologique et cellulaire

Colloques INSERM
ISSN 0768-3154

Other *Colloques* published as co-editions by John Libbey Eurotext and INSERM

133 Cardiovascular and Respiratory Physiology in the Fetus and Neonate. *Physiologie Cardiovasculaire et Respiratoire du Fœtus et du Nouveau-né.*
Scientific Committee : P. Karlberg,
A. Minkowski, W. Oh and L. Stern;
Managing Editor : M. Monset-Couchard.
ISBN : John Libbey Eurotext 0 86196 125 0
INSERM 2 85598 340 1

134 Porphyrins and Porphyrias. *Porphyrines et Porphyries.*
Edited by Y. Nordmann.
ISBN : John Libbey Eurotext 0 86196 087 4
INSERM 2 85598 281 2

137 Neo-Adjuvant Chemotherapy. *Chimiothérapie Néo-Adjuvante.*
Edited by C. Jacquillat, M. Weil and D. Khayat.
ISBN : John Libbey Eurotext 0 86196 125 0
INSERM 2 85598 340 1

139 Hormones and Cell Regulation (10th European Symposium). *Hormones et Régulation Cellulaire (10e Symposium Européen).*
Edited by J. Nunez, J.E. Dumont and R.J.B. King.
ISBN : John Libbey Eurotext 0 86196 125 0X
INSERM 2 85598 340 1

147 Modern Trends in Aging Research. *Nouvelles Perspectives de la Recherche sur le Vieillissement.*
Edited by Y. Courtois, B. Faucheux, B. Forette,
D.L. Knook and J.A. Tréton.
ISBN : John Libbey Eurotext 0 86196 126 0X
INSERM 2 85598 340 1

149 Binding Proteins of Steroid Hormones. *Protéines de liaison des Hormones Stéroïdes.*
Edited by M.G. Forest and M. Pugeat.
ISBN : John Libbey Eurotext 0 86196 125 0
INSERM 2 85598 340 1X

151 Control and Management of Parturition. *La Maîtrise de la Parturition.*
Edited by C. Sureau, P. Blot, D. Cabrol, F. Cavaillé and G. Germain.
ISBN : John Libbey Eurotext 0 86196 125 0
INSERM 2 85598 340 1

Suite page 299

Placental Communications: Biochemical, Morphological and Cellular Aspects

Communications placentaires: aspects biochimique, morphologique et cellulaire

Proceedings of the IIIrd Meeting of the European Placenta Group, Dourdan (France), September 27-30, 1989

Sponsored by the Institut National de la Santé et de la Recherche Médicale (INSERM) and the Groupe d'Etude et de Recherche sur le Placenta (GERP)

Edited by

L. Cedard
E. Alsat
J.-C. Challier
G. Chaouat
A. Malassiné

British Library Cataloguing in Publication Data
Placental communications
    1. Mammals. Placenta
I. Cedard, Lise
599.0334

ISBN 0 86196-227 3
ISSN 0768-3154

First published in 1990 by

Editions John Libbey Eurotext
6 rue Blanche, 92120 Montrouge, France. (1) 47 35 85 52
ISBN 0 86196 227 3

John Libbey & Company Ltd
13 Smiths Yard, Summerley Street, London SW18 4HR,
England.
(1) 947 27 77

Institut National de la Santé et de la Recherche Médicale
101 rue de Tolbiac, 75654 Paris Cedex 13, France.
(1) 45 84 14 41
ISBN 2 855 98 401 7

ISSN 0768-3154

© 1990 Colloques INSERM/John Libbey Eurotext Ltd,
All rights reserved
Unauthorised publication contravenes applicable laws

Preface

The third congress of the European Placenta Group was a great success from all points of view. It showed the increasing interest of researchers from Europe and other continents for our meetings. The congress was attended by 180 delegates from Europe, America, Asia and Africa. The atmosphere was warm and friendly and the discussions very lively.

We noticed a significant evolution in the research themes in placentology. Research in this area was first morphological and we were honoured by the presence of Dr Elisabeth Ramsey. The scientists more recently developed an interest in placental physiology. Two pioneers in this field, Maurice Panigel and Joseph Dancis, also attended the congress and their model of in vitro perfusions are still widely used in the study of materno-fetal transfers.

The endocrinology of the foeto-placental unit has been widely studied since E. Diczfaluzy started in this field. More recently, biochemists and molecular biologists have become interested in the trophoblast as a source of enzymes or DNA.

The placenta is a complex structure with many metabolic and hormonal functions ; it is thus an useful organ for the study of endocrine and paracrine regulation. Its limited life-span also makes it an interesting model for the study of growth and aging processes, and of cell division and differentiation.

The trophoblast which shares the fetus' genotype, is now widely used in antenatal diagnosis : caryotype and enzyme studies are carried out on trophoblast biopsies and hormones secreted by the trophoblast could be biochemical markers of chromosome anomalies such as Down's syndrome.

The placenta is the only normal organ disposable in large quantities, which justified the title of the symposium organized by Claude Villee and Peter Beaconfield "The placenta as a neglected experimental animal". It has been used in toxicology, as Richard Miller brilliantly demonstrated during the congress.

The immunological functions of the placenta are important, and its role in the implantation process is under current investigation. We were lucky to have experts in this field who discussed culture techniques, interactions between the trophoblast and the endometrial tissue, growth factors and differentiation, endocrinology and immunological receptors in lectures, workshops and poster presentations.

This volume contains documents which we hope will provide3 placentologists with updates and non-specialists with general ideas in the rapidly evolving field.

I would like to extend my thanks to the speakers and organizers, especially A. Malassiné, for their help and wish a prosperous future to the European Placenta Group.

Lise Cedard

Préface

Le IIIe Congrès de l'European Placenta Group fut un très grand succès à tous les points de vue. Il a, en effet, fait apparaître l'intérêt croissant des Européens et des chercheurs des autres continents pour nos réunions. 180 personnes ont assisté au Congrès, venues aussi bien d'Europe que d'outre-Atlantique, d'Asie ou d'Afrique. L'atmosphère fut chaleureuse et amicale et les discussions très animées.

Nous avons pu noter une évolution significative dans les thèmes de recherche en placentologie. Cette discipline était initialement essentiellement morphologique et ce fut un grand honneur pour nous d'avoir la présence d'Elisabeth Ramsey. Elle devint ensuite progressivement plus physiologique, et deux pionniers dans le domaine du transport transplacentaire, Maurice Panigel et Joseph Dancis, étaient également parmi nous.

Les modèles de perfusion *in vitro* qu'ils ont développés sont encore largement utilisés pour l'étude des transferts materno-fœtaux. L'endocrinologie de l'unité fœto-placentaire a été largement étudiée depuis les premiers travaux d'Egon Diczfaluzy. Nous avons, plus récemment, vu les biochimistes et les biologistes moléculaires s'intéresser au trophoblaste comme source d'enzymes ou de DNA.

Le placenta, dont la structure est très complexe, possède de si nombreuses fonctions métaboliques et hormonales qu'il est un organe privilégié pour les études de régulation endocrines et paracrines. Sa durée de vie limitée en fait aussi un modèle pour l'étude des phénomènes de vieillissement, aussi bien que pour la croissance, la multiplication et la différenciation cellulaires.

Le trophoblaste ayant le génome du fœtus est largement utilisé actuellement pour le diagnostic prénatal par biopsies villositaires (caryotype, enzymes), et des anomalies dans la synthèse hormonale pourraient être des marqueurs des anomalies chromosomiques telles que la trisomie 21.

Le placenta est le seul tissu normal disponible en grande quantité, ce qui a justifié le titre d'un Symposium organisé par Claude Villée et Peter Beaconfield, "The placenta as a neglected experimental animal" et son utilisation en toxicologie, comme l'a brillamment démontré Richard Miller au cours du Congrès.

Les fonctions immunologiques du placenta sont également importantes, et son rôle dans les processus d'implantation est à l'ordre du jour. Nous avons eu la chance d'avoir la présence d'experts dans ces différents domaines et de débattre au cours de lectures plénières, d'ateliers et de discussions des communications affichées, de différents problèmes très actuels : techniques de culture, interactions trophoblaste-endomètre, facteurs de croissance et différenciation, endocrinologie-récepteurs, immunologie, circulation, physiologie et transferts, structure et pathologie.

Ce volume contient un certain nombre de documents qui permettront, je l'espère, aux placentologues de faire le point sur certains sujets et aux non-spécialistes d'avoir une idée générale dans un domaine en pleine évolution.

Je remercie très vivement les orateurs et les organisateurs, plus particulièrement A. Malassiné, qui m'ont aidée si efficacement, et je souhaite longue vie à l'European Placenta Group.

Lise Cedard

Chers Placentologistes de l'Europe et des Etats-Unis,

Maintenant est un moment propice pour remercier le Dr Lise Cedard et ses collègues d'avoir organisé cet excellent symposium auquel nous avons participé ces trois derniers jours.

Durant les trois mois passés, nous avons tous entendu, vu et peut-être lu pas mal de choses concernant la Révolution française de 1789. Parmi cette panacée pour la presse, il y eut beaucoup d'articles sur la France depuis la Révolution, et un en particulier, qui a attiré mon attention, concernait les meilleures inventions provenant de France au dix-neuvième siècle.

Malheureusement, les noms cités ayant relation avec la médecine étaient lamentablement inadéquats : Laennec (stéthoscope, 1816), Braille (1829) et Pasteur (inoculation et pasteurisation, 1860) ont été bien entendu remarquablement accrédités, mais pas un seul physiologiste ne fut mentionné !

Claude Bernard, le fondateur de la médecine expérimentale, était un homme simple, pragmatique, qui ne se sera pas tourné dans sa tombe pour cette déplorable omission par la presse et les historiens. Mais je suis certaine qu'il applaudira la diversité de l'intérêt et de l'expertise qui fut démontrée durant cette conférence à un organisme qui n'échappa pas à son attention.

Je vous remercie, Lise, d'avoir rendu tout cela possible.

With compliments from

Maureen Young

Organizing Committee
L. Cedard (Président)
E. Alsat
J.-C. Challier
G. Chaouat
A. Firth
P. Kaufmann
R. Leiser
A. Malassiné (Secrétaire)
C. Nessman

List and address of participants
Liste et adresse des participants

Abramovich D.R., Department of Obstetrics/Gynaecology, University of Aberdeen, Foresterhill, Aberdeen, AB 2ZD, Ecosse, Royaume-Uni.

Ahmed M., U.L.K.C. School of Medicine (Obstetric/Gynaecology), 2411 Holmes, K.C., MO 64108-2792, Etats-Unis.

Alsat E., Laboratoire de physiopathologie et développement, CNRS-ENS, 46, rue d'Ulm, 75005 Paris, France.

Arnholdt H., Institut für Pathologie der Medizinische Universität, Ratzeburger Allee 160, D-2400 Lübeck 1, République fédérale d'Allemagne.

Atkinson D. E., Department of Child Health, Research Floor, St Mary's Hospital, Hathersage Road, Manchester, M13 OJH, Royaume-Uni.

Austgulen R., Children Department, Regional Hospital, 7006 Trondheim, Norvège.

Bambra C.S., Department of Reproduction, Institute of Primate Research, P.O. Box 24481, Karen, Nairobi, Kenya.

Barbera A., Via Washington, 106, 20146 Milano, Italie.

Barnea E.R., Feto-placental Endocrine Unit, Rappaport Institute, P.O. Box 9697, Rh. Haaliah, Haifa, Israël.

Beckers J.F., Service obstétrique, rue Vétérinaire, 45, B-1070 Bruxelles, Belgique.

Belisle D., Department Obstetrics/Gynaecology, Faculty of Medicine, Sherbrooke University Hospital, Sherbrooke, Québec JlH 5NA, Canada.

Benveniste R., Michael Reese Hospital, Medicine Center, Department of Obstetrics, University Chicago, Ill., Etats-Unis.

Bersinger N.A., University Hospital, Department of Obstetrics/Gynaecology, CH-8091 Zurich, Suisse.

Bielanska-Osuchowska Z., Department of Histology and Embryology, Faculty of Veterinary Medicine, Agricultural University, 02-766 Warszawa, Poland.

Bierings M.B., Department of Chemical Pathology, Medical Faculty, P.O. Box 1738, 3000 DR Rotterdam, Pays-Bas.

Bischof P., Department of Obstetrics/Gynaecology, University of Geneva, CH-1211 Geneva, 4, Suisse.

Boyd R. D.H., Department of Child Health, St Mary's Hospital, Manchester Ml3 OJH, Royaume-Uni.

Bruch F., Service d'Anatomie-Pathologie, Hôpital Antoine-Béclère, 92140 Clamart, France.

Bulfamante G.P., Cathedra Anatomia e Istologica-Patologia I, c/o H.S Paolo Via di Rudini 8, 20142 Milano, Italie.

Burton G.J., Department of Anatomy, University of Cambridge, Downing Street, Cambridge, CB2 3DY, Royaume-Uni.

Carter A.M., Department of Physiology University of Odense, Campusvej 55, DK-5230 Odense, Danemark.

Castellucci M., Department of Anatomy, RWTH Aachen, D-5100 Aachen, République fédérale d'Allemagne.

Cedard L., INSERM U.166, Clinique Beaudelocque, 123, bd de Port-Royal, 75014 Paris, France.

Čemerikić B., Institute of Endocrinology, Immunology and Nutrition NEP, Banatska 31B, 11080 Zemun, Belgrade, Yougoslavie.

Challier J.C., Biologie de la reproduction, Université Pierre et Marie Curie, 4 place Jussieu, 75005 Paris, France.

Chaouat G., INSERM U. 262, clinique Baudelocque, 123, bd de Port-Royal, 75014 Paris, France.

Chard T., Reproductive Physiology, Saint-Bartholomew's Hospital, London EC1A 7BE, Royaume-Uni.

Chou J.Y., NICHD, NIH, Building 10, Room 8C429, 9000 Rockville Pike, Bethesda, Maryland 20892, Etats-Unis.

Claben-Linke I., Department of Anatomy, RWTH Aachen, Wendlingweg 2, D-5100 Aachen, République fédérale d'Allemagne.

Dantzer V., Department of Anatomy, Royal Veterinary and Agricultural University, Bülowsvej 13, DK-1870 Frederiksberg C, Danemark.

Deleze J., Laboratoire de physiologie cellulaire, Faculté des Sciences, 86022 Poitiers, France.

Delmis J., Department of Obstetrics/Gynaecology, Medical School of Zagreb Meduliceva 40, 41000 Zagreb, Yougoslavie.

Desoye G., Department of Obstetrics/Gynaecology, University of Graz, A-8036 Graz, Autriche.

Dibbelt L., Institute of Biochemistry, Endocrinology, Medical University, Ratzburger Allee 160, D-2400 Lubeck l, République fédérale d'Allemagne.

Dohr G., Department of Histology and Embryology, University of Graz, Harrachgasse 21, 8010 Graz, Autriche.

Dumas J.-C., 20, rue de Brienne, 31520 Ramonville, France.

D'Souza S.W., Department of Child Health, St. Mary's Hospital, Hathersage Road, Manchester M13 OJH, Royaume-Uni.

Egeberg K., Unigen, Teknostallen, Prof. Brochsgt 6, 7030 Trondheim, Norvège.

Evain Brion D., Laboratoire de physiopathologie du développement, CNRS-ENS, 46, rue d'Ulm, 75005 Paris, France.

Evans J., Veterinary College, Ballsbridge, Dublin 4, Irlande.

Faure F., INSERM U. 168, CHU «La Grave», 31052 Toulouse Cedex, France.

Feige J.J., INSERM U.244, Biochimie, régulations cellulaires endocrines, LBIO/DRF, CENG 85 X 38041 Grenoble Cedex, France.

Fenichel P., INSERM U.210, Faculté de médecine, chemin de Valombrose, 06034 Nice Cedex, France.

Ferré F., INSERM U.166, clinique Baudelocque, 123, bd de Port-Royal, 75014 Paris, France.

Firth J.A., Department of Anatomy and Cell Biology, St-Mary's Hospital, Medical School, London W2 1PG, Royaume-Uni.

Fisher S.E., North Shore University Hospital, Cornell University, Medical College, 300 Community Drive, Manhasset, NY 11030, Etats-Unis.

Fondacci C., 9, rue Beautreillis, 75004 Paris, France.

Fox H., Department of Pathology, Stopford Building, University of Manchester M13 9PT, Royaume-Uni.

Funk A., Department of Obstetrics/Gynaecology, RWTH Aachen, Pauwelsstrasse 30, D-5100 Aachen, République fédérale d'Allemagne.

Genbačev O., Institute of Endocrinology, Immunology and Nutrition, INEP, Banatska 31B, 11080 Zemun, Yougoslavie.

Giroux M., INSERM U.168, CHU «La Grave», 31052 Toulouse Cedex.

Glasser S.R., Department of Cell Biology, Baylor College of Medicine, One Baylor Plaza, Houston, Texas 77030, Etats-Unis.

Grabowska A., Department of Pathology, University of Cambridge, Tennis Court Road, Cambridge CB2 lQP, Royaume-Uni.

Grudzinskas, Department of Obstetrics/Gynaecology, 4th Floor, Holland Wing, London Hospital Whitechapel, London El lBB, Royaume-Uni.

Grümmer R., Institut für Anatomie, Universitätsklinikum, Hufelandstrasse 55, D-4300 Essen, République fédérale d'Allemagne.

Hamelin B., INSERM U. 262, clinique Baudelocque, 123, bd de Port-Royal, 75014 Paris, France.

Hanssens M., Department of Obstetrics/Gynaecology, UI Saint-Raphaël, Gasthuisberg, B-3000 Leuven, Belgique.

Harbarth P., University Department of Biochemistry II, Im Neuenheimer Feld 328, D-6900 Heidelberg, République fédérale d'Allemagne.

Hartmann M., Department of Histology and Embryology, University of Graz, Harrachgasse 21, 8010 Graz, Autriche.

Hauguel de Mouzon S., CNRS, 9, rue Hetzel, 92190 Meudon, France.

Henrichs I., Department of Paediatrics 1, University of Ulm, Children's hospital, Prittwitzstrasse 43, D-7900 Ulm, République fédérale d'Allemagne.

Hess K., Laboratoire de biochimie médicale, faculté de médecine, BP 184, 54505 Vandœuvre-les-Nancy Cedex, France.

Hibbard J.U., Department of Obstetrics/Gynaecology, University of Chicago, 1169 So. Plymouth Ct 601, Chicago, Ill 60605, Etats-Unis.

Hitschold T., Dr. Horst-Schmidt-Klinik, Frauenklinik, Ludwig-Erhard-Strasse 100, D-6200 Wiesbaden, République fédérale d'Allemagne.

Hsi B.L., INSERM U.210, faculté de médecine (Pasteur), chemin de Vallombrose, 06034 Nice Cedex, France.

Hustin J., IMPL, allée des Templiers, 41, B-6280, Gerpines (Loverval), Belgique.

Iffy L., Department of Obstetrics/Gynaecology, UMDNJ-New Jersey Medical Rm. F. 246, 150, Bergen Street, Newark, NJ 07103, Etats-Unis.

Illsley N.P., Department of Obstetrics/Gynaecology and Reproduction, 1462 HSE, U.S.C.F., San Francisco, Cal. 94143, Etats-Unis.

Isaka K., Tokyo Medical College Hospital, 6-7-1, Nishishinjuku, Shinjuku-ku, Tokyo, Japan.

Jacobs R.A., Lawson Research Institute, St-Joseph's Hospital, 168 Grosvenor Street, London, Ontario N5 4223, Canada.

Jain S.K., National Institute of Immunology, Shahid Jeet Singh Marg., New Delhi 110067, Inde.

Jauniaux E., Obstetrics/Gynaecology, New Ward Block, King's College Hospital, Denmark Hill, London SE5 8RX, Royaume-Uni.

Jimenez E., Universitäts-Klinikum R. Virchow, Institut für Pathologie, Spandauer Damm 130, 1000 Berlin-west 19, République fédérale d'Allemagne.

Johnson P.M., Department of Immunology, University of Liverpool, PO Box 147, Liverpool L69 3BX, Royaume-Uni.

Jones C.J.P., Pathology Department, University of Manchester, Oxford Road, Manchester M13 9 PT, Royaume-Uni.

Karl P.I., North Shore University Hospital, Cornell University, Medical College, 300 Community Drive, Manhasset, NY 11030, Etats-Unis.

Kaufmann P., Department of Anatomy RWTH Aachen, Wendlingweg 2, D-5100 Aachen, République fédérale d'Allemagne.

Keramidas M., INSERM U.244, CENG 85 X, 38041 Grenoble Cedex, France.

King A., Department of Pathology, University of Cambridge, Tennis Court Road, Cambridge CB2 1QP, Royaume-Uni.

King B.F., Department of Human Anatomy, School of Medicine, University of California, Davis, Ca 95616, Etats-Unis.

Kingdom J.C.P., Department of Obstetrics/Gynaecology, Queen Mother's Hospital, Glasgow G3 85J, Ecosse, Royaume-Uni.

Kinskyi R., INSERM U.262, clinique Baudelocque, 123, bd de Port-Royal, 75014 Paris, France.

Klein, 27, rue des Ecoles, 75005 Paris, France.

Kliman H.J., Department of Pathology, 6 Founders, University of Pennsylvania, 3400 Spruce Street, Philadelphia, PA 19109-4283, Etats-Unis.

Kosanke G., Department of Anatomy RWTH Aachen, Wendlingweg 2, D-5100 Aachen, République fédérale d'Allemagne.

Kotowski A., Department of Anatomy, RWTH Aachen, Wendlingweg 2, D-5100 Aachen, République fédérale d'Allemagne.

Lalu K., Department of Pathology, University of Helsinki, Haartmaninkatu 3, Helsinki, Finlande.

Lampelo S., National Public Health Institute, Department of Environmental Hygiene, PO Box 95, SF - 70701 Kuopio, Finlande.

Laquerriere A., Laboratoire d'anatomie pathologique A, hôpital Charles-Nicolle, 1, rue de Germont, 76031 Rouen Cedex, France.

Laurini R.N., Institut de Pathologie, Rue de Bugnon 25, CH-1011 Lausanne, Suisse.

Leach I., Department of Anatomy and Cell Biology, Saint-Mary's Hospital, Medical School, Norfolk Place, London W2 1PG UK, Royaume-Uni.

Leichtweiss H.-P., Universitäts Frauenklinik Abt. für Exper. Medizin, Martinistrasse 52, D-2000 Hamburg 20, République fédérale d'Allemagne.

Leiser R., Institut für Tieranatomie, Universität Bern, Postfach 2735, CH-3001 Bern, Suisse.

Loke Y. W., Department of Pathology, University of Cambridge, Tennis Court Road, Cambridge CB2 1 QP, Royaume-Uni.

Luckhardt M., Universität Frauenklinik, Martinistrasse, 52, 2000 Hamburg, République fédérale d'Allemagne.

Maas, Logger 67, 1186 RN Amstelceen, Pays-Bas.

Maguiness S., London Hospital, Department of Obstetric/Gynecology, London El 1 BB, Royaume-Uni.

Malassiné A., Hôpital Robert Debré, Faculté des Savernes, 40, av. Recteur Pineau, bâtiment P, 86022 Poitiers, France.

Malek A., University of Arkansas for Medical Sciences, 4301 W. Maskham, Slot 518, Department of Obstetrics/Gynaecology, Little Rock, Ar 72205, Etats-Unis.

Martal J., Unité endocrinologie embryonnaire, Station de Physiologie Animale, INRA, 78350 Jouy-en-Josas, France.

Menu E., INSERM U. 262, clinique Baudelocque, 123, bd de Port-Royal, 75014 Paris, France.

Mermet I., INSERM U.80, CNRS UA 1177, UCBL Pavillon P, hôpital Edouard-Herriot, 69437 Lyon Cedex 03, France.

Merz W.E., University Department Biochemistry II, Im Neuenheimer Feld 328, D-6900 Heidelberg, République fédérale d'Allemagne.

Miller R.K., Department of Obstetrics/Gynaecology, University Medical Center, Box 668-601, Elmwood Av., Rochester, N.Y. 14642, Etats-Unis.

Mohammed T., Research Floor, Department of Child Health, St-Mary's Hospital, Hathersage Road, Manchester M130JH, Royaume-Uni.

Moirot H., Laboratoire d'anatomie pathologique. A, hôpital Charles-Nicolle, 1, rue Germont, 76031 Rouen Cedex, France.

Moll W., Institut für Physiologie, Universität Regensburg, Postfach 377, 8400 Regensburg, République fédérale d'Allemagne.

Mondon F., INSERM U.166, clinique Baudelocque, 123, bd de Port-Royal, 75014 Paris, France.

Moneghini I., Anatomia Pathologica, Ospedale San Paolo, Via Veglia 66, I - Milano 20159, Italie.

Morrish D.W., 7-117 Clinical Science Building, Department of Medicine, University of Alberta, Edmonton, Alberta T6G 2G3, Canada.

Mughal M.Z., Department of Child Health, St Mary's Hospital, Hathersage Road, Manchester M13 OJH, Royaume-Uni.

Mulholland J., Department of Cell Biology, Baylor College of Medicine, 1, Baylor Plazza, Houston, Texas 77030, Etats-Unis.

Nazzaro A., Department of Pathology, University of Cambridge, Tennis Court Road, Cambridge CB2 lQP, Royaume-Uni.

Nelson D.M., Department of Obstetrics/Gynaecology, Jewish Hospital of St-Louis, Washington University, School of Medicine, Saint-Louis, MO 63110, Etats-Unis.

Nessmann C., Hôpital Robert Debré, 48, bd Serurier, 75019 Paris, France.

Ockleford C.D., Department of Anatomy, University of Leicester Medical School, University Road, Leicester LE1 9 HN, Royaume-Uni.

Ogilvie S., Department of Pharmacology and Therapeutics, Box J-267 JHMHC, University of Florida, Gainesville, FL 32610, Etats-Unis.

Page K.R., University of Aberdeen, Department of Physiology, Marischal College, Aberdeen AB9 1 AS, Royaume-Uni.

Panigel M., Biologie de la reproduction, Université Pierre et Marie Curie, bâtiment A, 7e étage, 4, place Jussieu, 75252 Paris Cedex 05, France.

Phillippens L., Institut für Anatomie, Universitätsklinikum, Aufelandstrasse 55, D - 4300 Essen, République fédérale d'Allemagne.

Pierre F., Maternité du Beffroi, 23, avenue de Roubaix, 37100 Tours, France.

Pijnenborg R., Department of Obstetrics/Gynaecology, U.I. Saint-Rafaël, Gasthuisberg, B-3000 Leuven, Belgique.

Polliotti B., Unité de reproduction humaine, Hôpital Saint-Pierre, Université Libre de Bruxelles, Rue Haute 322, B-1000 Bruxelles, Belgique.

Pridjian G., Department of Obstetrics/Gynaecology, University of Chicago, 5841, Maryland Avenue, Chicago, Il 60637, Etats-Unis.

Radde I.G., Hospital for Sick Child, 555, University Avenue, Toronto, Ontario M5G 1X8, Canada.

Ramsey E.M., Research Department, Carnegie Institution, 3420 Que Street, NW, Washington, DC 20007, Etats-Unis.

Rao C.V., Department of Obstetrics/Gynaecology, 438 MDR Building, University of Louisville, School of Medicine, Louisville, Kentucky 40292, Etats-Unis.

Rebut-Bonneton C., INSERM U. 262, clinique Baudelocque, 123, bd de Port-Royal, 75014 Paris, France.

Reiber W., UniversitätsFrauenklinik Bachstrasse 18, DDR - 6900 Jena, République démocratique allemande.

Riemschneider R., Institut für Biochemie FH, Oldenbursallee 55, Postfach 1164, D-1000 Berlin, 19, République fédérale d'Allemagne.

Roeckelein G., Pathological Institute, University of Erlangen-Nürnberg, Krankenhausstrasse 8-10, 8520 Erlangen, République fédérale d'Allemagne.

Ross R., Division of Neonatology, University of Cincinnati, Medical Center, 231 Bethesda Avenue (ML 541), Cincinnati, Ohio 45267, Etats-Unis.

Scheffen I., Department of Anatomy, RWTH Aachen, Wendlingweg 2, D-5100 Aachen, République fédérale d'Allemagne.

Schmon B., Department of Obstetrics/Gynaecology, University of Graz, A-8036 Graz, Autriche.

Schneider H., Department of Obstretrics/Gynecology, University of Berne, Schanzeneckstrasse 1, CH - 3012 Berne, Suisse.

Schock C., Universitätsfrauenklinik, Martinistrasse 52, 2 - Hambourg 20, République fédérale d'Allemagne.

Schröder H.J., Universitäts Frauenklinik, Martinistrasse 52, 2 - Hamburg 20, République fédérale d'Allemagne.

Shaw A., Department of Child Health, St Mary's Hospital, Hathersage Road, Manchester M13 OJH, Royaume-Uni.

Sheppard B.L., Trinity College, Department of Obstetrics/Gynaecology, St James' Hospital, Dublin 8, Irlande.

Shiverick K.T., Department of Pharmacology and Therapeutics, University of Florida, College of Medicine, Box J-267 JHMHC, Gainesville, FL 32610, Etats-Unis.

Shorter S.C., Nuffield Department of Obstetrics/Gynaecology, John Radcliffe Hospital, Headington, Oxford OX3 9DU, Royaume-Uni.

Shrimanker K., London Hospital, Department of Obstetrics/Gynaecology, London El 1 BB, Royaume-Uni.

Sivonen P., Department of Pharmacology and Toxicology, University of Oulu, Kajaanintic 52 D, 90220 Oulu 22, Finlande.

Skeie Jensen T., Broegelmann Research Laboratory, Armauer Hansens Bldg, Haukeland Hospital, 5021 Bergen, Norvège.

Smith C.H., Washington University Medical School, St Louis Children's Hospital, 400, So. Kingshighway, St Louis, Missouri 63110, Etats-Unis.

Soma H., Tokyo Medical College, Department of Obstetrics/Gynaecology, 1-7, Nishishinjuku, 6, Shinjuku, Tokyo 160, Japon.

Stabile I., Flat 7 Rowan, 48-50 Muswell Road, Muswell Hill, London N10 2 BX, Royaume-Uni.

Starkey P.M., Nuffield Department of Obstetrics/Gynaecology, John Radcliffe Hospital, Oxford OX3 9DU, Royaume-Uni.

Stulc J., Department of Pharmacology, Faculty of Pediatrics, Albertov 4, 128 00 Prague 2, Tchécoslovaquie.

Sullivan M.H.F., Institute of Obstetrics/Gynaecology, Royal Postgraduate Medical School, Hammersmith Hospital, Du Cane Road, London W12 0NN Royaume-Uni.

Sunde L., Institute of Human Genetics, The Bartholin Bldg, University of Aarhus, DK-8000 Aarhus C, Danemark.

Suzuki Y., Department of Obstetrics/Gynaecology, Tokyo Medical College Hospital, 6-7-1, Nishi-Shinjuku, Shinjuku-ku, Tokyo 160, Japon.

Szulman A.E., Magee-Womens Hospital, Forbes Avenue and Halket Street, Pittsburgh, Pennsylvania 15213, Etats-Unis.

Takayama M., Department of Obstetrics/Gynaecology, Tokyo Medical College Hospital, 6-7-1, Nishi-Shinjuku, Shinjuku-ku, Tokyo 160, Japon.

Tardy M., IMEDEX, ZI des Troques, BP 38, 69630 Chaponost, France.

Tayot J.L., IMEDEX, ZI des Troques BP 38, 69630 Chaponost, France.

Thibault G., Laboratoire d'immunologie, Faculté de médecine 2 bis, bd Tonnellé, 37032 Tours Cedex, France.

Tiollier J., IMEDEX, ZI des Troques, BP 38, 69630 Chaponost, France.

Toder V., Department of Embryology/Teratology, Tel Aviv Univ. Med. School Tel Aviv 69978, Israël.

Tranter P.R., Institute of Obstetrics/Gynaecology, Royal Posgraduate Medical School, Hammersmith Hospital, Du Cane Road, London W12 OHS, Royaume-Uni.

Ugele B., Institut Frauenklinik der Universität, Maistrasse 11, D-8000 München 2, République fédérale d'Allemagne.

Unger M., Freie Universität, Institut für Pathologie, Klinikum «R. Vicklow», Spandauer Damm 130, 1000 Berlin 19, République fédérale d'Allemagne.

Verkeste C.M., Department of Obstretrics/Gynaecology, State University Limburg, PO Box 1918, 6201 BX Maastricht, Pays-Bas.

Vettenranta K., Children's Hospital, University of Helsinki, Stenbäckinkatu 11, SF-00290 Helsinki, Finlande.

Vičovac L., Institute of Endocrinology, Immunology and Nutrition, INEP, Banatska 31B, 11080 Zemun, Yougoslavie.

Vince G., Nuffield Department of Obstetrics/Gynaecology, John Radcliffe Hospital, Oxford OX3 9DU, Royaume-Uni.

Ward S., Department of Obstetrics/Gynaecology, St Mary's Hospital, Manchester M 13 0 JH, Royaume-Uni.

White T.E.K., Box 668, Department of Obstetrics/Gynaecology, University of Rochester Medical Center, 601 Elmwood Avenue, Rochester, NY 14642, Etats-Unis.

Williams J.M.A., University of Aberdeen, Department of Physiology, Marischal College, Aberdeen, Royaume-Uni.

Wells M., Honorary Consultation of Pathology, University of Leeds, Department of Pathology, Leeds L52 9JT, Royaume-Uni.

Young M., 4, Preston Close, Miller's Road, Toft, Cambridge, CB3 7RU, Royaume-Uni.

Contents
Sommaire

V Preface
VI *Préface*
IX List of participants
IX *Liste des participants*

I. TROPHOBLAST - ENDOMETRIUM INTERACTIONS
INTERACTIONS TROPHOBLASTE - ENDOMÈTRE

3   **H.J. Kliman, R.F. Feinberg, J.E. Haimowitz**
    Interactions between human term trophoblasts and endometrium *in vitro*
    Etude in vitro *de l'interaction entre le trophoblaste humain à terme et l'endomètre*

10 **Y.W. Loke**
    New developments in human trophoblast cell culture
    Nouveaux développements dans la culture des cellules trophoblastiques humaines

17 **S. Mani, J. Mulholland, J.A. Julian, S.R. Glasser**
    Implantation
    Implantation

22 **J. Hustin, J.G. Grudzinskas**
    Trophoblast-endometrium interaction : summary of the workshop
    Résumé de l'atelier : interaction trophoblaste-endomètre

30 **D.R. Abramovich, K.R. Page, C.K. Pearson**
    Does decidua influence 1st trimester trophoblast hCG secretion ?
    La sécrétion d'hCG par le trophoblaste du 1er trimestre est-elle influencée par la décidue ?

XIX

31 **R.F. Feinberg, J.E. Haimowitz, H.J. Kliman**
Human trophoblast interaction with the extracellular matrix : a histologic assay system for studying trophoblast invasion
Interaction entre le trophoblaste humain et la matrice extracellulaire : test histologique pour l'étude de l'invasion trophoblastique

32 **O. Genbačev, B. Cemerikic, V. Sulovic, L. Vičovac**
Long-term tissue culture of human first trimester villous trophoblast in collagen gel : evaluation of its possible use as a model system
Culture à long terme de trophoblaste du 1er trimestre sur gel de collagène : possibilité de son utilisation comme modèle expérimental

33 **R. Grümmer, H.W. Denker**
Investigations on the invasion of choriocarcinoma cells grown as spheroids
Etude des propriétés invasives des cellules de choriocarcinome en culture tridimensionnelle

34 **S. Mani, S. Lampelo, J.A. Julian, S.R. Glasser**
Biochemical indices of stromal cell differentiation are different following invasive vs superficial implantation
Les indices biochimiques de la différenciation des cellules du stroma sont différents après invasion ou implantation superficielle

35 **J. Mulholland, L. Hong, R.E. Kellems, S.R. Glasser**
Increased uterine adenosine deaminase activity is a function of decidualized stromal cells
L'augmentation d'activité de l'adénosine-déaminase de l'utérus est une fonction des cellules du stroma décidualisé

36 **A. Nazzaro, J.W. Fawcett, Y.W. Loke**
Interactions between trophoblast, choriocarcinoma, decidua and fibroblast cells *in vitro*
Interactions in vitro *entre les cellules du trophoblaste, de choriocarcinomes, de la decidue et les fibroblastes*

37 **D.M. Nelson, E.C. Crouch, E.C. Curran, D.R. Farmer**
Fibrin matrix modulates the phenotypic differentiation and proliferation of human placental trophoblast
La fibrine utilisée comme matrice module la différenciation phénotypique et la prolifération du trophoblaste humain en culture

38 **C.M. Salafia, N. Haynes, V.J. Merluzzi, C. Robiou, R. Rothlein**
The distribution of ICAM-l within decidua and placenta and gestational age associated changes

Distribution des ICAM-1 dans la décidue et le placenta, changements avec l'âge gestationnel

39 **L. Vićovac, N. Papić, O. Genbaćev**
Long-term tissue culture of decidual explants of different gestational age in collagen matrix
Culture à long terme d'explants de la décidue à différents stades gestationnels, sur matrice-collagène

40 **T.E.K. White, R.K. Miller, G. Grudzinskas, A. Lower, T. Fay**
Human endometrial cells grown on extracellular matrix (ECM) form an *in vivo* type morphology and produce the endometrial protein PP-14 ($\alpha 2$-PEG)
Les cellules endométriales humaines cultivées sur une matrice extracellulaire (ECM) ont une morphologie physiologique et produisent la protéine endométriale PP-14 ($\alpha 2$-PEG)

II. ENDOCRINOLOGY AND BIOCHEMISTY
ENDOCRINOLOGIE ET BIOCHIMIE

43 **S. Lampelo, K. Lalu**
Production of monoclonal antibody against maternal serum oxytocinase
Production d'un anticorps monoclonal contre l'ocytocinase du sérum maternel

50 **S. Belisle, G. Desoye**
Receptors and endocrine regulatory mechanisms : summary of the workshop
Résumé de l'atelier : récepteurs et mécanismes régulateurs endocriniens

63 **E. Alsat, A. Malassiné, R. Rebourcet, C. Besse, L. Cedard**
Characterization of high density lipoprotein (HDL_3) binding to human placenta : absence of internalization
Caractérisation d'une liaison des lipoprotéines de forte densité (HDL_3) dans le placenta humain : absence d'internalisation

64 **E.R. Barnea, D. Feldman, M. Kaplan**
The role of EGF in hCG secretion and trophoblast differentiation in the first trimester
Le rôle de l'EGF dans la sécrétion d'hCG et la différenciation trophoblastique au cours du 1er trimestre

65 **M.B. Bierings, M.R.M. Baert, H.G. van Eijk, J.P. van Dijk**
Regulatory signals in placental iron uptake
Signaux régulateurs dans la captation du fer par le placenta

66 S. Belisle, A. Petit, N. Gallo-Payet, J.G. Lehoux, D. Bellabarba, E. Esher, G. Guillon
Endocrine control of hPL and hCG production by the human placenta
Contrôle endocrinien de la production d'hPL et d'hCG par le placenta humain

67 B. Čemerikić, O. Genbačev, R. Beaconsfield
In vitro effect of dynorphine on hCG production and release by trophoblast tissue explants
Effet in vitro de la dynorphine sur la production d'hCG et sa libération par des explants de tissu placentaire

68 G. Desoye, B. Schmon, I. Andiel, R. Michlmayer, G. Dohr, M. Hartmann, A. Blaschitz
Antilipolytic hormones stimulate the secretion of hCG in cultivated trophoblasts via a pathway involving protein kinase C
Les hormones antilipolytiques stimulent la sécrétion d'hCG par des trophoblastes en culture, via un mécanisme faisant intervenir la protéine kinase C

69 L. Dibbelt, E. Kuss
On the reaction mechanism of human placental steryl-sulphatase
Sur le mécanisme de réaction de la stérol-sulfatase placentaire humaine

70 M. Dodeur, A. Malassiné, D. Bellet, A. Mensier, D. Evain-Brion
Characterization and differentiation of human first trimester placenta trophoblastic cells in culture
Caractérisation et différenciation des cellules trophoblastiques humaines au 1er trimestre, en culture

71 D. Evain-Brion, M. Dodeur, A. Mensier, E. Alsat, J.M. Bidart
Effect of parathyroid hormone on the endocrine functions of human trophoblastic cells of first trimester placenta
Effet de l'hormone parathyroïde sur les fonctions endocrines des cellules trophoblastiques humaines du 1er trimestre

72 I. Henrichs, G. Kreisel, G. Röckelein, W.M. Teller
Characteristics of carbohydrate metabolism in isolated cytotrophoblastic cells compared to whole placental tissue
Caractéristiques du métabolisme des hydrates de carbone par les cellules cytotrophoblastiques isolées comparées au tissu placentaire entier

73 K. Hess, H. Elouardirhi, D. Sekkat, F. Belleville, P. Nabet
A partially purified fraction of phosphorylated placental non-histone proteins which stimulates transcription in isolated nuclei

74 **K. Hess, H. Elouardirhi, D. Sekkat, F. Belleville, P. Nabet**
Purification partielle de protéines placentaires non-histone phosphorylées qui stimulent la transcription dans les noyaux isolés
An *in vitro* system to study the hPL genes expression in placental isolated nuclei
Système d'étude in vitro *de l'expression des gènes de l'hPL dans les noyaux placentaires isolés*

75 **N.P. Illsley, M.M. Jacobs**
Stimulation of the microvillous Na^+/H^+ exchanger by protein kinase C phosphorylation
Stimulation des échanges Na^+/H^+ microvillositaires par phosphorylation protéine kinase-C dépendante

76 **M.M. Jacobs, N.P. Illsley**
Is there dual regulation of syncytial adenylate cyclase ?
Existe-t-il une double régulation de l'adénylate cyclase syncytiale ?

77 **L. Kaplan, J.J. Lopez Costa, S.E. Carbone, O. Mastronardi, D.C. Rondina, J. Pecci Saavedra, J.A. Moguilevsky**
Neurotransmitters in human term placenta : biochemistry and immunochemistry
Neurotransmetteurs dans le placenta humain : biochimie et immunochimie

78 **P. Licht, W.E. Merz**
Evidence for a GABA-ergic modulation of hCG secretion by human first trimester placenta tissue
Evidence d'une modulation GABA-ergique de la sécrétion d'hCG par le tissu placentaire du 1er trimestre

79 **A. Malassiné, F. Mondon, J. Besson, M. Vial, G. Tanguy, W. Rostène, F. Ferré**
VIP receptors, positively coupled with adenylate cyclase activity on fetal vascularization of human placenta
Récepteurs au VIP positivement couplés à l'activité adénylate cyclase sur la vascularisation fœtale du placenta humain

80 **W.E. Merz, P. Licht, P. Harbarth**
Exogenous GnRH stimulates transcriptional and secretory rates as well as episodic secretion of human choriogonadotropin (hCG)
Le GnRH exogène stimule la sécrétion pulsatile de l'hormone gonadotrope chorionique humaine au niveau de sa synthèse et de sa sécrétion

81 **W. Moll, D. Scholl, H. Caffier, R. Götz**
Estradiol and estradiol receptors are present in and around placental arteries in guinea-pigs

Œstradiol et récepteurs à l'œstradiol sont présents dans et autour des artères placentaires chez le cobaye

82 **F. Mondon, C. Robaut, F. Ferré**
High-affinity specific binding for ^{125}I-ET-1 in membranes from human placenta vessels
Liaison spécifique et de forte affinité de l'^{125}I-ET-1 aux membranes des vaisseaux placentaires humains

83 **S. Ogilvie, L.H. Larkin, M.L. Duckworth, K.T. Shiverick**
Gestational profiles of the immunocytochemical localization of rat prolactin-like protein-B, rat placental lactogen II and pregnancy-specific β_1-glycoprotein in rat placenta
Profils gestationnels de la localisation immunohistochimique dans le placenta de rat, de la prolactine-like protéine B, de l'hormone lactogène placentaire II et de la β_1-glycoprotéine

85 **B. Polliotti, S. Meuris, P. Lebrun, C. Robyn**
Effect of extracellular calcium and potassium depolarization on the chorionic gonadotropin and placental lactogen releases by human placental explants
Effet du calcium extracellulaire et de la dépolarisation par le potassium sur la libération de l'hormone chorionique gonadotrope et de l'hormone lactogène placentaire par des explants placentaires humains

86 **Ch. V. Rao, Z.M. Lei**
The presence of gonadotropin receptors in human placenta, amnion, chorion and decidua
Présence de récepteurs aux hormones gonadotropes dans le placenta humain, l'amnios, le chorion et la décidue

87 **B. Ugele, L. Dibbelt, E. Kuss**
Sterylsulfatase activity of human placental cells in monolayer culture
Activité stérol-sulfatase dans les cellules placentaires humaines en culture monocouche

88 **K. Vettenranta, K.O. Raivio**
Extracellular adenine nucleotides in human trophoblastic purine nucleotide synthesis
Utilisation des adénines nucléotides extracellulaires pour la synthèse des nucléotides puriques dans le trophoblaste humain

III. GROWTH FACTORS AND DIFFERENTIATION
FACTEURS DE CROISSANCE ET DIFFÉRENCIATION

91 **D. Evain-Brion**
Growth factors and trophoblast differentiation
Facteurs de croissance et différenciation trophoblastique

98 **J. Tiollier, S. Uhlrich, V. Chirouze, M. Tardy, J.L. Tayot**
Biochemical and biological characterization of a crude growth factor extract (EAP) from placental tissue
Caractérisation biochimique et biologique d'un facteur de croissance brut (EAP) extrait du placenta humain

101 **D.W. Morrish, Ch. V. Rao**
Summary of the Workshop on growth factors and placental differentiation
Résumé de l'atelier : facteurs de croissance et différenciation placentaire

105 **E. Alsat, V. Mirlesse, M. Dodeur, D. Evain-Brion**
Regulation of epidermal growth factor receptors in human trophoblastic cells in culture : effect of parathyroid hormone
Régulation des récepteurs à l'EGF dans les cellules trophoblastiques humaines en culture : effet de l'hormone parathyroïde

106 **S.W. D'Souza, P. Ali, J.L. Smart**
Epidermal growth factor receptors in rat placenta, amnion and yolk sac : characteristics of specific binding are dependent on gestational age
Récepteurs à l'EGF dans le placenta, l'amnios et le sac vitellin chez le rat : les caractéristiques de la fixation spécifique sont dépendantes de l'âge gestationnel

107 **J.J. Feige, M. Keramidas, L. Multigner, J.J. Bourgarit, J.M. Saez, S. Uhlrich, E.M. Chambaz**
Anti-TGFβ antibodies as tools for placental TGFβ purification
Anticorps anti-TGFβ comme outils pour la purification du TGFβ placentaire

108 **S. Hauguel de Mouzon, D. Evain-Brion, M. Forestier, E. Alsat**
Expression of nuclear proto-oncogenes c-fos and c-myc messenger ribonucleic acids in human placenta
Expression des ARNm codant pour les proto-oncogènes nucléaires c-myc et c-fos dans le placenta humain

109 **D.W. Morrish, R. Sasi, C.C. Lin, D. Bhardwaj, S. Shiferaw**
Variable epidermal growth factor receptor regulation among normal trophoblast, fibroblast and tumor cells

Régulation différente de l'expression du récepteur à l'EGF dans le trophoblaste normal, les fibroblastes et les cellules tumorales

110 **J. Mühlhauser, C.A. Schroeter, P. Kaufmann, M. Castellucci**
Expression of C-erbB-2 protein product in human placental villi as compared to EGF-receptor
Comparaison entre la protéine produite par l'expression de C-erbB-2 dans le placenta humain et le récepteur à l'EGF

111 **V. Toder, E. Kochavi, H. Altaratz, N. Gleicher**
Growth factors as stimulators of placental cell growth
Stimulation de la croissance placentaire par les facteurs de croissance

112 **S. Uhlrich, J. Tiollier, M. Tardy, J.L. Tayot**
Biochemical and biological characterization of bFGF extracted from human placenta
Caractérisation biologique et biochimique du bFGF extrait du placenta humain

113 **M. Vučković-Tomanović, D.C. West, O. Genbačev**
Growth factors of the human placenta
Facteurs de croissance du placenta humain

IV. TROPHOBLAST ANTIGENS AND IMMUNOLOGICAL ASPECTS OF FETO-MATERNAL RELATIONSHIP
ANTIGÈNES TROPHOBLASTIQUES ET ASPECTS IMMUNOLOGIQUES DES RELATIONS FŒTO-MATERNELLES

117 **T. Chard**
Interferon as a fetoplacental signal in human pregnancy
L'interféron est un signal fœto-placentaire pendant la grossesse

121 **R. Kinsky, G. Delage, N. Rosin, M. Nguy Than, M. Hoffmann, G. Chaouat**
Modèle murin de résorption fœtale à médiation NK
Effect of Poly (1) (C12 U) on embryo resorption and implantation rates in mice

125 **J. Martal, N. Chene, M. Charlier, M. Guillomot, P. Reinaud, J. Bertin, G. Danet, K. Zouari, G. Charpigny**
Trophoblastin, oTP, embryonic interferons
Trophoblastine, OTP et interférons embryonnaires

132 **C. S. Bambra**
Effect of anti CG antibodies on baboon placental derived trophoblast *in vitro*

Effet des anticorps anti CG sur les cellules placentaires de babouin in vitro

133 **G. Chaouat, E. Menu, M. Hofmann, M. Dy, M. Minkowski, D.A. Clark, T.G. Wegmann**
Lymphokines at the feto-maternal interface affect fetal size and survival
Les lymphokines situées à l'interface fœto-maternelle affectent la taille du fœtus et sa survie

134 **P. Fenichel, C. Grivaux, G. Dohr, M. Samson, C. Milesi-Fluet, C.J.G. Yeh, B.L. Hsi**
Do human trophoblast, leukocyte and sperm share a common antigen recognized by GB 24 ?
Le trophoblaste humain, les leucocytes et les spermatozoïdes partagent-ils un antigène commun reconnu par GB 24 ?

135 **M. Garcia-Lloret, L. Guilbert, D.W. Morrish**
Functional expression of CSF-1 receptors on normal human trophoblast
Expression fonctionnelle des récepteurs au CSF-1 dans le trophoblaste humain normal

136 **A. Grabowska, G. Chumbley, Y.W. Loke**
Comparison of HLA class-I antigen expression in cultured human extravillous trophoblast, fetal skin and JEG-3 cells
Comparaison de l'expression de l'antigène HLA de classe 1 dans le trophoblaste extravilleux, dans les cellules de peau fœtale et les cellules JEG-3, en culture

137 **B. Hamelin, J. Demignon, E. Menu, G. Chaouat, C. Rebut-Bonneton**
Effect of 1,25 dihydroxycholecalciferol on IL-2 dependent CILL-2 proliferation and cAMP cell content. Role at the feto-placental interface
Effet du 1-25 dihydroxycholécalciférol sur la prolifération IL-2 dépendante des CILL-2 et leur contenu en cAMP : rôle à l'interface fœto-maternelle

138 **M. Hartmann, G. Dohr, G. Pilz, D. Ribitsch, G. Siwetz, G. Desoye**
Monoclonal antibodies GZ 100 and GZ 116 recognize different trophoblast antigens
Les anticorps monoclonaux GZ 100 et GZ 116 reconnaissent différents antigènes trophoblastiques

139 **T.S. Jensen, E. Ulvestad, R. Matre**
FCγ-receptors in sera from pregnant women
FCγ-récepteurs dans le sérum de femmes enceintes

140 **A. King, Y.W. Loke**
Interaction of decidual NK and LAK cells with human trophoblast *in vitro*
Interaction des cellules déciduales NK et LAK avec le trophoblaste humain in vitro

141 **E. Menu, D. Jankovic, J. Thèze, G. Chaouat**
Human placental supernatant and IL-2 and IL-4 dependent proliferation
Surnageants de placenta humain et prolifération dépendante de IL-2, IL-4

142 **P.M. Starkey, G.S. Vince, S.C. Shorter, I.L. Sargent, C.W.G. Redman**
The synthesis of tumour necrosis factor by human placental and decidual tissue
Synthèse de TNF par le placenta humain et la décidue

145 **S.C. Shorter, G.S. Vince, P.M. Starkey**
The identification of mRNA for the IL3-related cytokines in placental and decidual tissue
Identification d'ARNm pour les cytokines IL3 dans les tissus placentaire et décidual

146 **G. Thibault, D. Degenne, J.M. Guillaumin, A.C. Girard, P. Bardos**
Human syncytiotrophoblast plasma membranes (STPM) inhibit production of IL2 and expression of IL2 receptor in actived Jurkat cells
Les membranes plasmatiques du syncytiotrophoblaste humain inhibent la production d'IL2 et l'expression des récepteurs IL2 dans les cellules Jurkat activées

147 **D. Tornehave, B. Teisner, J. Chemnitz, J.G. Westergaard, H. Boye, J.G. Grudzinskas**
Immunocytochemical investigations of two «new» fetal antigens (FA-1 and FA-2) isolated from amniotic fluid
Etudes immunocytochimiques de deux nouveaux antigènes fœtaux (FA-1 et FA-2) isolés du liquide amniotique

148 **G.S. Vince, S.C. Shorter, P.M. Starkey, I.L. Sargent, C.W.G. Redman**
Cytokine production by human decidual macrophages and cytotrophoblast cells
Production de cytokine par les macrophages de la décidue humaine et les cellules cytotrophoblastiques

V. PLACENTAL CIRCULATION
CIRCULATION PLACENTAIRE

151 **B.L. Sheppard, C. Boyle, N. Gleeson, M. Jordan, L. Daly, J. Bonnar**
Plasminogen activator inhibitors of the placenta and placental bed in normotensive and hypertensive pregnancy
Inhibiteurs de l'activateur du plasminogène du placenta de grossesses normales et avec hypertension

160 **A.M. Carter, W. Moll**
Résumé de l'atelier : circulation placentaire
Placental circulation : summary of a workshop

171 **A.M. Carter, A. Detmer**
Fetal placental blood flow after uterine artery ligation in the guinea-pig
Flux sanguin fœto-placentaire après ligature de l'artère utérine chez le cobaye

172 **A. Funk, H. Fendel**
Trans-vaginal sonography of first trimester pregnancy : investigation on the nutritional value of the secondary yolk sac and on the maturation of chorionic vessels
Echographie transvaginale au 1er trimestre de la grossesse : étude de la valeur nutritionnelle du sac vitellin secondaire et de la maturation des vaisseaux chorioniques

173 **T. Hitschold, E. Weiss, H. Müntefering, P. Berle**
Dopplersonographic findings in umbilical chord anomalies : an *in vivo* model to show resistance parameters in fetoplacental circulation
Résultats de l'échographie Doppler dans les anomalies du cordon ombilical : un modèle in vivo *pour montrer les paramètres de résistance dans la circulation fœto-placentaire*

174 **T. Hitschold, E. Weiss, H. Müntefering, P. Berle**
Evaluation of placental risk using pulsed Doppler ultrasound of the umbilical arteries : a histometric-clinical investigation
Evaluation du risque placentaire par le Doppler pulsé des artères ombilicales : une investigation clinique histométrique !!!

175 **J.C.P. Kingdom, J. McQueen, A.G. Jardine, J.M.C. Connell, M.J. Whittle**
Alterations to vascular receptors for angiotensin II and atrial natriuretic peptide in placentae from pregnancies complicated by asymetrical growth-retardation

Modifications des récepteurs vasculaires pour l'angiotensine II et le peptide atrial natriurétique dans les placentas de grossesses compliquées de retard de croissance dysharmonieux

176 **J.C.P. Kingdom, A.G. Jardine, J.M.C. Connell, A. Templeton, M.J. Whittle**
Atrial natriuretic peptide opposes the vasoconstrictor effect of angiotensin II in a simple placental perfusion model
Le peptide atrial natriurétique s'oppose à l'effet vasoconstricteur de l'angiotensine II dans un modèle simple de perfusion du placenta

177 **I. Scheffen, L. Philippens, P. Kaufmann, R. Leiser, V. Mironov**
Maternal oxygen supply as a regulator of fetal placental capillarisation
L'oxygénation maternelle est un régulateur de la capillarisation placentaire fœtale

178 **M.H.F. Sullivan, L. Patel, M.G. Elder**
Anti-platelet activity of human trophoblast
Activité anti-plaquettaire du trophoblaste humain

179 **P.R. Tranter, M.H.F. Sullivan, M.G. Elder**
The effects of leukotrienes and their interaction with angiotensin II in the perfused human placenta
Effets des leucotriènes et leur interaction avec l'angiotensine II dans le placenta humain perfusé

180 **C.M. Verkeste, P.F. Boekkooi, L.L.H. Peeters**
The effect of reduced red cell deformability on placental perfusion in the awake late-pregnant guinea-pig
L'effet de la déformabilité réduite des globules rouges sur la perfusion placentaire chez le cobaye éveillé en fin de gestation

VI. PHYSIOLOGY AND PLACENTAL TRANSFERT
PHYSIOLOGIE ET TRANSFERT PLACENTAIRE

183 **C.H. Smith, J.K. Kelley**
Calcium dependent ATPases and nucleotide phosphates of the human placental basal plasma membrane
L'ATPase calcium-dépendante et les phosphatases nucléotidiques de la membrane plasmique basale du syncytiotrophoblaste humain

187 **J.C. Challier, H.J. Schroder**
Résumé de l'atelier : transport placentaire

Placental transport : summary of a workshop

192 **D.E. Atkinson, J.D. Glazier, C.P. Sibbey**
Gestational studies on the rat placental Na^+/H^+ exchanger
Variations de l'échange Na^+/H^+ durant la gestation chez le rat

193 **J.C. Challier, S. Goma, T. Bintein, G. Olive**
Epidermal growth factor, placental transfer and secretion in short term dual perfusion of human placental lobules
EGF, transfert et sécrétions au cours des perfusions en double circuit de courte durée des lobules placentaires humains

194 **J.U. Hibbard, G. Pridjian, P.F. Whitington**
Leucine and glutamic acid transport under diabetic conditions in the *in vitro* perfused human placenta
Transfert de la leucine et de l'acide glutamique dans le placenta humain perfusé in vitro *dans des conditions normales ou avec hyperglycémie et acidose*

195 **P.I. Karl, S.E. Fisher**
Na-dependent amino uptake by human placental microvillous membrane vesicles (MMV) : importance of storage conditions and preservation of cytoskeletal elements

196 **P.I. Karl, S.E. Fisher**
Taurine transport by microvillous membrane vesicles and the perfused cotyledon of the human placenta
Transport de la taurine par des membranes de vésicules microvillositaires et par le cotylédon perfusé du placenta humain

197 **M. Luckhardt, E. Aegerter, A. Malek, H. Schneider**
Transport and metabolism of glutamic acid studied in the human placenta by an *in vitro* perfusion method
Transport et métabolisme de l'acide glutamique étudiés dans le placenta humain par perfusion in vitro

199 **A. Malek, R.K. Miller, D.R. Mattison, R. Bryant, M. Panigel, L. Neth**
Energy production by the dually perfused term human placenta *in vitro* : ^{31}P nuclear magnetic resonance spectroscopy at 2.0 and 4.7 tesla
Production d'énergie par le placenta humain à terme perfusé en double circuit in vitro *: 31P-résonance magnétique nucléaire spectroscopie à 2.0 et 4.7 tesla*

201 **A. Malek, M.J. Meadows, F.C. Miller, E. Blann, D.R. Mattison**
Fetal to maternal transport of oxytocin across the dual perfused human placenta

Transport fœto-maternel de l'ocytocine par le placenta humain perfusé en double circuit

203 **A. Malek, M.J. Meadows, F.C. Miller, D.R. Mattison**
Continuous pO_2 monitoring during dual perfusion of the term human placenta
Monitorage continu de la pO_2 durant la perfusion en double circuit du placenta humain

205 **R.K. Miller, W. Faber, Y. Shah, P.J. Wier, R. Perez di Gregorio, A.A. Levin, P.A. Di Sant'Agnese, L. Neth, C. Eisenmann**
Placental toxicology : retinoids and cadmium
Toxicologie placentaire : rétinoïdes et cadmium

206 **T. Mohammed, J. Stulc, C.P. Sibbey, J. Glazier, R.D.H. Boyd**
Evidence for carrier-mediated potassium transfer across dually perfused rat placenta
Evidence pour un transfert du potassium médié par un transporteur à travers le placenta de rat perfusé en double circuit

207 **H.P. Leichtweiss, B. Mohar, S. Wohlers, H. Schröder**
The uptake of ascorbic acid and its oxidation products in the isolated cotyledon of the human term placenta
Captation de l'acide ascorbique et de ses dérivés oxydés dans le cotylédon isolé du placenta humain à terme

208 **K.R. Page, D.R. Abramovich, C.G. Dacke, T. Mayhew, J.M.A. Williams**
Inhibitor action on placental mineral metabolism
Action inhibitrice sur le métabolisme minéral placentaire

209 **G. Pridjian, A.H. Moawad, P.F. Whitington**
Transfer of betahydroxybutyrate (BOHB) by the *in vitro* perfused human placenta
Transfert du bétahydroxybutyrate (BOHB) par le placenta humain perfusé in vitro

210 **W. Reiber, H. Nöschel, S. Schröder, B. Müller**
Glucose and oxygen consumption, and lactate production of human placentae dually perfused *in vitro* under normoxia, hypoxia, and reoxygenation after a hypoxic period
Consommation de glucose et d'oxygène et production de lactate par le placenta humain perfusé in vitro en double circuit sous différentes conditions d'oxygénation

211 **C. Schoch, H. Schröder, H.P. Leichtweiß**
Uptake and transfer of amino acids in the artificially perfused guinea-pig yolk sac placenta

Captation et transfert d'acides aminés par la vésicule vitelline placentaire de cobaye artificiellement perfusée

213 **H. Schröder, W. Elwers, H.P. Leichtweiß**
D-glucose uptake and transfer in the artificially perfused guinea-pig yolk sac placenta
Captation et transfert de D-glucose par la vésicule vitelline placentaire de cobaye artificiellement perfusée

214 **A.J. Shaw, M.Z. Mughal, C.P. Sibley**
Magnesium transfer across the *in situ* perfused rat placenta
Transfert du magnésium à travers le placenta de rat perfusé in situ

215 **J. Štulc, B. Štulcovà, J. Švihovec, M. Břešťák**
Transcellular transport of Ca across the perfused human placental lobule
Transport transcellulaire du calcium à travers le lobule de placenta humain perfusé

VII. PLACENTAL STRUCTURE AND PATHOLOGY
STRUCTURE PLACENTAIRE ET PATHOLOGIE

219 **R. Leiser, T. Egloff**
Ultrastructural correlates of placental substance transfer : a state-of-art-review
Aspects ultrastructuraux du transport placentaire

239 **C.D. Ockleford, F.M. Bradbury, I. Indans**
Localization of cytoskeletal proteins in chorionic villi
Localisation des protéines du cytosquelette dans la villosité chorionique

251 **H. Arnholdt, U. Löhrs**
Proliferation and differentiation of Langhans cells
Prolifération et différenciation des cellules de Langhans

252 **Z. Bielanska-Osuchowska, A. Kunska**
Histochemical and ultrastructural research of the mesenchymal tissue in the placenta of pig
Etudes histochimique et ultrastructurale des tissus mésenchymateux dans le placenta de truie

253 **G.J. Burton, J.W. Thurley**
Correlative scanning and transmission electron microscopy of individual placental villi

Corrélation entre la microscopie électronique à balayage et par transmission de villosités placentaires individuelles

254 **V. Dantzer, M.H. Nielsen, S. Boisen, W. Prosbst**
Calcium in the pig placenta : Histochemistry and microanalysis
Calcium dans le placenta de truie : histochimie et microanalyse

255 **R. Demir, I. Ustünel**
Distribution of some enzymes in implantation site of pregnant rats
Distribution de quelques enzymes dans le site d'implantation chez la rate gravide

256 **R. Demir, I. Ustünel, N. Demir**
Light and electron microscopical observations on cellular interaction during initial stages of implantation and trophoblastic invasion in rats
Observation en microscopie optique et électronique d'interaction cellulaire pendant les stades précoces de l'implantation et de l'invasion trophoblastique chez le rat

257 **R.A. Jacobs, J. Oosterhuis, P. Libby, V. Han, J.R.G. Challis**
Localization of prostaglandin synthase in human and ovine placental membranes
Localisation de la prostaglandine synthétase dans les membranes placentaires humaines et ovines

258 **B.F. King, G.N. Fry, G.C. Douglas**
Morphological aspects of transferrin endocytosis by human trophoblast cells
Aspect morphologique de l'endocytose de la transferrine par les cellules trophoblastiques humaines

259 **L. Leach, B.M. Eaton, J.A. Firth, S.F. Contractor**
Localization of endogenous immunoglobulin-G in the human placenta
Localisation de l'immunoglobuline-G dans le placenta humain

260 **C.D. Ockleford**
Interference microscopic mass mapping of human first trimester chorionic villi
Cartographie quantitative par microscopie interférentielle des microvillosités choriales du 1er trimestre

261 **C.D. Ockleford, C. Barker, J. Griffiths, G. McTurk, R. Fisher, S. Lawler**
A «top stage» scanning ultrastructural study of healthy and premalignant trophoblast

Etude ultrastructurale à balayage de trophoblastes normal et précancéreux

262 **P. Ong, G.J. Burton**
The effects of hypoxia and reoxygenation on barrier thickness of placental villi maintained in organ culture
Les effets de l'hypoxie et de la réoxygénation sur l'épaisseur de la barrière placentaire des villosités maintenues en culture organotypique

VIII. PLACENTAL PATHOLOGY : CLINICAL ASPECTS
PATHOLOGIE PLACENTAIRE : ASPECTS CLINIQUES

265 **J.Y. Chou, S. Watanabe, T. Watanabe**
Induction of germ cell alkaline phosphatase gene expression in human malignant trophoblasts
Expression du gène de la phosphatase alcaline de cellules germinales dans les cellules de choriocarcinome humain

269 **E. Jimenez, M. Unger, F. Eitelbach, Z. Huang, G. Wagner, M. Vogel, I. Grosch-Wörner, A. Schäfer**
Demonstration of HIV-antigens in birth placentae and therapeutic abortions
Mise en évidence d'antigènes HIV dans les placentas à terme et dans des avortements thérapeutiques

276 **H. Fox, G. Roeckelein, M. Wells**
Placental structure and pathology, including trophoblastic neoplasia : summary of a workshop
Résumé de l'atelier : structure placentaire et pathologie, y compris les néoplasies trophoblastiques

278 **P. Bischof, T.M. Mignot, L. Cedard**
Are pregnancy-associated plasma protein-A (PAPP-A) and CA 125 measurements after IVF-ET possible predictors of early pregnancy wastage ?
Les mesures de la PAPP-A et du Ca125 sont-elles prédictives des pathologies précoces de la grossesse après FIVET ?

279 **J. Delmis**
Placental lipid content in preterm labor complicated by chorioamniotidis
Contenu lipidique du placenta dans l'acccouchement prématuré compliqué par une chorioamniotite

280 E. Jauniaux, G. Moscoso, S. Campbell, D. Gibb, K.H. Nicolaides
Ultrasound/pathologic correlations of placental anomalies associated with elevated maternal serum alpha-fetoprotein and a normal fetus
Corrélations entre les anomalies placentaires échographiques et anatomopathologiques associées à des taux élevés d'alpha-fœtoprotéine, dans le cas de fœtus normaux

281 H. Moirot, E. Thomine, M. Martin, G. Labadie, C. Fessard, P. Ensel, A. Pellerin, T. Ducastelle, J. Hemet, M.C. Bourreille, M.C. Boullie
Amniotic anomalies and lamellar ichthyosis. A case report
Anomalies amniotiques et ichthyose lamellaire : étude d'un cas

282 R. Pijnenborg, V. Ballegeer, D. Davey, M. Hanssens, B. Spitz, A. Tiltman, L. Vercruysse, A. Van Assche
The distribution of fibronectin and laminin in the placental bed of patients with different hypertensive disorders of pregnancy
Distribution de la laminine et de la fibronectine dans le lit vasculaire placentaire de patientes ayant différents types d'hypertension de la grossesse

283 R. Pijnenborg, A. Rees, J. Anthony, A. Tiltman, D. Davey, L. Vercruysse, A. Van Assche
Histological characteristics of placental bed spiral arteries related to the clinical classification of hypertensive disorders in human pregnancy
Caractéristiques histologiques des artères spiralées en relation avec la classification clinique des différents types d'hypertension de la grossesse

284 G. Roeckelein, J. Schroeder, R. Ulmer
Early spontaneous abortion with different karyotype : microscopic and morphometric investigations
Avortement spontané précoce avec différents caryotypes : études microscopique et morphométrique

286 C.M. Salafia, J.P. Burns, A.M. Vintzileos, L. Silberman
The correlation of placental and decidual histology with karyotype
Corrélation entre l'histologie du placenta et de la décidue et le caryotype

287 C.M. Salafia, A.M. Vintzileos, L. Silberman
Placental pathology of idiopathic intra-uterine growth retardation (IUGR) at term
Pathologie placentaire du retard de croissance intra-utérin idiopathique, à terme

288 **A.E. Szulman**
The placenta in human spontaneous abortion
Le placenta dans les avortements spontanés humains

289 **M. Takayama, K. Isaka, Y. Suzuki, H. Funayama, K. Akiya, H. Bohn**
Dynamics and synthesis of new placental tissue protein PP19
Dynamique et synthèse d'une nouvelle protéine placentaire : la PP19

CLOSING LECTURE
ALLOCUTION DE CLOTURE

293 **M. Panigel**
Non-invasive methods in placental research
Méthodes de recherche placentaire non invasives

1.
Trophoblast-endometrium interactions

Interactions trophoblaste-endomètre

Interactions between human term trophoblasts and endometrium *in vitro*

H. J. Kliman, R.F. Feinberg*, J.E. Haimowitz

*Departments of Pathology and Laboratory Medicine, and *Obstetrics and Gynaecology, University of Pennsylvania School of Medicine, 6 Founders Pavilion, Philadelphia, PA 19104, USA*

SUMMARY

Normal human implantation is a poorly understood process that is likely to be mediated by the cytotrophoblasts of the early conceptus. We utilized first, second, and third trimester purified cytotrophoblasts to develop an *in vitro* system for examining the interaction of trophoblasts with human endometrium. When endometrium was cultured alone it remained viable for up to three days. When trophoblasts were cultured alone, they formed small and large aggregates, and occasionally spherical shells with hollow centers—reminiscent of the outer trophoblastic layer of a blastocyst, the trophoblast shell. When trophoblasts and endometrium were cultured together, the trophoblasts adhered to the exposed stromal surfaces of the tissue fragments. The surface epithelium was rarely receptive to trophoblast attachment except in one experiment when day 19 endometrium was used for the co-incubation, suggesting that surface attachment is usually restricted. When proliferative endometrium was used in the co-culture, the trophoblasts induced a marked secretory effect on the endometrial glandular epithelium, suggesting a paracrine effect of trophoblast-secreted products. A common finding was the presence of an acellular zone in the endometrium only adjacent to the attached trophoblasts. We speculate that this zone may be caused by proteolysis and re-synthesis of extracellular matrix proteins by trophoblasts. Based on our results, this *in vitro* suspension system should prove useful for examining: 1) the specificity of trophoblast attachment to the epithelial surface of the endometrium; and 2) the biochemical and cellular processes involved in trophoblast invasion of the uterine extracellular matrix.

INTRODUCTION

Until recently, most studies examining implantation have been performed in the mouse, rat, rabbit, or monkey (1-4); and there is little information about implantation in the human (5-7). In addition, most of the these studies have been carried out using *in vivo* approaches which make it difficult to examine detailed cellular and biochemical interactions. Schlafke and Enders (8) performing electron microscopic studies on implantation sites in the rabbit proposed three possible mechanisms for implantation: 1) intrinsic implantation, which is characterized by trophoblast penetration between cells of the uterine epithelium and adhesion to the basal lamina; 2) fusion implantation, where syncytial trophoblasts fuse with uterine epithelial cells; and 3) displacement implantation, where trophoblasts dislodge the uterine epithelium from their basal lamina and replace it. Studies examining the interaction of mouse blastocysts with extracellular matrix (ECM) materials have been attempted (9), but no system exists at present for the *in vitro* examination of the interaction of human trophoblasts with endometrium. The unavailability of such *in vitro* systems has slowed the progress of our understanding of the cellular and biochemical events that take place during normal human implantation.

We present here an *in vitro* suspension model system which we have used to examine the interactions of purified human trophoblasts with explants of human endometria. Using this system we have been able to demonstrate that 1) trophoblasts adhere to human endometrial explants, 2) the trophoblasts produce a zone of necrosis at points of contact with the endometrium, and 3) attachment of trophoblasts to the surface epithelium appears restricted to day 19 endometrium.

RESULTS

Endometrium-trophoblast suspension co-culture

When endometrium was cultured alone in a 95% air - 5% CO_2 atmosphere it remained viable for up to three days. When tissue necrosis was present, it was mainly in the central portion of the tissue. Culturing endometrium in a 95% O_2 - 5% CO_2 atmosphere led to a slight improvement in tissue viability, but larger tissue fragments still, to a variable extent, exhibited central necrosis. When trophoblasts were cultured alone, they formed small and large aggregates. The large aggregates formed trabecular patterns and pseudo-villous structures. Occasional large aggregates formed spherical shells with hollow centers—reminiscent of the trophoblastic shell of an early blastocyst.

When trophoblasts and endometrium were incubated together in suspension, trophoblasts adhered firmly to the tissue fragments (Fig. 1). Utilizing antibodies against alpha-hCG, we confirmed by immunohistochemistry that these cells were trophoblasts, and not, for example, other minor cell types from the trophoblast preparation or break down products of the endometrium itself. In several co-incubation preparations, we also noted hCG-positive trophoblasts which had actually penetrated into the endometrial explants.

The concept of a "window" for implantation on days 19-20 has been discussed by a number of workers (10). We therefore wondered if we would see different patterns of trophoblast-endometrial interactions depending on the date of the endometrium. Trophoblast cells attached to the exposed endometrial stromal surfaces in 26 separate co-culture experiments. However, on one occasion when a day 19 endometrium was used, we observed trophoblast binding to the surface epithelium. We speculate that this association with the endometrial surface epithelium with day 19 endometrium may be related to the natural receptivity of the epithelium at this time of the cycle.

In 25 of 26 co-culture experiments the presence of "contact necrosis" was observed at the points where the trophoblasts made contact with the endometrial explants. Usually when endometrial explants were cultured alone in suspension they showed evidence of central necrosis, presumably because the central portions of the tissue were farthest away from the nutrients and oxygen in the media. When endometrium was co-cultured with trophoblasts, necrosis was usually not seen centrally, but peripherally, and always directly adjacent to the attachment sites of the trophoblasts. We speculate that this necrosis may be caused by the secretion of trophoblast proteases (11), although stimulation of endometrial proteases by the trophoblasts can not be ruled out. It is also possible that this zone contains newly synthesized ECM proteins (12) in addition to the breakdown products of the endometrium. This zone may have a counterpart *in vivo*—Nitabuch's layer (13). Nitabuch's layer is made up of eosinophilic fibrinoid material and separates the cytotrophoblastic cell columns from the decidua at the implantation site. The zone of necrosis we see *in vitro* may simply represent the first step in the formation of Nitabuch's layer since in the *in vitro* zone cellular remnants can be identified, while Nitabuch's layer appears to be largely acellular.

It has been suggested by others that only first trimester placental trophoblasts have the capacity to interact with and invade into endometrium (14). We have performed trophoblast-endometrium co-culture experiments using cytotrophoblasts derived from first trimester (8-12 weeks), second trimester (13 and 18 weeks) and term placentae (> 36 wks) without noting any substantial differences. In all three trimesters, the cytotrophoblasts bound to the cut surfaces of the endometrium (except on day 19, as noted above), produced contact necrosis, penetrated the endometrium to a limited extent in a two to three day period, and induced secretory endometrium if the starting material was proliferative endometrium. Therefore, the ability to bind and interact with endometrium was not limited to the first trimester, but was exhibited by trophoblasts from all three trimesters.

Paracrine effect of trophoblasts on proliferative endometrium

In addition to the observations noted above we also observed the induction of secretory changes on proliferative endometrium when co-cultured with trophoblasts. When proliferative phase endometrial explants were cultured for 24 h in the absence of exogenous steroids, little change was noted. As with the 0 h tissue, the 24 h explant showed a few subnuclear vacuoles in the glands, and the stromal cells showed only scant cytoplasm. When exogenous progesterone was added at physiologic levels (50 ng/ml), secretory phase endometrium consistent with day 17 (15) could be induced, as has been shown previously (16). In addition, the progesterone appeared to induce expansion of the stromal cytoplasm, consistent with early decidual changes. When this same proliferative endometrium was co-cultured in parallel with only cytotrophoblasts for as little as 24 h, secretory endometrium consistent with day 18-19 was induced. These secretory changes were noted in thirteen of fifteen proliferative endometrial co-culture experiments. We have shown previously (17, 18) that cultured cytotrophoblasts secrete progesterone, suggesting that trophoblast derived progesterone is active in stimulating the secretory changes we see in these co-culture experiments. These results suggest that localized hormonal effects of the implanting blastocyst may direct morphologic and biochemical changes in the specific area of the endometrium where trophoblast attachment and invasion is occurring.

Co-culture results as a function of tissue and cell type

We also investigated the specificity of our observations by co-culturing cytotrophoblasts with explants of breast tissue, fallopian tube, and umbilical cord; and incubating endometrium with human endothelial cells, JEG-3 choriocarcinoma cells, and malignant melanoma cells. When cytotrophoblasts were co-cultured with breast tissue they adhered to the cut surfaces, but showed no evidence of contact necrosis. When incubated with umbilical cord explants, the trophoblasts attached to the cut surfaces of umbilical cord, and again showed no evidence of contact necrosis. Co-culture experiments performed with JEG-3 cells (a malignant trophoblast cell line) revealed that they attached to endometrial cut surfaces, but not to the surface epithelium, and induced necrosis at the sites of attachment, much like normal trophoblasts. In the reverse experiments, endothelial cells co-cultured with endometrium exhibited attachment to the endometrial cut surfaces, but not the surface epithelium, and there was no evidence of contact necrosis. Malignant melanoma cells attached to the endometrial cut surfaces, did not attach to the surface epithelial layer, and did not induce necrosis at the sites of attachment. Interestingly, when these same melanoma cells were incubated with fallopian tube explants, the tumor cells again only attached to the cut surfaces. We conclude from these studies that attachment to the cut surfaces of tissue fragments, i.e., stromal extracellular matrix components, is not specific to trophoblasts, but is a behavior shared by a wide variety of normal and malignant cells. What appears to be specific is that: 1) it is the surface epithelium of both the endometrium and fallopian tube which allows or prevents cell attachment; 2) normal human trophoblasts bind to the endometrial surface epithelium only at a specific time in the menstrual cycle; and 3) when trophoblasts bind to endometrium, they induce a band of necrosis similar to the fibrinoid layer (Nitabuch's layer) noted at implantation sites *in vivo*.

DISCUSSION

To date, the greatest efforts in studying implantation have been in the use of animal blastocyst outgrowth models (2, 9). In these systems, mouse blastocysts are plated onto various surfaces where they proceed to hatch, attach and finally spread. This work has led to insights into the factors necessary for initial attachment, the factors necessary for outgrowth, and has permitted visualization of early developmental processes of peri-implantation blastocysts in the absence of endometrial influences. Since these systems do not offer the potential to study the effects of endometrial and hormonal factors, others have modified these outgrowth models by plating blastocysts onto endometrial monolayers in an attempt to examine the endometrial factor (19). These systems have offered insight into specific cell adhesion, but do not approximate the three dimensional relationships which exist in *in vivo* implantation. Recently Sengupta *et al.* (20) have attempted to improve on the monolayer co-culture model by proposing the use of mouse endometrial culture on floating collagen gels, and Glasser *et al.* have cultured mouse endometrial cells on Matrigel™ to induce columnar growth of endometrial cells (21).

Although these model systems have contributed to progress in animal implantation, the parallels to human implantation are unknown. Histologic studies of both animal and human implantation sites suggest that it is the trophoblastic shell that plays the critical role in the initiation and progression of implantation (22). Therefore, we speculated that purified cytotrophoblasts from placentae of different gestational ages could be co-cultured with endometria in order to elucidate implantation events in the human.

Our goals were to culture endometrial explants for several days and to observe specific attachment of trophoblasts to the endometrial explants. Utilizing a shaking suspension culture system, we were able to maintain viable endometria for at least three days. When trophoblasts purified from first, second or third trimester placentae were added to the proliferative or late secretory endometria, they attached to the cut surfaces of the explants. Alpha-hCG positive trophoblasts were seen to penetrate into the endometrial explants. One striking feature of these experiments was that trophoblasts did not attach to the epithelial surfaces of the explants. A noteworthy exception occurred when trophoblasts were co-cultured with endometria which was day 19 by histologic dating. In this experiment, the trophoblasts did bind to the surface epithelium, possibly because of the presence of menstrual cycle dependent cell adhesion molecules (23). These results suggest that it is the endometrium that determines whether trophoblast attachment will occur, and that the trophoblasts will bind to any surface that is permissive. Furthermore, it appears that the endometrial surface is generally not permissive, and may only promote trophoblast attachment during a relatively short time in the menstrual cycle. These results are consistent with clinical observations of *in vitro* fertilization patients, where it has been shown that pregnancies only occur when blastocysts are returned to the uterus on days 16-20, with no successful pregnancies occurring after day 20 (10). Even during this purported implantation window, successful implantation following *in vitro* fertilization probably only occurs 20-30 % of the time (24).

This model system, although not a duplicate of the *in situ* environment, does recapitulate a number of important physiological facets of nidation and placentation: 1) trophoblast association with endometrial surface epithelium occurs only on day 19; 2) at the trophoblast contact points with the endometrium, a zone of necrosis is noted, similar to the *in vivo* Nitabuch's layer; 3) occasional trophoblasts can be identified which appear to have penetrated into the endometrial explants, similar to the invading trophoblasts seen in the normal placental bed (25); and 4) the trophoblasts appear to induce a marked paracrine effect on the endometrium, inducing a conversion histologically from proliferative to secretory endometrium. We should, therefore, be able to utilize this model system to investigate the factors which induce endometrial permissiveness, mediate contact necrosis, promote a paracrine effect on the endometrium, and facilitate trophoblast invasiveness.

METHODS

Preparation and culture of human trophoblasts

The method of preparation and culture of human trophoblasts has been described previously (17). The procedure works well for first trimester through term placental tissue. Briefly, villous tissue was dispersed with trypsin and DNase. The dispersed cells were then purified on a Percoll gradient (5-70%). A middle band at density 1.040-1.060 g/ml is comprised of cytotrophoblasts. The cells isolated from this layer have the ultrastructural features of cytotrophoblasts, and contamination by macrophages (assessed by immunocytochemical localization of alpha$_1$-antichymotrypsin) and fibroblasts and endothelial cells (assessed by immunocytochemical detection of vimentin) is less than 5% (17). The cytotrophoblasts were cultured in Dulbecco's Modified Eagles' Medium (DMEM) containing 25 mM glucose and 25 or 50 mM HEPES (DMEM-HG) supplemented with gentamicin (50 µg/ml), glutamine (4 mM), and in some cases with 20% (v/v) heat-inactivated fetal calf serum (FCS).

Endometrial Explants

Endometrial tissue was collected in a sterile environment immediately after removal of the uterus or obtaining the specimen by D&C. The tissue was placed in warmed DMEM-HG supplemented with gentamicin (50 µg/ml). In the laboratory the tissue was minced into 1-2 mm cubes and transferred to DMEM-HG-20% FCS. A small portion of each specimen was fixed in Bouin's for histological dating. For endometrial co-culture experiments, 1 ml of a 1×10^6 cells/ml suspension was added to the endometrial explants and medium was changed every 24 h. Cells were cultured in sterile 17x100 mm polypropylene snap-top tubes at 37°C in an atmosphere of humidified 95% air - 5% CO_2 or 95% O_2 - 5% CO_2 while being gyrated on an angled (~30°) rotator (Red Rotor, Hoefer Scientific Instruments, San Francisco, CA) at a setting of five.

Preparation and culture of other tissues and cells
Endothelial cells were purified from human umbilical cords as previously described (26). 14-49B melanoma cells, derived from a metastasis of human melanoma, were generously supplied by Dr. Manard Herling of the Wistar Institute, Philadelphia, PA. JEG-3 choriocarcinoma cells were obtained from the American Type Culture Collection, Rockville, MD. Normal human breast tissue, fallopian tube, and umbilical cord were collected after clinical examination of residual tissue from the Surgical Pathology Laboratory of the Hospital of the University of Pennsylvania under the auspices of the Cooperative Human Tissue Network, Eastern Division. All specimen collection protocols were approved by the Institutional Review Board of the University of Pennsylvania.

Histology
Tissue was fixed in Bouin's, embedded in paraffin and sectioned at 5 µm intervals. Twenty to forty serial sections were made, with hematoxylin and eosin (H and E) staining of every other ten slides. H and E slides were assessed for the following: 1) trophoblast adhesion, including how many cells attached and whether they were attached to cut surfaces or surface epithelium; 2) tissue necrosis (contact necrosis or central necrosis); and 3) the presence and depth from the surface of trophoblast invasion.

Immunocytochemistry
Since trophoblast invasion can not be readily assessed using H and E staining alone, immunocytochemistry against alpha-hCG subunit was employed. Three to four slides per co-culture were stained for alpha-hCG as previously described (17). Rabbit anti-alpha-hCG was generously supplied by Dr. Steven Birken of Columbia University.

ACKNOWLEDGMENTS

The authors wish to thank Drs. G. Pietra and J. Strauss for helpful discussions and Dr. D. Cines for supplying us with the human endothelial cell cultures. We also wish to thank the histotechnologists of the Department of Pathology and Laboratory Medicine, and Ms. Melanie Minda and Mr. Lewis Johns of the Electron Microscopic Laboratory for helping to process the numerous samples required for this study. Our appreciation is also extended to the nurses and residents on the Labor and Delivery Suite of the Hospital of the University of Pennsylvania who helped us procure normal placentae for our trophoblast preparations. Finally, we wish to thank Dr. Virginia LiVolsi and Ms. Elizabeth Freeman, of the Cooperative Human Tissue Network, Eastern Division, who helped to procure the endometrial material used in these studies.

REFERENCES

1. Denker H-W: Basic aspects of ovoimplantation. Obstet Gynecol Annu 12:15-42, 1983
2. Enders AC, Chavez DJ, Schlafke S: Comparison of implantation in utero and in vitro. In: Cell and Mol Aspects of Implantation. SR Glasser and DW Bullock (eds). Plenum, p. 365-382, 1981
3. Enders AC, Hendricks AG and Schlafke S: Implantation in the Rhesus monkey: initial penetration of endometrium. Am J Anat 167:275- 298, 1983
4. Schlafke S, Welsh AO and Enders AC: Penetration of the basal lamina of the uterine luminal epithelium during implantation in the rat. Anat Record 212:47-56, 1985
5. Hata T, Ohkawa K and Uchida K: Contact patterns between cytotrophoblast and decidual cells in human implantation site. Acta Obst Gynaec Jpn 35:529-536, 1981
6. Hata T, Ohkawa K, Tomita M and Kishino M: Phagocytosis of human cytotrophoblast cell invading into decidual tissue in early stage of gestation. Acta Obst Gynaec Jpn 33:537-544, 1981

7. Glasser SR. Current concepts of implantation and decidualization. In: The Physiology and Biochemistry of the Uterus in Pregnancy and Labor. G Hoszar (ed) CRC Press, pp 127-154, 1986
8. Schlafke S and Enders AC: Cellular basis of interaction between trophoblast and uterus at implantation. Biol Reprod 12:41, 1975
9. Armant DR, Kaplan HA and Lennarz WJ: Fibronectin and laminin promote in vitro attachment and outgrowth of mouse blastocysts. Dev Biol 116:519-523, 1986
10. Navot D, Droesch K, Liu HC, Keriner D, Veeck L, Steingold K, Muasher S, Rosenwaks Z: Efficiency of human conception *in vitro* related to the window of implantation. Abst 419, 35th Annual Meeting of the Society for Gynecologic investigation, Baltimore, Maryland, 1988
11. Queenan JT Jr, Kao L-C, Arboleda CE, Ulloa-Aguirre A, Golos TG, Cines DB, and Strauss JF III: Regulation of urokinase-type plasminogen activator production by cultured human cytotrophoblasts. J Biol Chem 262:10903-10906, 1987
12. Ulloa-Aguirre A, August AM, Golos TG, Kao L-C, Sukuragi N, Kliman HJ, Strauss JF III: 8-Bromo-3'5' adenosine monophosphate regulates expression of chorionic gonadotropin and fibronectin in human cytotrophoblasts. J Clin End Metab 64:1002-1009, 1987
13. Boyd JD, Hamilton WJ: The human placenta, p 190. Cambridge, W. Heffer & Sons Ltd, 1970
14. Fisher SJ, Leitch MS, Kantor MS, Basbnaum CB, Kramer RH: Degradation of extracellular matrix by the trophoblastic cells of first-trimester human placentas. J Cell Biol 27:31-41, 1985
15. Blaustein A: Interpretation of endometrial biopsies. Second edition. Biopsy interpretation series, p 28. New York, Raven Press, 1985
16. Kohorn EI, Tchao RJ: Conversion of proliferative endometrium to secretory endometrium by progesterone in organ culture. J Endocrinol 45:401, 1969
17. Kliman HJ, Nestler JE, Sermasi E, Sanger JM, Strauss JF III: Purification, characterization, and *in vitro* differentiation of cytotrophoblasts from human term placentae. Endocrinology 118:1567-1582, 1986
18. Feinman MA, Kliman HJ, Caltabiano S, Strauss JF III: 8-Bromo-3'5'-adenosine monophosphate stimulates the endocrine activity of human cytotrophoblasts in culture. J Clin Endocrinol Metab 63:1211-1217, 1986
19. Glass RH, Spindle AI, Pedersen RA: Mouse embryo attachment to substratum and interaction of trophoblast with cultured cells. J Exp Zool 208:327, 1979
20. Sengupta J, Given RL, Carey JB, Weitlauf HM: Primary culture of mouse endometrium on floating collagen gels: a potential *in vitro* model for implantation. Ann NY Acad Sci 976:75-94, 1986
21. Glasser SR, Julian J, Decker GL, Tang JP, Carson DD: Development of morphological and functional polarity in primary cultures of immature rat uterine epithelial cells. J Cell Biol 107:2409-23,1988
22. Hertig AT, Rock J: A description of 34 human ova within the first 17 days of development. Amer J Anat 98:435-494, 1956
23. Nose A, Takeichi M: A novel cadherin cell adhesion molecule: Its expression patterns associated with implantation and organogenesis of mouse embryos. J Cell Biol 103:2649-2658, 1986
24. Laufer N, Tarlatzis BL, Naftolin F: In vitro fertilization state of the art. Semin Reprod Endocrinol 2:2, 1984
25. Feinberg RF, Kao L-C, Haimowitz JE, Queenan JT Jr, Wun T-C Strauss JF III, Kliman HJ: Plasminogen activator inhibitors types 1 and 2 in human trophoblasts: PAI-1 is a marker of invading trophoblasts. Lab Inv 61:20-26, 1989
26. Jaffe EA, Hoyer LW, Nachman RL: Synthesis of von Willebrand factor by cultured human endothelial cells. Proc Natl Acad Sci 71:1906-9, 1974

Résumé

INTERACTION AVEC LE TROPHOBLASTE HUMAIN À TERME ET L'ENDOMÈTRE ÉTUDIÉ IN VITRO

L'implantation humaine normale est mal comprise. Il est probable qu'elle est provoquée par le cytotrophoblaste précoce du conceptus au tout premier stage. Nous avons utilisé des cellules de cytotrophoblaste du 1er 2ème et 3ème trimestre pour étudier dans un système in-vitro l'interaction du trophoblaste et de l'endomètre humain. Quand l'endomètre est cultivé seul il reste viable pendant 3 jours. Quand les cellules du trophoblaste sont cultivées isolées elle se développent en agrégats de petite et grande taille, et parfois forment des amas sphéroides avec un centre creux. Cela rappelle la couche trophoblastique d'un blastocyste autrement dit l'enveloppe trophoblastique. Quand on a cultivé du trophoblaste et de l'endomètre en commun, le trophoblaste a adhéré à la partie stromale des tissus exposés parmi les fragments. L'épithélium de surface se trouvait rarement receptif pour les cellules trophoblastiques sauf dans le cas d'une expérience unique où l'endomètre du J19 a été utilisé pour les expériences de co-incubation. Cela suggère que l'attachement de surface est en général restreint quand l'endomètre prolifératif a été utilisé pour faire de la coculture, les cellules trophoblastiques ont induit une sécrétion importante de l'épithélium glandulaire secrétoire de l'endomètre suggérant que les produits secrétés par le trophoblaste ont un effet paracrine. Ila toujours été trouvé une zone acellulaire dans l'endomètre au seul point où il est adjacent au trophoblaste qui s'attache. Nous spéculons que cette zone pourrait être causée par la protéolyse et la resynthèse de protéines de matrice cellulaire par le trophoblaste. Nos résultats suggèrent que ce système de suspension in vitro devrait être utile pour comprendre (1) la spécificité de l'attachement des cellules trophoblastiques dans l'endomètre (2) les mécanismes biochimiques et cellulaires de l'invasion par le trophoblaste de la matrice extra-cellulaire de l'utérus.

New developments in human trophoblast cell culture

Y.W. Loke

Division of Cellular and Genetic Pathology, Department of Pathology, University of Cambridge, UK

The human placenta is a composite organ containing a variety of different cell types. To many investigators, the most interesting cell population is the trophoblast. For this reason, a great deal of effort has been devoted to developing techniques for the isolation and culture of these cells with a sufficient degree of viability and homogeneity in order to be used for experiments (see review by Loke 1983). Unfortunately, this is not easy to achieve, as all those who have tried will readily agree. Many problems remain and these were discussed in a recent Workshop (Loke and Hussa 1987). At present, there are three approaches to human trophoblast cell isolation which appear to show some promise. They are 1) density gradient separation, 2) immunological separation and 3) receptor-ligand interaction.

Density gradient separation
This method is based on the premise that the density of trophoblast cells is sufficiently different from that of other placental cells as to be separable by centrifugation over a Percoll gradient. An almost homogeneous population of trophoblast cells can be obtained from normal term placentae within the density band of 1.048-1.062g/ml (Kliman et al. 1987). When plated out in culture, many of these trophoblast cells are observed to be transformed into syncytiotrophoblasts which synthesise the usual placental proteins and hormones such as SP1, HCG and HPL. The cells isolated by this technique, therefore, are likely to be derived from villous cytotrophoblast since they appear to differentiate along the villous pathway to become syncytiotrophoblast *in vitro*. However, not all will agree that the cells isolated by this technique are necessarily trophoblast. For example, Bierings et al. (1988) using an identical protocol reported that as high as 40% of the cells obtained from the 1.048-1.062g/ml density band were CD14+ and DR+, both of which are macrophage markers. None of the cells stained for HCG, HPL or SP1. There is, therefore, still some doubt regarding the reproducibility of this isolation technique.

Immunological separation

This method is based on the use of appropriate monoclonal antibodies (Mabs) which can differentiate between trophoblast and other contaminating placental cells. Many such Mabs are now available (see WHO report by Anderson et al. 1987) so significant developments in this approach to trophoblast separation is anticipated within the next few years. Kawata et al. (1984) managed to obtain trophoblast from term placentae with a 74% purity using the fluorescence-activated cell sorter (FACS). More recently, Contractor and Soorana (1988) used a panning technique whereby cells labelled with an anti-trophoblast Mab are incubated on culture dishes coated with a goat anti-mouse IgG. The yield of trophoblast in the region of 50-60% by this panning technique appears to be inferior to that of using FACS but the fact that the former can be performed without the need for sophisticated instruments is a point in its favour. We ourselves have utilised Protein A conjugated to Sephadex microbeads as a means of pulling out antibody-coated trophoblast cells via binding to the Fc part of the immunoglobulin molecule (unpublished). At present, the results are still somewhat variable.

The anti-trophoblast Mabs to be used for this procedure must be selected with care. Not all of these are necessarily directed at surface antigens. Those antigens which are on the surface must be present in adequate density, otherwise there would be insufficient antibody bound to provide effective separation. Most importantly, a point which is not generally appreciated is that many trophoblast surface antigens are highly susceptible to degradation by proteolytic enzymes such as trypsin which, therefore, render many anti-trophoblast Mabs unsuitable if they happen to be directed at these epitopes since the initial dispersion of trophoblast cells from intact chorionic villi frequently involve enzyme digestion. Therefore, a systematic characterisation of the Mabs to be used and the nature of their trophoblast target antigens before embarking on isolation procedures could save a lot of wasted time.

Recently, Douglas and King (1989) described a negative selection procedure whereby non-trophoblast cells are removed by reacting them with a Class I HLA Mab and subsequent binding to immunogenetic spheres coated with an anti-mouse immunoglobulin. Recovery of trophoblast greater than 92% purity from term placentae was reported. The obvious constraint with this technique is that all extravillous trophoblast will also be lost because this subpopulation, unlike villous trophoblast, does express a Class I-like molecule. The principle of the method, however, is sound and no doubt other appropriate Mabs could eventually be found as a substitute for the presently used anti-Class I.

Receptor-ligand interaction

This is the method we have devised and is now in use in our laboratory. The rationale behind the development of this technique is the observation, in other culture systems, that epithelial cells frequently require a collagenous substrate for adhesion. We have observed that a similar requirement applies to trophoblast. From this, we described a method based on the use of a modified medium together with a substrate of extracellular matrix which consistently yields 80-90% trophoblast from first trimester placentae (Loke and Burland 1988). Since then, we have identified the component in extracellular matrix responsible for trophoblast adhesion as the protein laminin (Loke et al. 1989a). This protein is now used in our laboratory either as a substitute to coat culture dishes or via a more recent development whereby laminin is conjugated to magnetic beads (Dynal) which are then utilised to selectively isolate trophoblast cells bound to them (Loke et al. 1989b). Both methods are highly efficient, leading to trophoblast yields in excess of 80% purity.

Interestingly, the isolated cells are mainly extravillous trophoblast (Loke et al. 1989b), so this technique appears to be the first reported which is selective for this subpopulation. The cells are stained by Mabs reactive to extravillous trophoblast (e.g. 18A/C4 and CT.10.18), but not by those reactive only to villous trophoblast (e.g. 71.1). They also express a Class I HLA antigen recognisable by the Mab W6/32. This is characteristic of extravillous trophoblast while villous trophoblast is Class I negative. Furthermore, we now have the additional evidence that W6/32 immunoprecipitates a molecule from these isolated trophoblast cells which differ from a Class I molecule in having heavy chains associated with B_2M of 40KD rather than 45KD and that this trophoblast antigen is lacking in polymorphic domains (Grabowska et al. - submitted). These are similar to the unusual features described for the Class I antigen from term chorionic plate trophoblast (Ellis et al. 1986).

Unlike the trophoblast cells isolated by density gradient separation which are highly synthetic for placental products, the cells isolated by laminin-binding are relatively unproductive in culture. HPL and SP1 are localised to a few isolated cells while HCG is hardly ever observed. Even the multinucleated cells formed in culture do not contain any hormones which indicate they are more akin to placental bed giant cells than to syncytiotrophoblast (Loke 1988a). Thus, these trophoblast cells appear to be differentiating along the extravillous pathway.

We believe that the selection of extravillous trophoblast by laminin is due to the presence of receptors for this protein on the surface of these cells. This is supported by the observation of Wewer et al. (1987) that an antibody against the laminin receptor stained extravillous trophoblast intensely in histological sections of early decidua basalis while only faint immunoreactivity was detected along the basement membrane underlying trophoblast of the chorionic villi. We have already remarked on the possible relevance of this trophoblast laminin receptor expression to their _in vivo_ migration through decidua where individual stromal cells are surrounded by a thick pericellular layer of laminin thereby providing points of anchorage for trophoblast (Loke et al. 1989a).

Other methods for selective propagation of human trophoblast cells

Besides the aforementioned techniques, there are some others which should also be considered because the investigators using them have claimed a certain degree of success. Varying the enzymic digestion procedure or the type of enzyme used is said to affect the yield of trophoblast, the hypothesis being that if the underlying basement membrane of the chorionic villi is not broken, then only trophoblast cells are released without contamination by stromal cells of the villous mesenchyme. Thus, restricting the time of trypsin digestion to exactly 5 mins (Phillips et al. 1989) or using Dispase (Yeger et al. 1989) or Pancreatin (Bax et al. 1989) are said to fulfil the above conditions of selective villous disruption and trophoblast yields in excess of 95% purity have been reported.

Then, there are methods which do not use any enzymes at all. Yagel and his colleagues (1989) have reported that pure long-term trophoblast cultures can be established by the simple expedient of allowing these cells to grow out from explants. Because there has been no enzyme disruption of the villi, there is less chance for non-trophoblast cells to become attached to the culture dish and to prolfierate. With this method, the authors claim to have maintained trophoblast cells for as long as 8 months and 13 passages in culture. These investigators are not alone in successfully growing trophoblast in this way. Ungar and colleagues (1987) have already described a similar method earlier and it appears that pure trophoblast cells can also be cultured and maintained for several passages from the rat placenta using explants (Hunt et al. 1989). Therefore, there are sufficient successful results reported for this technique as to make it worthy of further investigation. Certainly, some cytotrophoblast during the early stages of gestation are highly proliferative as evidenced by the presence of mitotic figures as well as their expression of the transferrin receptor and the proliferation marker Ki67 (Bulmer et al. 1988). These cells could well be the progenitors of the trophoblast cells growing outwards from explants. Recently, the first reported development of a trophoblast cell line from a normal full term placenta which is highly tumorigenic when transplanted into nude mice came from investigators in Taiwan (Ho et. al. 1987). This probably represents an isolated case where some trophoblast cells are already transformed in vivo. According to Hertz (1978), neoplastic foci can often be detected in the normal human placenta so one begins to wonder how normal are the so-called 'normal' trophoblast cells which proliferate in culture. Perhaps all of these cells should be karyotyped in order to ascertain their normality. In mice, trophoblast cells established in culture from normal placentae will frequently induce tumours when injected into syngenic hosts (Log et al. 1981). Whether these are also instances of cells previously transformed in vivo or that trophoblast cells have a greater propensity to be transformed in vitro is not clear. The line drawn between what is considered as normal trophoblast behaviour and that of neoplastic cells has always been rather indistinct (Loke 1988b).

Finally, the theoretical possibility exists for manipulating the composition of the culture medium either to dissuade the growth of containments or to encourage that of trophoblast. Daniels-McQueen et al. (1987) and ourselves (Loke and Burland 1988) have addressed the former by substituting D-valine for L-valine since mesenchymal cells are said not to be able to utilise D-valine while cells of epithelial derivation such as trophoblast can. However, we have found that this regime does not result in complete stasis of contaminating cell growth. We, therefore, prefer to use our trophoblast cells in experiments within 2-3 days after seeding in culture during which period there is very little growth of contaminating cells anyway. As for encouraging trophoblast proliferation, hardly any systematic studies have been done to determine trophoblast cells' exact requirements. Ungar et al. (1987) observed that 8% of explants had trophoblast growths when cultured in medium supplemented with human pregnancy sera compared to only 19% in fetal calf sera. In contrast, we have found human pregnancy sera to be highly toxic to trophoblast cells (unpublished). We are now also investigating the effects of various cytokines in view of the report of Athanassakis et al. (1987) that murine trophoblast cell proliferation is enhanced by T-cell derived growth factors. It remains to be seen how significant a contribution such regimes would make to human trophoblast cell culture in future.

REFERENCES

Anderson, D.J., Johnson, P.M., Alexander, N.J., Jones, W.R., and Griffin, P.D. (1987): Monoclonal antibodies to human trophoblast and sperm antigens: Report of two WHO-sponsored workshops, Toronto, Canada. J. Reprod. Immunol. 10, 231-257.

Athanassakis, I., Bleackley, R.G., Paetkan, V., Guilbert, L., Barr, P.J., and Wegmann, T.G. (1987): The immunostimulatory effect of T cells and T cell lymphokines on murine fetally derived placental cells. J. Immunol. 138, 37-44.

Bax, C.M.R., Ryder, T.A., Mobberley, M.A., Tyms, A.S., Taylor, D.L., and Bloxam, D.L. (1989): Ultrastructural changes and immunocytochemical analysis of human placental trophoblast during short-term culture. Placenta 10, 179-194.

Bierings, M.B., Adriaansen, H.J., and Van Dijk, J.P. (1988): The appearance of transferrin receptors on cultured human cytotrophoblast and in vitro-formed syncytiotrophoblast. Placenta 9, 387-396.

Bulmer, J.N., Morrison, L., and Johnson, P.M. (1988): Expression of the proliferation markers Ki67 and transferrin receptor by human trophoblast populations. J. Reprod. Immunol. 14, 291-302.

Contractor, S.F., and Sooranna, S.R. (1988): Human placental cells in culture: a panning technique using a trophoblast-specific monoclonal antibody for cell separation. J. Devel. Physiol. 10, 47-51.

Daniels-McQueen, S., Knichevsky, A., and Boime, I. (1987): Isolation and characterisation of human cytotrophoblast cells. Trophoblast Res. 2, 423-445.

Douglas, G.C., and King, B.F. (1989): Isolation of pure villous cytotrophoblast from term human placenta using immunomagnetic microspheres. J. Immunol. Methods 119, 259-268.

Ellis, S.A., Sargent, I.L., Redman, C.W.G., and McMichael, A.J. (1986): Evidence for a novel HLA antigen found on human extravillous trophoblast and a choriocarcinoma cell line. Immunology 59, 595-601.

Grabowska, A., Carter, N., and Loke, Y.W. Human trophoblast cells in culture express an unusual major histocompatibility complex class-I like antigen. Submitted to Am. J. Reprod. Immunol.

Hertz, R. (1978): Spectrum of gestational trophoblast diseases. In Choriocarcinoma and Related Gestational Trophoblastic Tumours in Women, ed. R. Hertz, pp.23-43. Raven Press, New York.

Ho, C-K., Chiang, H., Li, S-Y., Yuan, C-C., and Ng, H-T. (1987): Establishment and characterization of a tumorigenic trophoblast-like cell line from a human placenta. Cancer Res. 12, 3220-3224.

Hunt, J.S., Deb, S., Faria, T.N., Wheaton, D., and Soares, M.J. (1989): Isolation of phenotypically distinct trophoblast cell lines from normal rat chorioallantoic placentas. Placenta 10, 161-177.

Kawata, M., Parnes, J.R., and Herzenberg, L.A. (1984): Transcriptional control of HLA-A,B,C antigen in human placental cytotrophoblast isolated using trophoblast- and HLA-specific monoclonal antibodies and the fluorescence-activated cell sorter. J. Exp. Med. 160, 633-651.

Kliman, H.J., Feinman, M.A., and and Strauss, J.F. (1987): Differentiation of human cytotrophoblast into syncytiotrophoblast in culture. Trophoblast Res. 2, 407-421.

Log, T., Chang, K.S.S., and Hsu, Y.C. (1981): Carcinomas induced by cell lines cultivated from normal mouse placentas. Int. J. Cancer 27, 365-372.

Loke, Y.W. (1983): Human trophoblast in culture. In Biology of Trophoblast, eds Y.W. Loke and A. Whyte, pp.663-701, Elsevier/North Holland.

Loke, Y.W. (1988a) Immunocytochemical characterisation of human trophoblast. In Placental Protein Hormones, eds M. Mochizuki and R. Hussa, pp.19-31. Elsevier Science Publishers.

Loke, Y.W. (1988b): The human placenta: allograft or cancer? Cambridge Medicine, Issue for Autumn, 19-21.

Loke, Y.W., and Burland, K. (1988): Human trophoblast cells cultured in modified medium and supported by extracellular matrix. Placenta 9, 73-182.

Loke, Y.W., and Hussa, R.O. (1987): Trophoblast cell culture. A workshop report. Trophoblast Res. 1, 461-464.

Loke, Y.W., Gardner, L., Burland, K., and King, A. (1989a): Laminin in human trophoblast-decidua interaction. Human Reprod. 4, 457-463.

Loke, Y.W., Gardner, L., and Grabowska, A. (1989b): Isolation of human extravillous trophoblast cells by attachment to laminin-coated magnetic beads. Placenta 10, 407-415.

Phillips, C.N., McCue, P.A., Whitsett, C.F., and Priest, J.H. (1989): Differentiation in human chorionic villus cultures: hCG and HLA expression. Prenatal Diagnosis 9, 227-242.

Ungar, L., Csanka, E., Kazy, Z., Siklos, P., and Hercz, P. (1987): The use of pregnancy serum to obtain trophoblastic cell cultures. <u>Placenta</u> 8, 639-646.

Wewer, U.M., Taraboletti, G., Sobel, M.E., Albrechtsen, R., and Liotta, L.A. (1987): Role of laminin receptor in tumor cell migration. <u>Cancer Res.</u> 47, 5691-5698.

Yagel, S., Casper, R.F., Powell, W., Parhar, R.S., and Lala, P.K. (1989): Characterization of pure human first-trimester cytotrophoblast cells in long-term culture: growth pattern, markers and hormone production. <u>Am. J. Obstet. Gynecol.</u> 160, 938-945.

Yeger, H., Lines, L.D., Wong, P-Y., and Silver, M.M. (1989): Enzymatic isolation of human trophoblast and culture on various substrates: comparison of first trimester with term trophoblast. <u>Placenta</u> 10, 137-151.

Implantation

S. Mani, J. Mulholland, J.A. Julian, S.R. Glasser

Department of Cell Biology, Baylor College of Medicine, Houston, Texas, USA

Each year the literature devoted to implantation increases. In three quarters of this century we have added to this library in every language and dialect that scientists have mastered. Most recently we have attempted to use the ciphers of cell and molecular biology to analyze the mechanisms which regulate the transition of a non-receptive uterus to a uterus which is competent, albeit transiently, to interact with its blastocyst.

It is an impressive library. Regrettably, it is a library devoid of information which can be used predictably to control the implantation process. Our very modest success with embryo transfer in domestic animals and particularly in humans (1) underscores the limited utility of our knowledge. In large part our failure to define, in any species, the mechanisms which control the events which characterize the peri-implantation uterus can be attributed to the lack of an experimental cell model that can produce the types of data required. Possibly we have been so concerned with obvious differences between species (temporal, hormonal regulation, depth of attachment, etc.) and even intra-species differences between the various components of the attachment process (uterine epithelial (UE) and stromal (US) cells, trophectoderm (TE)) that we have failed to recognize important common elements. Attention to such commonalities could prove more instructive than comparative differences in providing new insights.

It makes more sense to regard attachment, per se, rather than the mode of attachment, as part of a developmentally coordinated system. Attachment in this sense is not an independent process. Rather, attachment initiates the integrative series of events between UE, US and TE that establishes a definitive placenta. For this reason we ask what is important about attachment itself rather than what is different between interstitial versus superficial implantation. Is it more important that one type of attachment occurs superficially in one species and interstitially in another? Or is it more important that UE cells, in whatever species, become receptive at some species specific time? As a consequence of this receptivity events also occur at the UE basal surface. Attachment assures a critical sequence of events follows, each occurring at a specific place, a specific time that establishes and maintains a pregnancy. We suggest that is is more important that there is a change in the secretory phenotype of UE and US cells. Changes in biochemical profiles are relevant because they are indices of such phenotypic changes.

It is difficult to address these questions. Existing models are not adequate. They are not readily available to experimental or analytical manipulation. They do not recognize the stringent conditions required by secretory epithelia in order to express their unique properties. This is particularly evident in assessing changes occurring in the sub-epithelial stromal compartment which are initiated by attachment. Similarly we are not able to analyze the nature of signals arising from the US cells of the post-neonatal animal that could mediate the endocrine and paracrine responses of the UE cell.

Our efforts to resolve such issues has led us to develop a cell culture system which establishes the polarity of UE cells (2,3). Polarity of epithelial cells is necessary for such cells to express their special functions (secretory, transport, absorption). We proposed that if we could polarize UE cells they would retain their ability to respond to hormonal signals. In response to these hormonal signals UE cells would synthesize stage specific directories of proteins and glycoproteins and sort them to the appropriate (apical, basal) plasma membrane domain. This would be consonant with the idea that the transition from a non-receptive to a receptive uterus is a reflection of the profile of hormonally directed proteins and glycoproteins disposed on the apical surface and in its secretory compartment.

This proves to be the case. UE cells isolated from immature rats and cultured on matrix-impregnated semi-permeable filters proliferate to confluence and establish morphological and functional polarity. Among the indices of functional polarity is the maintenance of hormone responsiveness and the preferential apical secretion of marker proteins and glycoproteins stimulated in vivo by estradiol (2,3,4,5). For the first time it has been possible to regulate an in vitro implantation system. Polarized UE cells, in response to estradiol, are non-receptive to blastocysts (Table 1). Non-attached blastocysts are functionally viable because they will attach and grow out, in the presence of estrogen, if they are transferred to non-cellular matrices or to US cells. These experiments validate the biological relevance of this experimental cell system.

TABLE 1
IN VITRO ATTACHMENT OF RAT BLASTOCYSTS TO VARIOUS SUBSTRATES

Substratum	% Attachment at			
	48h	72h	90h	120h
(A) Bare plastic wells	90	94	94	94
Bare CM filters	0	0	0	0
CM + EHS [a]	92	94	96	98
CM + EHS + UE [b]	0	0	0	0
(B) Transfer from bare CM filters to				
CM + EHS (48h)	–	83	–	96
CM + EHS + UE (48h)	–	0	0	0
(C) Transfer from CM + EHS + UE to				
CM + EHS (48h)	–	82	88	90
CM + EHS + US[c] (48h)	–	95	95	98

[a] Engelbreth-Holm-Swarm tumor matrix
[b] Primary cultures of immature uterine epithelial cells
[c] Primary cultures of immature uterine stromal cells

Our studies with polarized UE cells have been extended by our ability to culture these cells in serum-free, phenol red-free defined medium (0.2% serum extender, Collaborative Research, Bedford, MA). These studies have yielded rather provocative results:

(1) The apical and basal UE cell surfaces are differentially responsive to E_2 in a dose dependent manner (Table 2).

TABLE 2

RESPONSE OF POLARIZED UTERINE EPITHELIAL CELLS TO ESTRADIOL-17β*

		0.2% + ESTRADIOL-17β				
	0	1×10^{-13} M	1×10^{-12} M	1×10^{-11} M	1×10^{-10} M	1×10^{-9} M
TOTAL SECRETION**	4.8	3.7	3.8	3.9	7.2	10.0
APICAL SECRETION**	4.3	3.4	3.5	3.8	7.0	9.9
BASAL SECRETION**	.45	.32	.32	.09	.19	.12
APICAL/BASAL	9.6	10.6	10.9	42.2	36.8	82.5

* Immature rat uterine epithelial cells cultured (7d) on EHS matrix impregnated semipermeable filters in serum-free, phenol red-free defined media (0.2% S-SX)
** Expressed as a % of total incorporation of ^{35}S-methionine, 8h incubation
Underlined figures designate dose dependent responses to E_2

(2) Polarized UE cells are directly responsive to E and P. This response is not dependent in an obligatory manner on the presence of mesenchymally derived cells. It is provocative that P does not down-regulate E *in vitro* as it does *in vivo* (Table 3).

IMMATURE RAT UTERINE EPITHELIAL CELLS CULTURED IN SERUM-FREE, PHENOL-RED-FREE, STRIPPED SERUM EXTENDER (0.2% S-SX), 0-7d.

	SECRETION (% TOTAL INCORPORATION*)				
	TOTAL	APICAL	BASAL	A/B	Cells/mm^2
0.2% S-SX	1.8	1.6	.21	7	1075
(1.7)**	(1.5)	(.20)	(7.5)		
0.2% S-SX*** $+1 \times 10^{-9}$ ME_2	3.9 (2.9)	3.8 (2.8)	.05 (.04)	74 (70)	1352
0.2% S-SX $+1 \times 10^{-9}$ ME_2 $+1 \times 10^{-6}$ MP	5.5 (4.5)	5.4 (4.4)	.05 (.04)	108 (110)	1231
S-SX+E_2 vs. S-SX	+72%	+89%	-80%	+842%	
S-SX+P vs. S-SX	+165%	+195%	-80%	+1364%	
S-SX+E_2 vs. S-SX+P	+55%	+56%	-	+56%	

* ^{35}S-methionine (100 μCi, in methionine-free, phenol-red free medium; 8h)
** Normalized per 1000 cells/mm^2

*** Steroids added d.1 - d.7

(3) The polarized epithelial cell system can be applied to the study of rabbit UE cells. Rabbit UE cells proliferate to confluence and develop accepted indices of morphological and functional polarity including hormonal responsiveness (Table 4). It is now feasible to compare superficial versus interstitial attachment and their sequelae in order to test our hypothesis regarding the common nature of certain processes.

TABLE 4
POLARIZED IMMATURE RABBIT UTERINE EPITHELIAL CELLS CULTURED IN SERUM-FREE, PHENOL RED-FREE STRIPPED SERUM EXTENDER (0.2% S-SX), 0-7d*

	SECRETION % TOTAL INCORPORATION**			
	TOTAL	APICAL	BASAL	A/B
0.2% S-SX	1.8	1.68	.11	15.3
0.2% S-SX +1×10^{-9} ME_2***	1.7	1.59	.11	14.5
0.2% S-SX +1×10^{-7} MP	3.9	3.48	.41	8.5
S-SX + E_2 VS. S-SX	-5.5%	-5.4%	-0-	-5.5%
S-SX + P VS. S-SX	+116.7%	+107.2%	+2727%	-44.4%
S-SX + P VS. S-SX + E	+129.4%	+118.8%	+2727%	-41.3%

* Confluent cultures
** ^{35}S-methionine (100 µCi in methionine-free, phenol-red free, defined medium; 8h)
*** Steroids added d.1-d.7

It is of considerable interest that relative to the process associated with superficial implantation the basal surface of the rabbit UE cell is particularly sensitive to the program of steroid hormone treatment used to mimic implantation.

This research was supported by NIH grants HD 25189, HD-13663 and HD 07495.

REFERENCES

1. Lindenberg, S., Hyttel, P., Sojgren, A. and Greve, T. (1989) A comparative study of attachment of human, bovine and mouse blastocysts to uterine epithelial monolayer. Hum. Reprod. 4:446.

2. Glasser, S.R., Julian, J.A., Decker, G.L., Tang, J.-Y. and Carson, D.D. (1988) Development of morphological and functional polarity in primary cultures of rat uterine epithelial cells. J. Cell Biol.107:2409.

3. Carson, D.D., Tang, J.-Y., Julian, J.A. and Glasser, S.R. (1988) Vectorial secretion of proteoglycans by polarized uterine epithelial cells. J. Cell Biol. 107:2425.

4. Wheeler, C., Komm, B.S. and Lyttle, C.R. (1987) Estrogen regulation of protein synthesis in the immature rat uterus. The effect of progesterone during *in vitro* incubation. Endocrinology 120:910.

5. Takeda, A., Takahashi, N. and Shimazu, S. (1988) Identification and characterization of an estrogen inducable glycoprotein (VSP-1) synthesized and secreted by rat uterine epithelial cells. Endocrinology 122:105.

Trophoblast-endometrium interaction : summary of a workshop

J. Hustin, J.G. Grudzinskas*

Institut de Morphologie Pathologique, Allée des Templiers 41, 6280 Gerpinnes (Loverval), Belgique
*Department of Obstretics/Gynaecology, 4th Floor, Holland Wing, London Hospital whitechapel, London E1 1BB, Royaume-Uni

Implantation of a fertilized egg in a suitably prepared uterine mucosa is a pre-requisite for the establishment of normal pregnancy and thus for the survival of mankind. The moment during which the blastocyst can implant is very limited and depends on several factors which are the hormonal maternal milieu and the number and intensity of signals produced by the 'travelling' blastocyst. The nidation phenomenon is unique in that cells of different origins come into close contact and eventually intermingle at the so-called site of implantation. In recent years there has been a growing interest on this unique situation. Any study dealing with this question should address one or more of the following:
- Which are the cells present in the maternal-trophoblastic interface?
- Do they have a special behaviour? What are their biological (biochemical, endocrinological) properties?
- What is the control of trophoblastic penetration within the decidua?
- Do trophoblastic cells influence maternal cells and vice versa?
- Can we obtain some explanation for the immunologic tolerance of the embryo as a graft?

The Workshop No.I of the III European Placenta Group Meeting in Dourdan addressed these matters in stimulating and comprehensive discussions.

J HUSTIN (Loverval - Belgium) discussed the very early stages of pregnancy and introduced the subject of endometrial preparation for implantation of the blastocyst.

During the secretory period of the menstrual cycle, three phases can be delineated which are biochemically different from each other: the early secretory phase, the implantation phase and the late secretory phase. It was emphasized that the surface epithelium of the endometrium plays a major role. During the implantation phase, the surface epithelium expressed important beta-L-Fucose binding while the expression of sialic acid and 3-Fucosyl-N-acetyl-lactosamine is strongly reduced. During this period, gland secretions begin (mainly PP5 and PP12, and gamma-glutamyltranspeptidase) while protein production by stromal cells is still largely non-existent. During the secretion phase, there is a tremendous increase of gland secretions (PP14, prostaglandins) while protein production by stroma (i.e. decidual cells) becomes established (PP12, Prolactin ...).

Lastly it was stressed that the normal lymphoid population of the endometrium increases tremendously with an increased T4/T8 ratio and an inhibition by stroma granulocytes of suppressor-cytotoxic T-cells.

T WHITE (Rochester, NY - USA) discussed the production of endometrial proteins namely PP14 by endometrial cells on an extracellular matrix (ECM). The cells were grown for 16 days, they remained rounded, migrated to form mounds after 1-2 days and continued to grow in this formation. They divided (as judged by tritiated thymidine incorporation) throughout the culture period. Histologically, the mounds contained either simple or branching glandular structures, which secreted mucin, and a stromal component and in many cases, a simple columnar epithelial covering. To test the secretory ability of endometrial cells on ECM, the culture media were measured for PP14 by RIA, using ^{125}I-labelled PP14 and a monospecific antibody (rabbit) to human PP14. All ECM cultures secreted detectable levels of PP14, with levels as high as 21.86 μg/48 hr (e.g. in 4 different endometria, the average PP14 level in ECM cultures by 1-3 days (d) was 2.23 μg/1/48 hr (range = 0 - 16.9), by 8-9d: 10.93 μg (8.3 - 18.2) by 14-16 d: 6.4 μg (0 - 15.40), while corresponding monolayer cultures showed either non- or barely-detectable levels of PP14. Thus, endometrial cells grown on ECM divide, form glands, epithelium and stroma, and secrete PP14 (also demonstrated by immunohistochemistry in their epithelial compartment), suggesting that this ECM culture is appropriate for studying endometrial structure and function in vitro. The maximum production was obtained in late secretory endometria. Further increase was observed by coculture with trophoblast.

P STARKEY (Oxford - UK) described studies on the bone marrow derived endometrial stroma cells which account for 28% of large granular lymphocytes, 8% T-cells and 19% macrophages. At term the bone marrow derived cells present 53% of all stromal cells with only 4% of larger granular lymphocytes, 8% T-cells, 19% macrophages. Other LCA factors, including cytokines, exist which may play a role in the trophoblast maternal cell-interaction. The tumour necrosis factor (TNF) is produced in decidual and villous tissue, TNF, mRNA apparently being confined to the macrophage population. Proposed functions are inhibition of trophoblast growth and modulation of maternal immune cells. Platelet Derived Growth Factor

(PDGF) is produced by macrophages and cytotrophoblasts and stimulates trophoblastic growth. Granulocyte Colony Forming Stimulating Factor is found only in association with macrophages and stimulates trophoblastic growth.

The subject of trophoblast invasion within the decidua was addressed by __R GRUMMER__ (Essen - FRG) with the use of multicellular aggregates of human choriocarcinoma cell lines. These spheroids (SPHs) maintain several morphological and functional characteristics of the normal trophoblast.

Invasiveness of three choriocarcinoma cell lines grown as SPHs (BeWo, Jeg-3, JAr) was tested in the Mareel-assay. Tumour cell SPHs and chicken heart fragments (PHF) were precultured separately for 3 days. They were then placed on semi-solid agar in petri dishes for 24 h and subsequently cocultured on a gyratory shaker at 37°C, 5% CO_2 in air for up to 14 days. All three cell lines were shown to invade the heart fragments. After 3 days of coculture, evidence of infiltration could already be observed. Whereas invasion of BeWo and Jeg-3 cells into PHFs progressed relatively slowly, invasion proceeded faster with JAr cells. With the latter, PHFs were greatly replaced by cancer cells after 14 days of coculture, while major remnants of PHFs were still detectable in case of the other two cells lines. Human endometrium of the early secretory phase was also used as a host tissue in a pilot series of experiments. The endometrium was precultured with E^2 and progesterone for 3 days in roller jars. In confrontation culture evidence of adhesion and invasion of choriocarcinoma cells was observed.

__P BISCHOF__ and __A CAMPANA__ (Geneva - Switz) discussed the origin and secretion of CA 125 in the human endometrium, CA 125 , originally described as an ovarian surface epithelial tumour associated antigen, has been localized in epithelia of Mullerian origin. The antigenic determinant is found on a high molecular weight ($\pm$ 700,000) glycoprotein complexes present on the cell surface, in serum and in serous and amniotic fluids. CA 125 is currently detected by a monoclonal antibody (OC 125) of the IgG class. The concentration of CA 125 in serum is widely used as a marker of epithelial ovarian cancer. The presence of CA 125 is not specific for ovarian carcinoma in that increased circulating levels of CA 125 have also been found in women with endometriosis, ovarian hyperstimulation, pelvic inflammatory disease or pregnancy.

The source of serum CA 125 in healthy pregnant or cycling women has not been definitely established. While it has recently been proposed that the level of CA 125 in serum reflects the growth of the dominant follicle and is dependent on cyclic changes in the female genital tract, the exact site of synthesis and secretion has not yet been described.

In an attempt to resolve this problem a study was undertaken which included immunohistochemistry, cell and tissue cultures and measurements of CA 125 in different biological fluids. The results suggest that an ovarian origin of CA 125 could be excluded since:

a. peripheral CA 125 levels were not increased by treatment which induces ovulation;
b. serum CA 125 levels did not change during the cycle;
b. peripheral and ovarian vein levels of CA 125 were similar.

An endometrial origin of CA 125 is, however, possible since:

a. menstrual blood levels of CA 125 were significantly higher than peripheral levels;
b. washings of the uterine cavity contain high concentrations of CA 125 which vary with the stages of the cycle;

c. by immunochemistry, CA 125 is localized in the glandular epithelium of the endometrium and decidua, being maximum in the early secretory phase.

When cultured in vitro, decidual explants or endometrial cells, but not trophoblast explants, release CA 125. Endometrial epithelial cells from the proliferative phase, when cultured in presence of oestradiol and progesterone, produced significantly more CA 125 than the same cells cultured in absence of hormones. Progesterone, however, when added to cells obtained from a secretory endometrium, exerts an inhibitory effect. It was concluded that CA 125 is an exocrine product of endometrial glands and it is hypothesized that CA 125 appears in the peripheral circulation only when the basement membrane of the uterine epithelium is broken, i.e. during menses, at implantation and with endometrial carcinoma. Furthermore, the endometrium is not the only source of CA 125 as shown by the unchanged levels of CA 125 after hysterectomy. The most probably source of CA 125 in the non-pregnant state is the peritoneum.

J G GRUDZINSKAS (London - UK). In the systemic study of endocrine and metabolic events occurring in relation to ovulation, conception and implantation and the first trimester of pregnancy, the endometrium before and after fertilisation may be an important active participant rather than just a target organ responding to gonadal signals. Studies on the two major secretory proteins of the endometrium, insulin-like growth factor binding protein (IGF-bp), also known as PP12) and progesterone-dependent endometrial protein (PEP, also known as PP14, alpha-2-PEG, beta-lactoglobulin endometrial homologue) may lead to the development of the first non-invasive indices of endometrial and decidual cell function.

IGF-bp (or PP12) is derived from a population of stromal cells, the trigger for synthesis apparently being related to implantation. The trends in levels in peripheral blood are highest in the second trimester of pregnancy and parallel the growth of decidual tissue,. High levels of IGF-bp are also seen in pre-ovulatory ovarian follicular fluid, amniotic fluid and fetal blood. The biological activity of IGF-bp suggests an important role in paracrine events in relation to ovulation, implantation and fetal growth.

Studies on PP14 (or PEP) have now established that the principal site of origin is the glandular epithelium of the endometrium. The highest concentrations of PP14 are seen in the first trimester of pregnancy, but it is also found in amniotic fluid and uterine washings. The precise controlling mechanisms are still unclear, but progesterone is likely to have an important trophic influence. In the non-pregnant state, circulating PP14 levels are highest in the perimenstrual phase of an ovulatory cycle. Since circulating PP14 is absent or present in low levels in anovulatory cycles, it is likely that high levels during the menstrual phase are indicative of ovulation. In complications of early pregnancy, PP14 synthesis is apparently normal in anembryonic pregnancy up to 8 weeks gestation, but very low serum levels are seen in ectopic gestation.

It was concluded that IGF-bp and PP14 may be useful markers of glandular and stromal cell populations for the study of endometrium and decidua in vitro and in early human pregnancy,.

T CHARD (London - UK) discussed studies on IGF-bp (or PP12) with reference to the rapid expansion of information on the possible paracrine role of this substance. He described depressed serum levels of PP12 in association with intrauterine growth retardation, and other observations in diabetes mellitus, polycystic ovarian disease, and puberty. The importance of knowledge of growth factors and their metabolism was emphasised.

However, the observations that serum IGF-bp had a diurnal variation must indicate that earlier studies on IGF-bp (PP12) need to be reviewed to determine what effect, if any, the diurnal rhythm may have influenced the conclusions.

O GENBACEV (Zemun - Yu) discussed the interaction between trophoblast and decidua in tissue culture conditions. Culture of human villous trophoblast up to seven days in a three dimensional collagen gel was achieved with satisfactory viability and obvious hCG (beta-subunit) production. Similar conditions were applied to decidua, coculture of decidua with trophoblast tissue explants resulted in the significant increase of prolactin production. Decidual prolactin was considered to be under trophoblastic control probably through hCG regulation,

CONCLUSIONS

The Workshop addressed a variety of questions of considerable importance in the understanding of events occurring at implantation, in early pregnancy and tumour behaviour. The critical and lively discussion following each presentation emphasised the benefits of considering these matters in a forum of cell biologists, physiologists, biochemists, endocrinologists and clinical scientists. Whereas the answers to the particular questions were not necessarily conclusive, it was generally agreed to pursue the answers to these questions in a multi-disciplinary strategy.

Does decidua influence 1st trimester trophoblast hCG secretion ?

D.R. Abramovich, K.R. Page*, C.K. Pearson**

*Departments of Obstetrics and Gynaecology, * Physiology and ** Biochemistry, University of Aberdeen, UK*

The controlling factors of human chorionic gonadotrophin (hCG) secretion by trophoblast cells are unknown though second messengers (e.g. cyclic AMP) are known to be implicated. A knowledge of the intracellular biochemistry may help in the understanding of the implantation of the fertilized ovum and of causes of early and recurrent abortion.

The specific object of this work is to see whether hCG production by early trophoblast is influenced by the decidua.

In our early experiments individual trophoblast cells were prepared from 9-11 week placentae. We obtain between 400-850 X 10^5 cells/g wet weight of placenta and incubate 150 000 cells/ml in each well. These cells were incubated in triplicate for 72 h with 50 per cent decidual conditioned medium (d.c.m.) from first trimester pregnancies.

Preliminary results show that the greater the concentration of d.c.m. used (25 per cent versus 50 per cent) the greater the suppression of hCG production and the 11 week gestational age d.c.m. suppressed hCG production more than 8 week d.c.m.

Decidua thus appears to influence hCG trophoblast production but whether by its prolactin or alpha-1 or -2 PEG content is unknown.

Human trophoblast interaction with the extracellular matrix : a histologic assay system for studying trophoblast invasion

R.F. Feinberg, J.E. Haimowitz*, H.J. Kliman*

Departments of Obstetrics and Gynaecology and* Pathology and Laboratory Medicine, University of Pennsylvania School of Medicine, 6 Founders Pavilion, Philadelphia, PA 19104, USA

Trophoblast invasion of the uterus is critical for human implantation. The cellular basis for this process is likely to depend on specific interactions between trophoblasts and extracellular matrix (ECM) components. In order to dissect the complex cellular properties which are required for trophoblast invasion of the ECM, we have developed an *in vitro* assay system for studying the histologic features of trophoblast invasion. Human cytotrophoblasts were isolated from term placentae (Kliman et al., **Endocrinology 118**:1567, 1986), and either: 1) maintained for 24 h in suspension culture to allow for aggregation and placed on a Millicell™ filter layered with a 0.9 mm thick bed of undiluted Matrigel™ (a mixture of ECM proteins including type IV collagen and laminin); or 2) co-incubated with Matrigel™ fragments in suspension culture. Histologic sections at 18-24 h demonstrated trophoblast attachment and adherence to the surface of the Matrigel™. Sections examined after 48 h of incubation revealed that aggregates of trophoblasts appeared to initiate erosion of the Matrigel™ surface. A thin zone of clearing was also noted around many cells, possibly indicating trophoblast lysis of Matrigel™ components. By 120 h, invasion into the Matrigel™ by trophoblasts was seen on multiple histologic sections, with a depth of invasion ranging from 10 µm to 50 µm. Immunohistochemical analysis of the trophoblasts within these Matrigel™ sections revealed positive staining for alpha-hCG, suggesting that the cells remained viable during the invasive process. Interestingly, very little immunostaining for plasminogen activator inhibitor type 1 (PAI-1) within actively invading trophoblasts was found. In contrast, trophoblasts which have invaded the placental bed *in vivo* display prominent PAI-1 immunoreactivity (Feinberg et al., **Lab. Invest 61**:20-26, 1989). This implies that actively invading trophoblasts may suppress PAI-1 synthesis, whereas cells that are no longer invading produce increased amounts of this protease inhibitor. Our results demonstrate that: 1) trophoblast-ECM interactions can be studied histologically by culturing human cytotrophoblasts in the presence of Matrigel™; 2) human cytotrophoblast aggregates attach to and invade Matrigel™ *in vitro*, a process that may recapitulate some features of blastocyst implantation *in vivo*; and 3) this *in vitro* system should permit direct examination of specific cell adhesion molecules, proteases, and protease inhibitors involved in trophoblast invasion. Supported by the March of Dimes (RFF) and USPHS grant HD00715 (HJK).

Long-term tissue culture of human first trimester villous trophoblast in collagen gel : evaluation of its possible use as a model system

O. Genbačev, B. Čemerikić, V. Šulovic, L. Vićovac

Institute of Endocrinology, Immunology and Nutrition, INEP, Zemun, Belgrade, Yugoslavia

The selection of tissue culture technique depends on what one intends to study. To establish maximal duration of experiment and to understand what kind of a model we are offering, it is necessary to operate strict criteria for checking tissue viability.

We have cultured first trimester human villous trophoblast up to seven days in three-dimensional collagen gel. As vitality control glucose consumption, lactate and LDH production, ^{14}C-leucine incorporation and morphological analysis were followed in a 24-hour interval. HCG and HPL production rates, and progesterone concentration in the medium were studied as specific markers of the trophoblast functional integrity.

Effects of the defined media, the media supplemented with fetal cord and first trimester pregnancy serum, and decidual-conditioned media on the monitored parameters were compared and discussed from the aspect of the use of trophoblast tissue culture in the study of: (A) trophoblast tissue ability for protein synthesis and its regulation; (B) drug testing; (C) trophoblast insufficiency and its correction.

Investigations on the invasion of choriocarcinoma cells grown as spheroids

R. Grümmer, H.-W. Denker

Institut für Anatomie, Universitätsklinikum, D-4300 Essen, Federal Republic of Germany

For studies of invasion in vitro, multicellular threedimensional systems have been found to give more reliable data than monolayer systems (MAREEL 1980).
Therefore we are using multicellular aggregates of human choriocarcinoma cell lines. These spheroids (SPHs) maintain several morphological and functional characteristics of the normal trophoblast (GRÜMMER et al., Trophoblast Res. 4, in press).
Invasiveness of three choriocarcinoma cell lines grown as SPHs (BeWo, Jeg-3, JAr) was tested in the Mareel-assay (MAREEL et al. 1979). Tumor cell SPHs and chicken heart fragments (PHF) were precultured separately for 3 d. They were then confronted on semi solid agar in petri dishes for 24 h and subsequently cocultured on a gyratory shaker at 37°C, 5% CO_2 in air for up to 14 d. All three cell lines were proven to be invasive in this assay. After 3 d of coculture beginning of invasion could already be observed. Whereas invasion of BeWo and Jeg-3 cells into PHFs progressed relatively slowly, invasion proceeded faster with JAr cells. With the latter, PHFs were greatly replaced by cancer cells after 14 d of coculture, while major remnants of PHFs were still detectable in case of the other two cell lines.
Human endometrium of the early secretory phase was also used as a host tissue in a pilot series of experiments. The endometrium was precultured with E_2 and progesterone for 3 d. In confrontation culture adhesion and invasion of choriocarcinoma cells was observed. Existing morphology leaves it open so far whether invasion proceeded always through an intact regenerated epithelium or through minor defects in the epithelium.
This three-dimensional organ culture system may be an interesting model for the study of trophoblast invasion.

(Supported by the Minister f. Wissenschaft und Forschung NRW; Proj. Nr. 500 019 88).

Biochemical indices of stromal cell differentiation are different following invasive vs superficial implantation

S. Mani, S. Lampelo*, J.A. Julian, S.R. Glasser

Department of Cell Biology, Baylor College of Medicine, Houston, TX 77030, USA
*Department of Environmental Toxicology, National Public Health Institute, Kuopio, Finland

Rat and human uterine stromal (US) cells decidualizing *in vivo* are characterized by elevated or *de novo* expression of cell-specific marker proteins, i.e., desmin, an intermediate filament protein, and laminin, an extracellular matrix protein. These proteins are negligibly expressed in non-decidualized US cells. *In vitro*, decidualizing rat US cells also express these markers implying that the program of stromal cell differentiation is similar *in vivo* and *in vitro*. Induction and maintenance of decidualization *in vivo* are progesterone (P) dependent whereas *in vitro* decidualization is not dependent on hormonal status at the time of US cell isolation. Neither is P required by US cells decidualizing *in vitro*. The US cells from immature, castrate or cycling rats decidualize *in vitro* as validated by desmin and laminin expression. These data prompted the question: Are the biochemical indices which characterize invasive implantation (human, rat) common to different implantation strategies, i.e., superficial (rabbit)? *In vivo* post-attachment rabbit US cells express laminin but not desmin. Desmin also could not be detected in cultured rabbit cells. These data suggest that species specific *in vivo* programs of stromal cell differentiation are reflected *in vitro*. We propose (1) different implantation programs (invasive vs. superficial) are characterized by different biochemical profiles. (2) US cells *in vitro* express the same program of differentiation as they do *in vivo* but the *in vitro* environment itself is not a stimulus to decidual cell differentiation, and (3) *in vitro* stromal cell differentiation (rats, rabbits) is hormonally independent.

(Supported by U.S. National Institute of Health grants HD-13663, HD-22785 and HD-07495.)

Increased uterine adenosine deaminase activity is a function of decidualized stromal cells

J. Mulholland, L. Hong*, R.E. Kellems*, S.R. Glasser

Department of Cell Biology and Department of Biochemisry, Baylor College of Medicine, Houston, TX 77030, USA*

Adenosine deaminase (ADA) is a significant enzyme in the uterus because it irreversibly deaminates adenosine and 2'deoxyadenosine. These natural ADA substrates are active growth inhibitors which could interfere with decidualization and the maintenance of pregnancy. The enzyme was localized in maternal tissues at the maternal-fetal interface during formation of the hemochorial placenta which is stringently regulated by steroid hormones. To identify the specific uterine cells which express ADA, the enzyme was examined by histochemical, immunocytochemical, and biochemical methods in estrogen-primed, progesterone-treated ovariectomized rats. Decidualization (DCR) was induced in one uterine horn and ADA specific activity was analyzed at 24h intervals in control and decidualized endometrial tissue from DCR + 1 through DCR + 11 days. The decidual response was monitored by the weight of the decidualized horn which increased from days + 1-3 (80-120% per day vs. control) plateaued at DCR + 5-6 (600-800%), began to regress (DCR + 8-9), and returned to control levels (DCR + 11-12). An increase in the specific activity of ADA was not detected until DCR + 3. Activity rose markedly from 30-40 nmol/min/mg protein to 196 nmol/min/mg on DCR + 7-9 and regressed to 295 nmol/min/mg on DCR + 11-12.

These studies establish the decidual cell as the source of ADA during early placental development in the rat. The increase and regression in decidualization and ADA specific activity describe parallel curves separated by 24 hours. These data suggest that ADA activity is one expression of the differentiated phenotype of the decidual cell.

(Supported by U.S. National Institutes of Health grants HD-22785 and HD-07495.)

Interactions between trophoblast, choriocarcinoma, decidua and fibroblast cells *in vitro*

A. Nazzaro, J.W. Fawcett, Y.W. Loke

Departments of Pathology and Physiology, Cambridge University, UK

We have performed time-lapse video microscopy to examine the behaviour of choriocarcinoma (JAR and JEG) cells, and trophoblast cells derived from first trimester elective terminations. JEG and JAR cells show similar behaviour when cultured alone in tissue culture plastic: the cells have a highly motile border, which is constantly ruffling and putting out filopodia. However, the cells seldom move around the dish individually; rather they remain stationary, attached to a cell clump, which gradually increases in size as the cells divide. First trimester trophoblast cells, cultured according to the method of Loke et al. (1988), have a very similar appearance to JEG and JAR cells. However, trophoblast cells are rather more motile than the cell lines; they still tend to remain in clumps, but individual cells move around within the clump, leading to frequent changes of cell position.

We have examined preparations in which cultures of JEG and JAR cells have been apposed to cultures of a fibroblast cell line (Rat1). Under these conditions the choriocarcinoma cells quite rapidly invade the fibroblasts, often as single cells. It is possible that the relatively high motility we see in trophoblast cultures is in part due to the presence of approximately 5-10% of fibroblasts.

We have also made three dimensional cultures of trophoblast and decidual cells, by packing cells into porous cellulose ester tubes. When a trophoblast containing tube is apposed to a decidua containing tube, cytokeratin containing trophoblast cells rapidly invade the decidua.

Fibrin matrix modulates the phenotypic differentiation and proliferation of human placental trophoblast

D.M. Nelson, E.C. Crouch, E.C. Curran, D.R. Farmer

Department of Obstetrics and Gynaecology, and Pathology, Jewish Hospital of St Louis, Washington University School of Medicine, St. Louis, MO, USA

We used immunocytochemistry, morphometric analysis, and electron microscopy to study perivillous fibrin deposits on normal term human placental villi, and we examined the response of cultured cellular trophoblast to fibrin matrix in vitro.

Histologically, the perivillous deposits were hypocellular, eosinophilic masses immunoreactive for the B-β monomer of fibrin II. Of 3477 villi examined, 258 villous profiles had denudations of the syncytial trophoblast and 191 (74.4 per cent) of these, or 5.5 per cent of the total villi, had associated fibrin deposits. Ultrastructurally, damage to the syncytial trophoblast was apparent at the edge of some deposits, where syncytial denudation was accompanied by a fibrin coating of residual cellular trophoblast and trophoblastic basal lamina. Other deposits were surfaced by syncytial trophoblast with underlying cellular trophoblast grown on a fibrin matrix, but not on uncoated plastic, differentiated cytologically and histologically into a trophoblast layer like that on term villi. Trophoblast cultures grown 72 h on fibrin had medium levels of hCG (225 mIU/ml), oestradiol (1511 pg/ml) and progesterone (45 ng/ml) comparable to cultures growth without matrix. However, the labelling index of cells grown 24 h on fibrin (0.49 per cent $\pm$ 0.57 per cent) was 12-fold lower than cells grown on plastic without matrix (5.82 per cent $\pm$ 1.85 per cent). We suggest that re-epithelialization of perivillous fibrin deposits is a form of villous repair and that trophoblast-fibrin interactions modulate trophoblastic differentiation and proliferation. (Supported by NICHHD Grant 22913 to DMN).

The distribution of ICAM-1 within decidua and placenta and gestational age associated changes

C.M. Salafia, N. Haynes, V.J. Merluzzi, C. Robiou, R. Rothlein

Danbury Hospital, Danbury, Connecticut, USA and Boehringer-Ingelheim Pharmaceuticals, Inc., Ridgefield, Connecticut, USA

Intercellular adhesion molecule (ICAM-1) is a ligand of leukocyte functional antigen (LFA-1) present on many cells, including monocyte/macrophages. It is considered to play an important role in the induction and/or maintenance of inflammatory responses, permitting leukocyte adhesion. Its expression is inducible in epithelial cells exposed to antigenic stimuli (preceding expression of histocompatible locus antigen (HLA D-DR) and is maturation dependent in certain cell lines. The distribution of ICAM-1 in decidua and placenta was evaluated using monoclonal antibodies and peroxidase-antiperoxidase immunohistochemistry. In decidua of first and third trimesters, scattered ICAM-1 (+) cells were observed. In placentas of first and third trimesters, all types of trophoblast were ICAM-1 (-). Prior to 10 weeks gestation the villous stroma was uniformly ICAM-1 and HLA D-DR (-). Beginning in the chorionic plate at approximately 10 weeks, scattered ICAM-1 (+) stromal cells were observed, while stromal cells of the terminal villi were ICAM-1 (-). By 14-16 weeks, approximately 40-50% of the villous stromal cells which share other immunohistochemical markers (such as EB-11) with monocyte/macrophages. The lack of functional maturation of the villous stromal macrophage may explain the rarity of chronic villitis early in gestation.

Long-term tissue culture of decidual explants of different gestational age in collagen matrix

Ljiljana Vićovac, Nadežda Papić*, Olga Genbačev

Institute of Endocrinology, Immunology and Nutrition, INEP,
*Zvezdara Clinical Hospital Centre, Department of Obstetrics and Gynaecology, Belgrade, Yugoslavia

The in vitro model for decidual explant long-term culture has been developed using three-dimensional gel of collagen with the aim to standardise a model for studying paracrine and autocrine regulation of decidual protein production, interaction of decidual and trophoblast tissues in vitro, and possible drug effects.

Decidual tissue from 1st trimester pregnancy (6-10 weeks), 2nd trimester and term was cultivated in collagen gel for at least 14 days. Standard conditions included Eagle's MEM supplemented with 5% FCS. The morphology, alkaline phosphatase, LDH, glucose consumption, ^{14}C-leucine incorporation and PRL production were monitored. The obtained results indicate that decidual tissue of all gestational ages can be maintained in collagen gel tissue culture for at least 14 days and preserve the morphological integrity of decidual cells.

Prolactin production rate followed a different pattern in 1st trimester decidua. Co-culture with trophoblast tissue explants resulted in the significant increase of prolactin production between the 7th and 14th day in all gestational ages.

Human endometrial cells grown on extracellular matrix (ECM) form an *in vivo* type morphology and produce the endometrial protein PP-14 (α2-PEG)

T.E.K. White, R.K. Miller, G. Grudzinskas*, A. Lower*, T. Fay*

Department of Obstetrics/Gynaecology and Toxicology Division, University of Rochester, Rochester, New York, USA
* *Academic Unit of Obstetrics and Gynaecology, The London Hospital (Whitechapel), London, UK*

Human, premenopausal, nonpregnant endometria were studied in culture by growing isolated cells in monolayers or on ECM (Matrigel) (*Biol. Reprod.* 38 (Suppl. 1):132, 1988). Cells were grown for 16 days, and the media was collected every other day. Monolayers, grown to and maintained at confluency, showed flat stromal and epithelial cells, but no formation of multiple cell layers or differentiated structures. Conversely, cells on ECM remained rounded, migrated to form mounds after 1-2 days, and continued to grow in this formation. ^{3}H-Thymidine uptake showed that the cells on ECM divided throughout the culture period. Histologically, most mounds contained either simple or branching glandular structures, which secreted mucin; a stromal component; and in many cases, a simple columnar epithelial covering. To test the secretory ability of endometrial cells on ECM, the culture media were measured for PP-14 by RIA, using ^{125}I-labeled PP-14, and a monospecific antibody (rabbit) to human PP-14. All ECM cultures secreted detectable levels of PP-14, with levels as high as 21.86 µg/L/48hr (e.g., in 4 different endometria, the average PP-14 level in ECM cultures by 1-3 days (d) was 8.23 µg/L/48hr (range=0-16.9), by 8-9 d: 10.93 (8.3-18.2), by 14-16 d: 6.40 (0-15.4)), while corresponding monolayer cultures showed either non- or barely-detectable levels of PP-14. Thus, endometrial cells grown on ECM divide, form glands, epithelia and stroma, and secrete PP-14, suggesting that this ECM culture is appropriate for studying endometrial structure and function in vitro. Future studies will investigate *de novo* synthesis of PP-14 and other endometrial proteins (e.g., PP-12) using radio-line immunoelectrophoresis. (Supported by NIH ES 02774, ES 07026, USA; CRC-SP1828, UK).

2.
Endocrinology and biochemistry
Endocrinologie et biochimie

Production of monoclonal antibody against maternal serum oxytocinase

S. Lampelo, K. Lalu

National Public Health Institute, P.O.B 95, SF-70101 Kuopio, Finland and Department of Anatomy, University of Kuopio, P.O.B. 6, 70211 Kuopio, Finland

INTRODUCTION

Serum aminopeptidase activity is known to increase during pregnancy. A high-molecular-weight cystine aminopeptidase appears in maternal circulation and its serum level rises throughout normal pregnancy reaching the maximum in the last trimester (Lampelo & Vanha-Perttula, 1980a; Vanha-Perttula et al., 1988). Cystine aminopeptidase and oxytocinase are regarded as the same enzyme which in maternal serum hydrolyses oxytocin (Ryden, 1966). Oxytocinase has been purified from human placenta (Lampelo et al., 1982) and from maternal serum (Lalu et al., 1985). A comprehensive analysis of cystine aminopeptidase in maternal serum and human placenta has suggested that the placenta is the source of oxytocinase in maternal serum (Lampelo & Vanha-Perttula, 1980b; Lalu et al., 1985).

Cystine aminopeptidase is probably the only serum aminopeptidase which is able to cleave Cys-Tyr bond in oxytocin. The most obvious assumption of its function has been that it serves to regulate the oxytocin level in maternal serum and, thus, may prevent a premature labour. Due to the wide spectrum of synthetic amino acid derivatives hydrolysed by cystine aminopeptidase (Lampelo et al., 1982; Lalu et al., 1985) the enzyme may also have other functions in maternal circulation and placental tissue. Besides oxytocinase both serum and the placental tissue contain several other aminopeptidases (Lampelo, 1982; Lalu et al., 1986) with marked overlapping substrate specificities. Serum oxytocinase activities have been found to correlate with some clinical and pathological findings in placental function (Spellacy et al., 1977). However, the lack of a specific method for the assay of oxytocinase has prevented its wider use in clinical diagnostics. The monoclonal antibodies would offer a useful tool in the specific determination of oxytocinase concentrations in maternal serum for monitoring the placental function. Such specific antibodies are also applicable for immunohistochemical localization of the enzyme in normal and pathological placental tissue to disclose its physiological role and importance in diagnostic pathology.

METHODS

Purification of maternal serum oxytocinase

The oxytocinase from pooled maternal serum was purified utilizing the methods described by Lalu et al. (1986). The purification steps are shown in Scheme 1.

Scheme 1. Purification of oxytocinase from maternal serum.

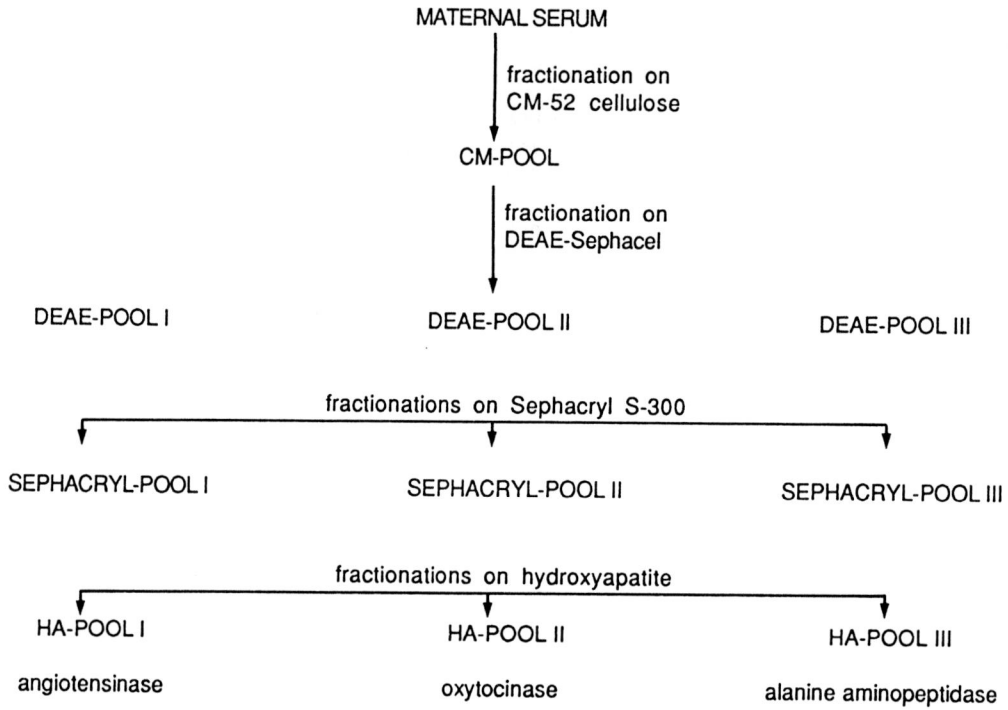

Production of monoclonal antibodies

A myeloma cell line X63-Ag8/653 was selected for cell fusion. The cells were treated with 8-azaguanine two weeks before the cell fusion. The cells were grown in RPMI-1640 medium supplemented with 10 % fetal calf serum, glutamine and penicillin/streptomycin (regular medium).

The purified oxytocinase was suspended in complete Freund's adjuvant (1:1) and drawn through a 21-gauge needle into a syringe. BALB/c mice were injected intraperitoneally with 0.5 ml of the oxytocinase suspension on three occasions at 3-4-week intervals. A blood sample was withdrawn periodically and tested for the presence of a immunoresponse by immuno dotblot assay as described below. The spleen of a mouse with a high immunoresponse was removed 3 days after the last booster and used for fusion in supplemented RPMI-1640 medium containing hypoxanthine, aminopterin and thymidine (HAT medium).

The fusate was distributed in portions of 0.1 ml to wells of 96-well plates containing a feeder layer of mouse macrophages prepared one day earlier. After one week the hybridomas were fed with the HT medium and subsequently with the regular medium.

Antibody screening by immuno dotblot

The 96-well dotblot apparatus of BioRad was used. A 0.1-ml sample of diluted maternal or control serum was pipetted on a wetted nitrocellulose membrane. After a 1-hr incubation the dots were blocked with 0.1 % bovine serum albumin (BSA) in TBS (0.02 M Tris-HCl, pH 7.4, containing 0.15 M NaCl). The hybridoma media were added for 1 hr. After washings with TBS with 0.05 % Tween included, a horseradish peroxidase-linked rabbit anti-mouse immunoglobulin (Dakopatts) was added as the secondary antibody. It was diluted 1:100 with the washing buffer. Peroxidase was indicated with diaminobenzidine as substrate.

Localization of oxytocinase in human placental tissue

Pieces of human term placenta embedded in O.C.T. were frozen in hexane cooled with liquid nitrogen. Sections of 4-6 µm were cut at -20 °C and transferred to glass slides, which had been pretreated with 2 % HCl in 70 % ethanol and stored at -20 °C. Sections were thawed briefly and fixed in methanol for 10 min. After fixation the morphology of the sections was monitored after staining with Mayer's hematoxylin-eosin.

The endogenous peroxidase was inhibited by adding 1.5 ml hydrogen peroxide into 50 ml of the fixative. After rehydration in TBS, the non-specific protein binding was blocked by incubation in TBS with 1 % BSA for 1 hr. Hybridoma medium, diluted mouse serum or regular medium was placed over the tissue sections. The incubation proceeded at 37 °C in a moist chamber for 30 min. The slides were rinsed well with TBS and the secondary antibody in 1:100 dilution was applied to each section for an incubation of 30 min. The slides were rinsed well and incubated in the peroxidase substrate solution. The sections were washed and mounted.

The Ig class specificity of the monoclonal antibody was tested by the Ouchterlony immunodiffusion technique.

RESULTS

Scheme 1 outlines the principles of the purification procedure used for the maternal serum oxytocinase. Three aminopeptidases were partially separated and numbered according to their elution from the DEAE-Sephacel column (Fig. 1). The dialyzed HA II pool was the final purified oxytocinase. SDS-PAGE (polyacrylamide gel electrophoresis) of the oxytocinase subunits purified from maternal serum and placenta revealed identical subunits. The calculation of the molecular weight of the subunit gave an approximate value of 145 000 for both enzymes (Fig. 2; Lalu et al., 1988).

The first visible cell colonies were seen within 7 days after fusion. After 12 days of culture the screening of the hybridome media from the wells was started. Each well, which recognized the maternal serum, was subcloned into 96 wells. In the next screening all hybridoma media from the wells with cell colonies were analyzed using the maternal or control sera as antigens. Those wells, which recognized the maternal serum but showed no or very low affinity to the nonpregnant serum, were further subcloned. After the third subcloning nearly all wells were positive (Fig. 3), while only a few wells contained nonproducers. The positive cell lines were expanded for storage in liquid nitrogen and for antibody production.

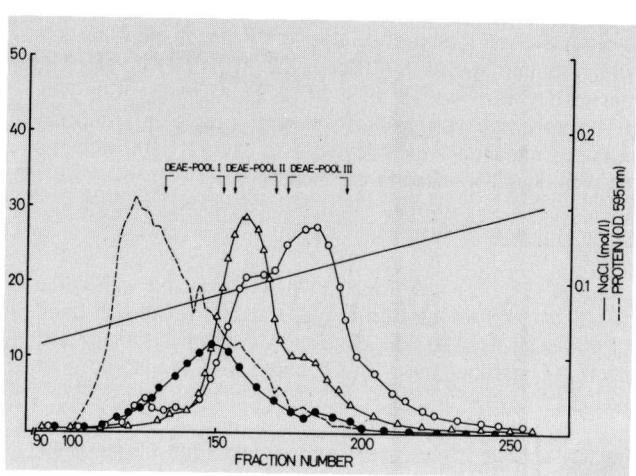

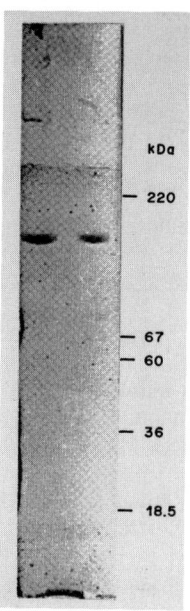

Fig. 1. Separation of the main aminopeptidase activities from the maternal serum CM-pool in DEAE-Sephacel chromatography for further purification.

Fig. 2. SDS-PAGE of the purified oxytocinase from maternal serum (left) and human placenta (right).

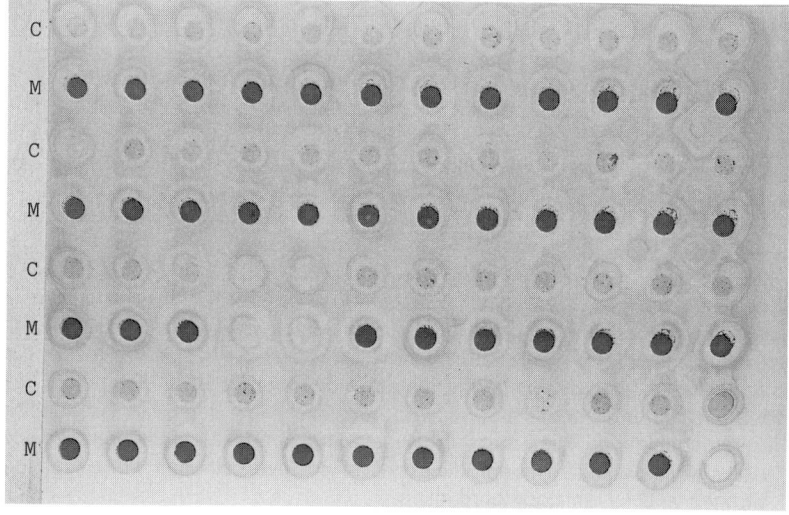

Fig. 3. Immuno dotblot assay of L4HB8 hybridoma media after the third subcloning. The non-pregnant (C) and maternal (M) sera were used as antigens in parallel for all media.

The staining pattern of a frozen placental section is shown in Fig. 4. An intense staining of the syncytium is evident. The fibrinoid was also occasionally stained, but its stainability was not consistent in all areas.

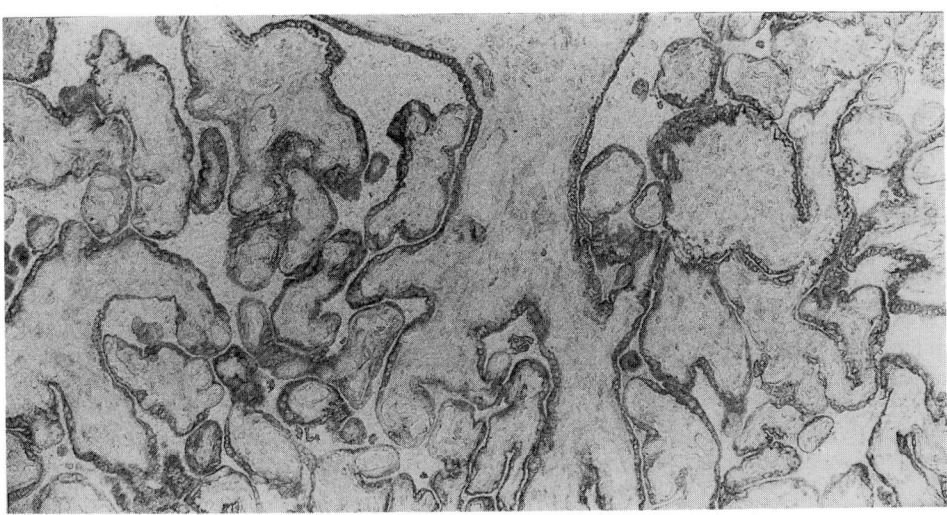

Fig. 4. Immunohistochemistry of frozen sections of human term placenta with anti-oxytocinase.

A single precipitation line was obtained against the anti-mouse IgM antiserum in the Ouchterlony immunodiffusion method.

DISCUSSION

The human placenta has been found to contain at least six distinct aminopeptidases, of which oxytocinase (cystine aminopeptidase; EC 3.4.11.3) has been suggested to be secreted by syncytial cells into the maternal circulation. An increase of serum aminopeptidase activity has been demonstrated during pregnancy with several synthetic substrates (Lampelo & Vanha-Perttula, 1980a). The relative increase in specific activity has been variable with different substrates, but the most apparent increase has been shown with cystine derivatives (Lampelo & Vanha-Perttula, 1980a; Vanha-Perttula et al., 1988). Three main aminopeptidases have been purified and characterized from the maternal serum and compared with two aminopeptidases purified from the control serum (Lalu et al., 1986). It has been confirmed that the biochemical and physical characteristics of the maternal serum oxytocinase, absent in nonpregnant serum, corresponded closely to those of the purified placental enzyme (Lalu, 1988). Some obvious structural differences were, however, obtained in studies with various lectins. In particular, the subunits of the purified oxytocinase preparations from maternal serum and placenta displayed differences in their affinity to certain lectins. These indicated a higher sialic acid content in the serum enzyme (Lalu et al., 1988). The higher sialic acid content of serum oxytocinase may protect the enzyme in the circulation.

By immunization of mice with the purified serum oxytocinase we have produced hybridomas secreting antibodies which recognized oxytocinase in the maternal serum. There was no cross-reactivity with proteins of the non-pregnant serum in the dotplot assay. Our preliminary studies have shown that the antibody does not work in the solid-phase Elisa-assay with the antigen bound to a plastic microtitre plate. Because the concentration of oxytocinase has been very low in the protein-rich serum, there might be a competition in binding to the plastic well. It is also possible that the antigen may leak off from the plate during subsequent incubation with the antibody. It also might be due to the Fc-fragment of the immunoglobulins, which reacts with rheumatoid factors in serum. Fragmentation of antibody molecules and purification of Fab-fragments may solve this problem.

In this study we cloned a cell line that produces an antibody with high affinity for an antigen present in syncytial cells of the human term placenta. Since an anti-oxytocinase raised against the maternal serum enzyme reacts with the syncytial cells, it confirms the suggestion that the oxytocinase in maternal serum is of placental origin.

At present we do not know, whether the histochemical reaction in the placental fibrinoid is due to a specific or nonspecific binding. The origin of the fibrinoid as well as its function have remained obscure. It is possible that oxytocinase is captured by the fibrinoid or it may even form this material at the end of pregnancy. This may result in a gradual decrease of serum oxytocinase at the end of pregnancy and indirectly initiate the labour.

The evidence that the antibody was monoclonal came from the observation that the hydridoma medium was produced by a cell line that was cloned three times by tranferring each time one cell into one well, in which a single colony started to grow. The antigen gave the same staining pattern in immunohistochemistry and in the dotblot assay. It precipitated in the Ouchterlony gel with anti IgM but not with other anti-Ig molecule classes.

At present we do not know whether this antibody recognizes its antigen in EM sections as well. In general, the high molecular weight of the IgM molecule prevents its use in electronmicroscopic immunohistochemistry..

Anti-oxytocinase produced in this study could be a useful tool in developing a specific immunological assay for routine clinical diagnostics to monitor the placental function and fetal wellbeing. This monoclonal antibody might also provide a specific immunohistochemical assay to disclose the intracellular localization of the enzyme, to visualize the mechanism of its secretion into maternal circulation, to analyse its metabolism and in applications for diagnosis of placental pathology.

Acknowledgements - The authors wish to thank professor Tapani Vanha-Perttula for providing the necessary facilities at the Department of Anatomy and for his invaluable support. We are very grateful to professor Jouko Tuomisto for his positive attitude toward our work. We also thank professor Peter Kaufmann for his valuable advice. Miss Arja Venäläinen and Miss Teija Koponen are acknowledged for their technical assistance.

REFERENCES

Lalu, K. (1988): Human serum aminopeptidases during pregnancy: Purification and characterization. Publ. Univ. Kuopio, Medicine 7: 1-70.
Lalu, K., Lampelo, S., and Vanha-Perttula, T. (1985): Purification of three aminopeptidases from human maternal serum. Int. J. Biochem. 17: 1227-1235.
Lalu, K., Lampelo, S., and Vanha-Perttula, T. (1986): Characterization of three aminopeptidases purified from maternal serum. Biochem. Biophys. Acta 873: 190-197.
Lampelo, S. (1982): Partial purification and characterization of four aminopeptidases from the soluble fraction of human placenta. Med. Biol. 60: 278-281.

Lampelo, S., and Vanha-Perttula, T. (1980a): Human placental aminopeptidases. In *The Human Placenta*, eds A. Klopper, A. Genazzini and P.G. Crosignani, pp. 283-291. London and New York: Academic Press.

Lampelo, S., and Vanha-Perttula, T. (1980b): Fractionation and characterization of cystine aminopeptidase (oxytocinase) and arylamidase of human serum during pregnancy. *J. Reprod. Fert. 58*: 225-235.

Ryden, G. (1966): Cystine aminopeptidase and oxytocinase activity in pregnancy. *Acta Obstet. Gynecol. Scand. 45*: suppl. 3: 1-105.

Spellacy, W., Usategui-Gomez, M., and Fernandez-deCastro, A. (1977): Plasma human placental lactogen, oxytocinase and placental phosphatase in normal and toxemic pregnancies. *Am. J. Obstet. Gynecol. 127*: 10-16.

Vanha-Perttula, T., Lampelo, S., Lalu, K., and Saarikoski, S. (1988): Secretory and nonsecretory exopeptidases of the human placenta. In *Placental and Endometrial Proteins: Basic and Clinical Aspects*, eds Y. Tomoda, S. Mizutani, O. Narita and A. Klopper, pp. 273-281. Utrecht: VSP BV

Receptors and endocrine regulatory mechanisms : summary of a workshop

S. Belisle, G. Desoye*

*University of Sherbrooke, Quebec, Canada, * University of Graz, A-8036 Graz, Austria*

Placental receptors are known to regulate endocrine functions of the placenta by endo-, para-, auto- and intracrine mechanisms. Therefore it comes as no surprise that placental receptors are receiving increasing interest among placentologists. This is reflected by the number of articles published about receptors relative to all papers pertinent to placental/trophoblast. This portion was only 1.4% from 1977-1979 and increased to 8.3% for 1986-1988.

New binding sites are identified and characterized, post-receptor binding cellular events are investigated, and the ultimate physiological responses mediated by these receptors are sought for.

The workshop on 'Placental receptors and endocrine regulatory mechanisms' attempted to review the current understanding of general receptor linked signalling mechanisms in the perspective of endocrine control of placental hormone production.

I S. Belisle (Univ. Sherbrooke, Canada) reviewed the signalling mechanisms of hormones which have been identified so far in human placental cells. Generally, there are three different pathways, by which cell surface receptors generate intracellular signals. The first group of hormones (i.e. LHRH) affects (activates/inactivates) a plasma membrane bound enzyme - adenylate cyclase. However, activated receptors do not modulate adenylate cyclase directly. A third membrane protein - a GTP binding protein (G protein), serves as link between receptors and adenylate cyclase. Binding of a ligand to its receptor results in the binding of GTP by the regulatory component G_s, which in turn triggers its binding to the catalytic component of adenylate cyclase and its dissociation from receptor. After binding of the stimulated regulatory component R the affinity of R for its substrate ATP is increased and the actual second messenger cAMP is generated. Another set of hormones (e.g. alpha-adrenergic agonists, opioid peptides, adenosine, prostaglandins) does not activate G_s but rather an inhibitory component G_i which can produce a GTP dependent inhibition of adenylate cyclase, thereby lowering intracellular cAMP.

cAMP itself activates a group of enzymes known as cAMP dependent protein kinases ('kinases A') which are either soluble or membrane bound. These enzymes activate or inactivate a series of endogenous proteins by phosphorylation, thus giving rise to the ultimate intracellular effects of an extracellularly bound hormone.

A second group of hormones (e.g. angiotensin II) which act via intracellular messengers after binding to their specific and distinct receptors, activate a membrane phospholipase C, which in turn cleaves membrane phosphoinositides to produce diacylglycerol. Diacylglycerol then binds to an enzyme, protein kinase C, and activates it, a process requiring low intracellular concentrations of phospholipid and Ca^{++}. Protein kinase C, like kinases A, is a serine kinase and activates/inactivates other intracellular proteins by phosphorylation.

A third group of hormones (e.g. insulin, EGF, IGF-I, PDGF) does not have to use second messengers or phosphatidylinositol breakdown for activation of intracellular kinases, but acts via receptors with an intrinsic kinase activity. Binding of hormones to the extracellular moiety of the receptor leads to activation, likely by a conformational change of the receptor, of an intrinsic kinase localized on the intracellular moiety of the receptor. These kinases are, unlike kinases A and C, tyrosine specific. They phosphorylate membrane bound and cytosolic proteins, thereby activating physiological effectors. The kinases can also phosphorylate kinases A and protein kinase C, which in turn by serine phosphorylation of the receptor could inhibit receptor kinase activity. This would link two different signalling pathways and generate a feedback control of receptor/effector coupling.

A last group of hormones functions via ion coupled receptors (e.g.ACTH). Typically, binding of hormones to the receptors results in an increasing flux of Ca^{++} into the cell, thus increasing intracellular Ca^{++}. Ca^{++} exerts its effects by binding to specific Ca^{++}binding proteins, the best characterized being calmodulin. These Ca^{++}binding proteins by interaction with enzymes or other effector proteins activate biochemical pathways that lead ultimately to physiological responses.

II In his lecture M. S. Ahmed (Univ. Kansas City, USA) reviewed the current knowledge about the placental opioid system and opioid receptors as an example for second messenger (cAMP) coupled receptors.

1. The Receptor

A 63,000 dalton glycoprotein has been purified from villous tissue. The purified protein retains its high affinity saturable and stereospecific binding to kappa-ligands. The binding of this protein to bremazocine and etorphine is close to the theoretical value for a molecule of this size assuming one binding site.

2. The Natural Ligand

Opiate peptides endogenous to the placental tissue have been identified. These are methionine and leucine enkephalins, ß-endorphin and the dynorphins. Dynorphin 1-8 is the predominat opiate peptide in placental tissue. The ocatapeptide shows high affinity binding to the membrane bound receptors. Its affinity to the purified receptor is less, however, its k_d is the lowest when compared to other peptides. A similar decrease in affinity of opiate peptides to their purified receptors has been observed in other tissues.

3. Relation of the Receptor and Peptide Levels to Mode of Delivery

a. Receptor levels: The number of villous tissue membrane binding sites is affected by the route of delivery. The number of binding sites in placenta obtained abdominally are higher than those from vaginal deliveries.

b. Opiate peptide levels: The levels of placental opiate peptides extracted from villous tissue are affected by the route of delivery. Peptide levels are higher in vaginal delivery than those obtained abdominally.

These observations in addition to those of others showing increased maternal serum levels of ß-endorphin with labor as well as the absence of a correlation between maternal and cord or fetal levels of opiate peptides suggest:

a) a role for these peptides in labor and or delivery; and, b) their fetal or placental and not maternal origin. The inverse proportion observed for peptide and receptor levels could be explained by receptor down regulation during labor. If this is true, then higher peptide levels in vaginal deliveries would correspond to lower receptor levels, i.e., the results obtained. This is further substantiated in the following:

4. The Relation of Receptor Levels to Drug Abuse During Pregnancy

Patients who used the opiate derivative talwin (pentazocine) during pregnancy have little or no detectable opiate binding sites in their placental villous tissue membranes. This supports the assumption of receptor down regulation due to increased agonist levels because of chronic illicit opiate use during pregnancy.

5. In vitro Modulation of Acetylcholine Release from Villous Tissue by Opioids

a. Role of opioids: Opiates were shown to modulate acetylcholine release in certain cholinergic synapses. In absence of neurons, i.e., in placental villous tissue, a similar observation has been made. Kappa-opiate agonists are more potent than u-agonists in inhibiting acetylcholine release. Kappa- and u-antagonists are able to reverse the inhibitory action of their corresponding agonists. In addition, the antagonists alone cause stimulation of acetylcholine release probably due to reversal of the autoinhibitory action of endogenous opiate peptides. Several roles for acetylcholine in placental tissue have been proposed of which regulation of amino acid transport and vascular contractility are examples.

b. Role of calcium: Presynaptic calcium uptake is required for acetylcholine release in certain cholinergic neurons. Similarly, placental villous tissue requires the presence of calcium in organ baths for acetylcholine release to occur. The calcium channel blocker, diltiazem, inhibits acetylcholine release and the inhibition is potentiated by morphine.

Other functions for the placental opiate system have been proposed. These are regulation of LHRH and hCG release. Ahmed pointed out that further work is needed to establish the physiologic role(s) of the opiate system in human placenta and its effect on fetal development.

III EGF receptors like insulin receptors are present in the microvillous membrane of the syncytiotrophoblast. It has been shown that insulin receptor sites are also found on intracellular organels in other tissues, predominantly on nuclear membranes. One of the fundamentally important unanswered questions regarding the action of EGF and the action of peptide regulatory agents in general is, whether receptors for these agents are also present in the intracellular organelles and whether they play any role in signal transduction mechanisms. Human placenta, because of its easy availability, abundance and being the richest source of EGF receptors, allowed Ch.V. Rao (University of Louisville, USA) to address these issues for EGF.

EGF receptors are present not only in plasma membranes but also in lysosomes, rough and smooth ER, Golgi and nuclei of human placental trophoblast cells. Some of the intracellular organelle receptor properties are similar, while others are different from those in the plasma membranes. The intracellular organelle receptors are functional in that they have kinase activity which can autophosphorylate the receptors. On a per mg protein basis, some of intracellular organelles contain as much as 75 % of the receptors found in plasma membranes.

The availability of monoclonal receptor antibodies specific to different receptor regions allowed Rao to independently verify the biochemical results using immunogold electron microscopy. In agreement with biochemical results, EGF receptors were also found in Severalintracellular organelles including nuclei. The nuclear EGF receptors were present in nuclear membranes, condensed chromatin and perimeter of condensed chromatin, where transcriptionally active genes are believed to be present. Exogenous EGF can internalize and associate with several intracellular organelles. Nuclear association was hightest among the intracellular organelles.

In conclusion, studies from their laboratory during the past 10 years have shown that EGF receptors are also present in several intracellular organelles, including nuclei, of human placental trophoblast cells. This strongly suggests that intracellular organelle receptors, particularly those in nuclei, may participate in signal transduction mechanisms of EGF and that EGF may belong to the very recently known group of hormones exerting also intracrine effects.

IV One of the long standing, most widely discussed and still not fully solved problems in placental endocrine regulation is the question, how placenta specific hormones, like hCG and hPL, are biologically synthesized and how their biosynthesis is regulated. During the last decade enormous progress has been made in unravelling the various steps of hCG synthesis, which were reviewed by W.E. Merz (Univ. Heidelberg, FRG).

Human chorionic gonadotropin (hCG) is composed of two noncovalently linked dissimilar subunits alpha and beta. The alpha-subunit is encoded by a single gene located at chromosome 6, whereas a cluster of seven genes or pseudogenes for the beta-subunit is present on chromosome 19. Among those, beta-gene 5 and to a minor extent beta-gene 3 are expressed in the placenta. The subunit genes are independently expressed. The fact that hCG biosynthesis in placenta tissue as well as in tumor cell lines can be stimulated by cAMP was often used to study the regulation of the subunit gene expression. Gene transcription of the alpha-subunit is stimulated by well defined transcriptional cAMP-responsive enhancer elements (CRE). In the 5´ region of the alpha-gene several cis-acting DNA sequences are located containing two tandemly linked 18 bp CRE (comprising the 8 bp palindrom 5´-

TGACGTCA) flanked by upstream and downstream elements. In contrast, a TATA-box (highly conserved consensus of sequences in the promotor region of eucaryotic genes important for positioning of RNA polymerase II) and the cAMP-responsive 8 bp palindrom is lacking in the case of the beta-gene 5. As shown by Fenstermaker et al. (Mol. Endocrinol. 3, 1070, 1989) in the BeWo choriocarcinoma cell line the CRE of the beta-subunit gene comprises most probably multiple domains. The enhancer-promotor combination seems to be responsible for the cell-specific expression of the alpha-gene (Jameson et al. Mol. Endocrinol. 3, 763, 1989). CAMP seems to increase also the half-lives of alpha- and beta-mRNAs (Fuh et al. Mol. Endocrinol. 3, 1148, 1989).

Whereas a lot of information is available about the action of cAMP on transcription little is known about possible effects of cAMP on translation and post-translational modification. The biologically active hormone requires subunit assembly and glycosylation. Therefore the question arises whether the glycosylation is a prerequisite for subunit assembly and to what extent the biological properties of hCG may depend on the degree of glycosylation. It was shown by Hilf and Merz (Mol. Cell. Endocrinol. 39, 151, 1985) that hCG which was synthesized by first trimester placental tissue in the presence of cAMP showed a significant increase of receptor binding activity as well as of biologic activity which coincided with a decrease in the isoelectric points of the microheterogeneous hCG forms. This hints at a direct influence of cAMP on the glycosylation since the microheterogeneity of hCG is caused by the carbohydrate part. Closer inspection of the carbohydrate processing of placenta-hCG and the influence of cAMP by means of pulse-chase experiments (Hilf and Merz, Biochem. Biophys. Res. Commun. 159, 26, 1989) showed three intracellular subunit precursors (11 kDa (non-glycosylated), 16.5 kDa (one N-linked carbohydrate residue), 19.5 kDa form (two N-linked carbohydrate chains)) and two extracellular alpha-subunit forms (alpha-subunit contained in hCG (20.6 kDa form), free alpha-subunit (higher carbohydrate content, 23.4 kDa)). Studies of the kinetics of these different forms showed that the trimming of the carbohydrate chain by alpha-mannosidase II in the Golgi seems to be the rate-limiting step of the glycosylation of the alpha-subunit and possible a signal for subunit assembly. The hCG synthesized in vitro as well as in vivo by placental tissue in part contains an alpha-subunit with an uncompleted N-linked carbohydrate part ot the hybrid type (Hilf and Merz, Acta endocr. Suppl. 287, 69, 1988). CAMP seems to influence the processing of the alpha-subunit carbohydrate part since it accelerates glycosylation and causes the terminal steps (including sialylation) to be performed more efficiently. The uncompleted N-linked carbohydrate structures are lacking in hCG secreted from cAMP-treated cultures. These investigations demonstrate an influence of cAMP on glycosylation possibly by interfering with the intra-Golgi vesicle transport as suggested by Camilli et al. (J. Cell Biol. 103, 189, 1986). Despite the clear effect of cAMP on gene activity of hCG subunits as well as its influence on glycosylation the physiological role of cAMP in pregnancy is still unclear. There are other suggestions for a physiological regulation of hCG biosynthesis one of these concerning the GnRH-induced stimulation of hCG-biosynthesis seems to be especially an attractive hypothesis. GnRH may be synthetized by the cytotrophoblast. It was supposed to stimulate hCG biosynthesis by the syncytiotrophoblast. In superfusion cultures using first trimester placenta tissue it was shown that exogenous GnRH induces a rapid transient increase of the hCG secretion rate and exerts a long-standing increase of the alpha-mRNA level as well as of hCG secretion. Furthermore, it increases frequency and amplitude of episodic hCG secretion in vitro (Merz et al., this volume). In the hypothalamic-hypophyseal axis the GnRH-induced stimulation of biosynthesis of gonadotropic hormones is controlled by GABA. The presence of a similar system in the placenta is suggested by investigations of Licht and Merz (this volume) which have shown that GABA stimulates hCG-biosynthesis and secretion. The GABA effect is probably mediated by $GABA_A$ - receptors since the $GABA_A$-agonist muscimol is a potent stimulator of hCG-biosynthesis. The $GABA_A$ antagonist bicuculline decreases hCG secretion and depletes the episodic hCG secretion pattern. GABA is able to abolish the effect of bicuculline. These experiments show that a considerable analogy in the regulation of gonadotropine biosynthesis and secretion seems to exist between the placenta and the hypothalamic-hypophyseal axis.

V E. Barnea (Rappaport Inst. Haifa, Israel) further discussed the effects of steroids on placental protein hormone secretion.

The role of steroids in controlling the HPO (hypothalamic-pituitary-ovary) and HPA (hypothalamic-pituitary-adrenal) axis is already well established. In the placenta however, both steroid and peptide hormone are produced in close vicinity unabling the establishment of a clear endocrine axis. Most available data on steroid effect on placental peptide production is based on in vitro observations. Until now, the most investigated hormone, is HCG, while the information on other peptides is scarse.

Steroids act through receptors. The placenta contains both estrogen and corticoid binding sites. The data about progesterone receptors is controversial and the presence of androgen receptors was not documented. Sources of steroids reaching the placenta are: maternal, fetal as well as locally produced ones. The placenta is an incomplete steroid producing organ, namely it is unable to convert progesterone to androgens and it requires androgen substrates from mother and fetus for estrogen production. Although the presence of progesterone receptors is still debated, progesterone was shown to inhibit HCG secretion in the early placenta. This block was achieved by reducing the free beta HCG sub-unit mRNA production. In vivo, however, progesterone increases HCG secretion, perhaps indirectly, through its trophic effect on the endometrium. Using a novel superfusion model Barnea & Kaplan (J.C.E.M. 89) have shown that progesterone has a rapid inhibitory effect upon pulsatile HCG secretion, decreasing both pulse amplitude and frequency. This inhibition, which is not dependent on protein synthesis, may be exerted by the membrane stabilizing effect of progesterone, which blocks the secretion of HCG stored in secretory granules. Other hormones than progesterone, 20-alpha-OH-progesterone and pregnenolone are known to inhibit HCG secretion. In contrast, 17-OH-progesterone stimulates HCG secretion.

Corticosteroids stimulate HCG secretion while androgens and estrogens have no such effect. In conclusion, a number of unresolved questions remain regarding the role of steroids in placental peptide hormone production: 1) auto/paracrine effects, 2) in vitro/in vivo results, 3) early/late placenta differ in effects, 4) are steroid receptors present in placental villous tissue, 5) short/long term effects, and 6) the role of the human embryo.

VI The human placenta produces a large amount of steroid hormones throughout pregnancy. It was calculated that a placenta at term synthesizes about 250 mg progesterone per day. For a long time it was unclear if the placenta de novo synthesizes the cholesterol as precursor for steroid production or if the cholesterol is provided by the maternal circulation. Meanwhile receptors for low-density lipoproteins (LDL) on the placenta were identified by E. Alsat, A. Malassine and L. Cedard. A. Malassine Poitiers, France) emphasized that human placental progesterone synthesis is essentially (up to 90%) dependent on the maternal cholesterol delivered from plasma LDL. Utilization of lipoprotein cholesterol involves a receptor-mediated endocytosis process, as it does for fibroblasts and steroidogenic cells.

Saturable binding sites specific for LDL have been characterized on microvillous placental membranes, as early as the 6 th week of gestation. Their presence has been confirmed by ultrastructural studies of placental villi incubated with ferritin-LDL or gold-LDL.

Furthermore, microvilli possess distinct receptors for Acetyl-LDL (AcLDL) suggesting the existence in the placenta of a "Scavenger Pathway". The ^{125}I-LDL and the ^{125}I-AcLDL binding are inhibited only by the unlabeled homologus ligand and two distinct receptor proteins have been detected by ligand-immunoblotting. The number of binding sites for AcLDL is higher than that for native LDL. Acetyl-like modified LDL are present in human placental blood and can be used as biological natural ligand.

Native or AcLDL were conjugated to colloidal gold, to visualize their route for internalization in human syncytiotrophoblast in culture. Cells obtained from placental villi after cesarean section by a standard trypsin-DNase dispersion method followed by a Percoll gradient centrifugation step, were cultured during 3 days to obtain syncytio-trophoblast-like cells. After 24h in medium without serum, the cells are incubated with gold native LDL or AcLDL, at 37 C, for various times. Both ligands are internalized by multinucleated cells following the classical receptor-mediated endocytosis process: coated pits, coates vesicles, endosomes and lastly lysosome-like dense bodies.

Besides, isolated placental microvilli display a saturable and temperature dependent binding for high density lipoprotein, devoid of Apolipoprotein E (HDL_3). Nevertheless, the

HDL_3 binding is not specific, since HDL_3 can not be replaced by native and AcLDL. Furthermore, ultrastructural study confirms that gold-HDL_3 bind to the trophoblast cell surface but without subsequent internalization process.

The existence of various microvillous binding sites for different types of lipoproteins, with or without internalization pathway may be related to the high amount of cholesterol indispensable to the progesterone biosynthesis and cellular growth of the placenta. Interestingly, other studies showed that HDL stimulates the release of hPL from placental and trophoblastic cells in culture.

Conclusions

Two theories currently prevail on the control of hormone production by the human placenta: A) An autonomous regulation which depends on the changing ratio of cyto- and syncytiotrophoblast during pregnancy, B) an endocrine regulation comparable to that observed in any mature endocrine cell. By far, existing evidences favor the last hypothesis. Thus neurohormones from the mother, fetus or produced in situ in the cytotrophoblast do bind to specific membrane receptors in the syncytiotrophoblast. This hormone-receptor complex then activates biochemical and biophysical cellular events to stimulate or inhibit synthesis and secretion of placental hormones.

Characterization of high density lipoprotein (HDL_3) binding to human placenta : absence of internalization

E. Alsat, A. Malassiné*, R. Rebourcet, C. Besse*, L. Cedard

INSERM U. 166, Maternité Baudelocque, 75014 Paris, France
* Laboratoire de Physiologie Cellulaire UFR Sciences, Poitiers, France

Studies on cultured cells demonstrated that for progesterone biosynthesis, trophoblast utilizes the cholesterol carried by maternal plasma low density lipoprotein (LDL). Specific receptor and internalization process for LDL and modified LDL (acetyl-LDL) have been well characterized. Winkel et al. (1980) found that HDL may stimulate progesterone production by trophoblastic cells in culture when added in high concentration. So, it is of interest to study the HDL binding and its eventual subsequent internalization by purified placental preparations.

Isolated term placental microvilli were used for binding of ^{125}I-HDL_3 (devoid of Apoprotein E). HDL_3 were conjugated to colloïdal gold for ultrastructural visualization of binding and internalization in syncytiotrophoblast in culture.

Saturable receptor sites specific for HDL_3 were identified. Scatchard analysis revealed a K_D value of 24.2 ± 8.0 µg HDL_3 protein/ml and a maximum binding capacity at 4°C of 128.2 ± 54.5 µg HDL_3 protein/mg of membrane protein. These sites have broad specificity : both LDL and Acetyl-LDL were able to partially inhibit the HDL_3 binding.

Ultrastructural study confirms that gold-HDL_3 bind specifically to syncytiotrophoblast membrane. However, after incubation at 37°C, an internalization process similar to those described for gold LDL and gold acetyl-LDL was not observed for gold HDL_3.

These results demonstrate specific HDL_3 binding without internalization. The physiological significance of an HDL_3 membranous interaction and the placental steroidogenesis remains to be established.

The role of EGF in hCG secretion and trophoblast differentiation in the first trimester

E.R. Barnea, D. Feldman, M. Kaplan

Feto-Placental Endocrine Unit, Rappaport Institute, Technion, Haifa, Israel

The role of EGF in the placenta has been previously addressed : at term and midterm it stimulated hCG secretion and induced cellular differentiation. The information on the role of EGF in the first trimester placenta (FTP) is, however, scant.

Here we investigated the effect of EGF on FTP in three different in vitro models. Placental tissue obtained from elective pregnancy terminations (7-10 weeks) was cleared of blood with 0.9% NaCl and rinsed with DMEM containing 1% antibiotics. Cells were disloged with trypsin/DNAse and cultured with DMEM, 5% FCS for 1-50 days with or without 100-200 ng/ml EGF. The media was changed frequently and saved for assay. In addition, cells were examined for morphological changes. In isolated cells following exposure to EGF, hCG secretion, as measured by RIA (MAIA clone, Serono) increased significantly, p less than 0.05, however it had no effect on differentiation. Further, in placental explants (where cell to cell integrity is maintained) incubated for 24 h 100 ng/ml EGF caused a significant 157% increase in hCG secretion (p less than 0.05).

Finally, using a novel multichannel superfusion apparatus (Accusyst, Endotronics) we recently reported that hCG secretion is pulsatile (Barnea, Kaplan, JCEM, 1989) one minute pulses of EGF caused a significant increase (calculated as area under the curve) in pulsatile hCG secretion, at 50 ng/ml.

In conclusion, the results obtained in static (cells and explants) and dynamic (superfusion) cultures point out to the stimulatory role that EGF has on placental hCG secretion. In contrast, no associated cellular differentiation was noted.

Regulatory signals in placental iron uptake

M.B. Bierings, M.R.M. Baert, H.G. van Eijk, J.P. van Dijk

Department of Chemical Pathology, Medical Faculty, Erasmus University P.O. Box 1738, 3000 DR Rotterdam, The Netherlands

Transferrin receptors (TfR) on the apical side of syncytiotrophoblast mediate placental iron uptake. Regulation of TfR - expression may be an important mechanism to control trans-placental iron transport. We use an *in vitro* model of cultured human term cytotrophoblasts to study TfR - expression regulation. Receptor-ligand interactions are studied using ^{125}I - transferrin. Kinetics of receptor-mediated endocytosis will be discussed.

The level of TfR - expression is highly dependent on cellular density in the culture dishes. High cellular density results in low TfR - levels, and vice-versa. Low cellular density does not represent a proliferative trigger for term cytotrophoblasts in vitro. Mean cellular protein of around 100 μg per 35 mm dish leads to receptor numbers of $3 - 4 \times 10^4$ per cell surface. This is considerably lower than TfR expression in malignant trophoblast cell lines.

Cellular differentiation, induced by 8-bromo-cAMP or prolonged culture periods does not influence either the level of TfR expression, or the affinity of the ligand for its receptor.

High availability of iron in the culture medium, especially if added in the physiological diferri-transferrin form, reduces the level of TfR's on cytotrophoblasts.

The lack of effect of insulin-addition and serum-withdrawal on the level of surface TfR's in cultured term cytotrophoblasts is probably related to their being not-proliferative *in vitro*.

Endocrine control of hPL and hCG production by the human placenta

S. Belisle, A. Petit, N. Gallo-Payet, J.G. Lehoux, D. Bellabarba, E. Esher, G. Guillon*

Faculté de Médecine, Université de Sherbrooke, Québec, Canada
* INSERM U. 264, Centre CNRS-INSERM de Pharmacologie-Endocrinologie, 34000 Montpellier, France

The regulatory mechanism(s) for the production of human placental peptide hormones are poorly understood and two theories currently prevail: (1) an autonomous production which varies according to the changing ratio of cyto- to syncytiotrophoblasts throughout pregnancy (2) an autocrine-paracrine control mediated through receptor binding events. We have favored the latter hypothesis and postulated a functional analogy between the regulation of peptide hormone production by the pituitary with that by human placenta. Trophoblastic cells obtained by trypsin digestion of placentas from mid-term (13-18 weeks) and term (38-41 weeks) normal gestations were rested for 36-48 hrs prior to study A- hCG studies. We observed the presence of specific, moderate affinity (Ka 10^{-8}M) low capacity LHRH receptor sites which could be up-and down-regulated depending on the endocrine milieu in vitro. Photo- affinity labelling studies showed that the molecular weight of this placental receptor for LHRH was comparable to that observed in rat pituitary (60 kd) and structure-activity relationships indicated that it bound both agonists and antagonists of LHRH. The concentrations of these receptor sites decreased four-fold from mid-gestation to term which was paralleled by a decreased production of bioactive hCG by agonists of LHRH. Ca^{2+} and cAMP, but not membrane lipid hydrolysis, mediated this hCG production by LHRH in human placentas. B- hPL studies. We also found that its production by human placental cells was regulated by cellular events comparable to those responsable for Prolactin production by the pituitary. Specific opioids and angiotensin II (AII) receptor sites were observed in human placentas with characteristics of high affinity (Ka 10^{-10}M) and low capacity for EKG and AII. The concentrations of these receptors increased from mid-pregnancy to term and was manifested by an increased ability to release hPL which was stimulated 2-fold by 10^{-8}M AII and 2.2⁻ fold by 10^{-7}M EKC. Dopamine (DA), naloxone and AII antagonist (SAR-ALA II) completely inhibited this hPL production (>85%). AII stimulated inositol phosphate (IP) production whereas EKC had no effect. DA was without effect on basal IP production but inhibited the AII-induced accumulation of IP. The production of cAMP was inhibited by EKC (55%), stimulated by DA (600%) but unaffected by AII. Extracellular Ca^{2+} did not influence basal or AII stimulated hPL release but was essential for DA effects. We conclude that the placental production of hCG and hPL is under the control of paracrine influences. (Supported by MRC Canada).

In vitro effect of dynorphine on hCG production and release by trophoblast tissue explants

B. Čemerikić, O. Genbačev, R. Beaconsfield*

Institute of Endocrinology, Immunology and Nutrition, INEP, Yugoslavia
* SCIP Research Centre, University of London, UK

We have studied the in vitro effect of dynorphine on hCG production and release by first trimester trophoblast tissue explants using two different protocols: (A) short-term incubation up to six hours, and (B) tissue culture up to seven days.

The results of the short-term experiments show that dynorphine (1 nM) consistently increases the release of hCG in first trimester placentae. The effect varies between individual placentae and is antagonised in the presence of naloxone. The stimulatory effect of dynorphine on hCG is less pronounced with advancing gestational age of the placenta.

In Long-term experiments, the stimulatory effect of dynorphine on hCG was seen again over the first two days of exposure. The highest increase in hCG production rate was observed during the first 24 hours.

As the dose of dynorphine used in these experiments falls within the range of Kd values for opiates, it is tempting to speculate that our results support a role of opioid peptides in the autocrine/paracrine regulation of hCG synthesis and release by first trimester trophoblast.

Antilipolytic hormones stimulate the secretion of hCG in cultivated trophoblasts via a pathway involving protein kinase C

G. Desoye, B. Schmon, I. Andiel, R. Michlmayer, G. Dohr*, M. Hartmann*, A. Blaschitz*

*Dept. Obstetrics and Gynaecology and * Institute of Histology and Embryology, University of Graz, A-8036 Graz, Austria*

The present study was undertaken in order to investigate if insulin (I) and nicotinic acid (NA) affect the secretion of hCG and if this occurs via a common pathway involving protein kinase C (pkC).

Cytotrophoblasts were isolated from human term placenta following established protocols. Every 24 hours the culture media (D-MEM) - which had or had not been supplemented with the reagents under study - were changed during 8 days and the concentrations of alpha-hCG (a-hCG), hCG and hPL were measured by RIA. Viability of cells was assessed by dye exclusion and glucose consumption.

Both, I and NA increased the total amount of secreted a-hCG and hCG after 8 days in a concentration dependent manner. hPL was unaffected by I and decreased by NA, respectively. Monoclonal anti-insulin-receptor antibodies mimicked I effects, suggesting that I effects are mediated by I receptors. Supplementation of maximally stimulating concentrations of either hormones plus variing concentrations of an inhibitor of pkC reduced the secretion of a-hCG and hCG to as low as 30% of control (D-MEM). hPL levels were higher (I) or equal (NA) compared to control. Phorboldibutyrate, a stimulator of pkC, increased in a concentration dependent manner a-hCG and hCG concentrations by up to 50% and 40%, respectively, whereas it did not alter hPL levels. Generally, the time course of hormone secretion and of the formation of syncytia was unaffected.

The data indicate: 1) I and NA stimulate hCG secretion by a common mechanism involving protein kinase C; 2) this mechanism is not specific for I and NA but may also apply to other stimulators of protein kinase C, like adenosin, which is generally supplemented to D-MEM; 3) hCG and hPL secretion are differently regulated.

On the reaction mechanism of human placental steryl-sulphatase

L. Dibbelt, E. Kuss

Inst. Biochem. Endokrinol., Med. Univ., Lübeck, D-2400 Lübeck I and I. Frauenklinik, University of München, D-8000 München 2, Federal Republic of Germany

Sterylsulphatases (EC 3.1.6.2) catalyse the hydrolysis of sulphuric acid esters of certain neutral steroids such as cholesterol or dehydroepiandrosterone. The enzymatic activity exists in several species and organs in a strongly membrane-bound insoluble form which has many kinetic and physicochemical properties in common with the type I arylsulphatase, arylsulphatase C, but appears to differ markedly from the soluble type II arylsulphatases A and B. In the human species, the placenta is the richest sources of sterylsulphatase activity among both adult and fetal tissues. Recently, we solubilized and purified the enzyme from human placental cellular membranes : according to the results of gel electrophoresis and N-terminal amino acid sequencing, the preparation turned out to be homogenous. We performed kinetic experiments with the isolated sterylsulphatase applying a number of potential substrates and inhibitors of this enzyme.

Besides neutral steroid sulphates, the sulphates of phenolic steroids (e.g. oestrone) and even of non-steroidal phenols (e.g. p-nitrophenol) efficiently were hydrolyzed by the sterylsulphatase. The enzyme activity was strongly affected by vanadate, which structurally resembles sulphur trioxide and therefore may act as a transition state analogue. Diethylpyrocarbonate severely inactivated the sulphatase indicating that histidyl residues are essential for the enzyme activity. Taken together, our findings reveal a considerable similarity between the reaction mechanism of human placental sterylsulphatase and the one proposed previously by other authors for the type II arylsulphatases.

Characterization and differentiation of human first trimester placenta trophoblastic cells in culture

M. Dodeur, A. Malassiné*, D. Bellet**, A. Mensier, D. Evain-Brion

Laboratoire de Physiopathologie du Développement, CNRS-ENS, 46 rue d'Ulm, 75005 Paris,
* UER des sciences, 86022 Poitiers, ** Institut Gustave-Roussy, 94805, Villejuif, France

A preparation of highly enriched isolated cytotrophoblasts was obtained from human first trimester placenta using Dispase incubation of villous tissue at 4°C, followed by a spontaneous cell release at 37°C. After 24 hours of culture, 90 to 95% of cells were immunostained by anticytokeratin antibody and by anti α subunit antibody, showing their epithelial characteristic and their ability to synthesize a subunit of human chorionic gonadotropin (hCG). After 48 hours of culture, these cells differentiated into syncytiotrophoblast as shown by electron microscopic study and by their capacity to synthesize and to secrete human chorionic sommatomammotropin (hCS). Moreover we studied the secretion of hCG, and of its free α and β subunits and the secretion of hCS as a function of cell culture time. While the level of secreted hCG (40ng/ml) and of its free α subunit (30ng/ml) was stable during 72 hours of culture, the hCS level was undetectable during the first 48 hours of culture increasing continuously afterwards. Addition of dibutyryl cAMP(10^{-3}) in continuous or after 96 hours of cell culture induced an increase of secreted hCG, and its free subunits and stimulated also the secretion of hCS. After 96 hours of culture in presence of cAMP, the level of secreted hCG was increased by ten fold as compared to control and secretion of free α and β subunits was increased by seven fold.

This suggests that these cells possess the capacity to respond to stimuli which increase intracellular cAMP level. Such a cell culture will be of interest to further determine at the early gestational term the mecanisms implicated in the differentiation and the growth of placental cytotrophoblasts, and in the regulation of their endocrine functions.

Effect of parathyroid hormone on the endocrine functions of human trophoblastic cells of first trimester placenta

D. Evain-Brion, M. Dodeur, A. Mensier, E. Alsat, J.M. Bidart*

*Laboratoire de physiopathologie du développement, CNRS-ENS, 46 rue d'Ulm, 75005 Paris, * Institut Gustave-Roussy, 94805 Villejuif, France*

Our previous study in teratocarcinoma cells suggested a role of human parathyroid hormone (hPTH) in the early development of placenta.

The purpose of this study was to evaluate the possible role of hPTH on the functions of first trimester trophoblast cells. Adenylate cyclase activity in crude membranes from first trimester human placental villous tissue is stimulated 2 fold by hPTH (1-34) (10^{-6} M) from 265 ± 32 to 532 ± 80 pmoles of cAMP/mg of protein/15min. A similar stimulation of adenylate cyclase is observed in human term placental villous tissue but not in two different choriocarcinoma cell lines.

In order to evaluate the possible role of hPTH on the functions of first trimester human trophoblast cells, cells were isolated by enzymatic digestion, purified on percoll gradient and cultured (2 X 10^5 cells per plate) in DMEM supplemented with 20 % fetal calf serum. On day three of culture h PTH stimulated in a dose dependent manner cAMP production by these cells (from 0.34 to 2.40 pmoles/10^6 cells/60min.) and increased by 2 fold human chorionic gonadotropin (hCG) secretion as compared to control (35 ng/ml/10^6 cells). hPTH stimulated the secretion of free α subunit of hCG as measured by specific IRMA assay from 35ng to 70ng /ml/10^6 cells but did not change significantly the secretion of free b subunit. hPTH did not modify the secretion of human chorionic somatomammotropin of these cells.

In conclusion human trophoblastic cells are target cells for hPTH which regulates some of their endocrine activities such as hCG secretion.

Characteristics of carbohydrate metabolism in isolated cytotrophoblastic cells compared to whole placental tissue

I. Henrichs, G. Kreisel, G. Röckelein*, W.M. Teller

*Dept. of Paediatrics I, University of Ulm, and * Pathological Institute, University of Erlangen-Nürnberg, Federal Republic of Germany*

Introduction: The carbohydrate metabolism of human placenta is the main source of the fuel supply for the placenta itself. The borderline cell layers to maternal blood are the syncytiotrophoblast and cytotrophoblast surfacing the villi. Therefore, we studied the gross and the free glucose utilization (U_G), L-lactate production (P_L) and pyruvate production (P_P) of **placental explants** in comparison to **isolated cytotrophoblastic cells**. - **Methods:** Incubation studies (for 120 min with glucose concentration of 100 mg/dl) were made with a) placental tissue explants (20 term placentae), 4-6 pieces (60 mg) of each placenta; b) purified syncytio- and cytotrophoblastic cells (9.9 x 10^6 mononuclear cells/ml) freshly isolated from 30 g placenta tissue by enzymatic digestion and density discontinuous gradient centrifugation in 3-4 incubation samples (17 term placentae). The separated mononuclear trophoblastic cells were characterized immunocytochemically. In some experiments (n= 12) with isolated cells, tracing tests by D-(U-^{14}C)glucose were performed and the metabolized lactate and pyruvate were analyzed by HPLC preparation. The determination of glucose, L-lactate and pyruvate concentration were done by adapted enzymatic methods. The computed U_G and P_L of each experiment (in nmol/10^6 cells/60 min or µmol/g tissue weight/60 min) were analyzed as ratio of P_L/U_G in explants tissue (R^E) or in isolated cytotrophoblastic cells (R^C). **Results: 1.a)** The absolute gross metabolic rates for isolated cytotrophoblastic cells were: $U_G{}^C$= 30.95 (± 9.51) (mean ± S.D.); $P_L{}^C$=14.89 (± 6.77). b) The absolute gross metabolic rates for human placental explants were: $U_G{}^E$=3.38 (± 0.85), $P_L{}^E$=4.81 (± 1.27). **2.** The $P_L{}^C$ is (related to absolut values) about a half of $U_G{}^C$: i.e. mean R^C was 0.578 (± 0.212). But the $P_L{}^E$ is about 1 1/2 times higher than $U_G{}^E$, i.e. mean R^E was 1.51 (± 0.15). This difference is highly significant (p 0.01). **3.** The percentual part of $U_G{}^C$ derived from free glucose was 4.85%/10^7 cells/60 min (± 1.86). The lactate production ($P_L{}^C$) out of free glucose was 8.41 %/10^7 cells/60 min (± 3.25). **Conclusions: 1.** The preparation of isolated trophoblastic cells of term placenta shows higher metabolic activity in glucose utilization ($U_G{}^C$) calculated to values of placental explants. For we computed that only 6.55 x 10^7 cells would achieve a $U_G{}^E$ of 600 mg placental explant. - **2.** The $R^C=P_L{}^C/U_G{}^C$ is significantly lowered compared to R^E (placental explants). This result could present a more active citric acid cycle with more gained energy in isolated epithelial trophoblastic cells.-**3.** In isolated cells the glycolytic pathway seems to be used more by free glucose than represented in the gross glucose utilisation: They exemplifie a ratio P_L/U_G=1.70 (± 0.09).-**4.** We speculate that different enzymatic and hormonal properties of carbohydrate metabolism exists in trophoblastic cells related to whole placental tissue.

Supported by Deutsche Forschungsgemeinschaft (He 1107/2-3)

A partially purified fraction of phosphorylated placental non-histone proteins which stimulates transcription in isolated nuclei

K. Hess, H. Elouardirhi, D. Sekkat, F. Belleville, P. Nabet

Laboratoire de biochimie médicale I, Faculté de médecine, BP 184, 54505 Vandœuvre-les-Nancy Cedex, France

The chromatin non-histone proteins (NHP) are particularly interesting because of their expected roles in gene regulation. Difficulties in the purification of non-histone chromosomal proteins arise because most of them display a strong tendency to aggregate and also by a lack of discernable biological activities that can be utilized to monitor their presence during purification schemes.

Phosphoproteins extractible from chromatin have been studied. It has been found that chromatin proteins, especially those engaged in the transcriptional machinery are phosphoproteins. We partially purified a phosphorylated non-histone protein by a simple procedure involving a non-denaturing gel electrophoresis and extraction from the gel. We tested its effect on RNA synthesis.

Nuclear proteins extracted from isolated placental nuclei by a high ionic strengh buffer (buffer A : 2 M NaCl ; 5 M urea ; 1 mM KH_2PO_4, K_2HPO_4 ; 0,1 mM PmSF ; 2 mM Tris pH 6.8) are chromatographed on an hydroxyapatite ultrogel (HA-U) column. They were step-wise eluted by buffer A with increasing amounts of phosphates (1, 50, 100 mM). Three peaks were obtained called respectively HAP_1, 2 and 3. HAP_2 contained most of the phosphorylated NHP. HAP_2 were treated by BioRex 70 (Na^+) which retained histones and chromatographed on a calcium phosphate gel which selectively retained the phosphorylated NHP. Amino acid composition of crude phosphorylated NHP (P_2) shows a content of 12.6 % serine and 17.6 % glutamic + aspartic. This fraction P_2 was analysed by one dimensional gel electrophoresis in a non-denaturing condition. The apparent molecular weights of the protein bands were estimated at 56 000, 32 000 and 10 000. The largest band (PM 56 000) was extracted from the gel and migrated on dissociating SDS-PAGE electrophoresis. The result showed a single electrophoretic band which confirmed that this protein is neither an aggregate nor an oligomer.

To monitor its presence during purification, we used a transcription system of isolated nuclei. The usefulness of a given in vitro transcription assay system for studies of gene regulation is highly dependent on its ability to transcribe relevant genes in a manner reflecting the in vivo situation with preserving in vivo site(s) of regulation. The nuclei elongated previously initiated RNA chains and initiate new chains.

When purified fraction NHP 56 000 were added to nuclei transcription system, a stimulation of RNA synthesis was observed. Among the RNA synthesized, the hPL species are increased as shown by Northen blot and after in vitro translation, by immunoblot.

An *in vitro* system to study the hPL genes expression in placental isolated nuclei

K. Hess, H. Elouardirhi, D. Sekkat, F. Belleville, P. Nabet

Laboratoire de biochimie médicale, Faculté de médecine, BP 184, 54505 Vandœuvre-les-Nancy Cedex, France

The placenta provides a ready source of large amounts of human tissue for nuclei preparation. The availability of purified nuclei that structurally and functionally resemble conditions in vivo is important for investigating the synthesis and processing of RNA.

Morphologically, nuclei were studied by optical and electron microscopy. Such preparation is able to incorporate labeled |^{3}H|-UTP into neosynthesized RNA and provides a system where nuclei are permeable to nucleotides and macromolecules allowing the study of RNA synthesis regulations. 80 % of RNA polymerisation was inhibited by 4 µl/ml of α amanitin showing that it is mostly the effect of polymerase II. High salt concentration is optimal for this synthesis (0.4 M $(NH_4)_2SO_4$; 4 mM Mn^{2+} ; 4 mM Mg^{2+}). Polyacrylamide gel electrophoresis indicated that synthesized RNA was heterogenous in size having a distribution from 5 to 60 S with a significant fraction migrating as 10-20S. Approximately 7 % of total RNA was retained on an oligo (dT) cellulose or poly U Sepharose-4 B-column suggesting that some RNA poly-A processing occured.

A crucial question in the evolution of the in vitro transcription system is how accurately it reflects the physiological transcription initiation in corresponding intact cells. It seems unlikely that some loosely bound transcription factors, indispensable for accurate initiation, could have completely leached out of the nuclei and were subsequently last in the course of nuclear preparation ; however, we cannot exclude such a possibility. Thus our major argument in favor of the transcription initiation is based on the response of the transcriptional machinery to the action of heparin. Heparin effectively inhibits binding of RNA polymerase to DNA and thus inhibits initiation of RNA synthesis. The activity of free polymerase was reduced (37 %) by the addition (2 mg/ml) of the initiation inhibitor heparin.

A technical problem associated with these systems is the difficulty to differentiate previous in vivo transcripts from in vitro neosynthesized RNA chains. An approach to circumvent this problem is to use 5-mercury-UTP (Hg UTP) in the transcription system. Newly synthesized nascent RNA transcripts, which contained Hg-UMP were fractionated from the excess endogenous nuclear RNA by agarose-ethane-thiol affinity chromatography. Hg-UTP did not markedly affect transcription rate. These RNA are translated in an optimized wheat germ cell-free system. HPL synthesized in vitro as shown by radioimmunoassay accounts for 20 % of total proteins. This system should permit us to study the regulation of hPL genes expression in their environment.

Stimulation of the microvillous Na$^+$/H$^+$ exchanger by protein kinase C phosphorylation

N.P. Illsley, M.M. Jacobs

Department of Obstetrics, Gynaecology and Reproductive Sciences, University of California, San Francisco, CA 94143, USA

The plasma membrane Na$^+$/H$^+$ exchanger is a ubiquitous transporter that serves a variety of functions including intracellular pH regulation and cellular volume control. In polarized epithelial tissues such as the placental syncytiotrophoblast this transporter can also serve specialized functions including transepithelial pH regulation and salt transport. In a number of tissues, transporter activity is modulated by intracellular and extracellular factors which promote protein kinase C- or tyrosine kinase-mediated phosphorylation. We asked whether the microvillous Na$^+$/H$^+$ exchanger is regulated by protein kinase C phosphorylation. Microvillous membranes (MVM) were prepared from term human placenta; alkaline phosphatase activity (MVM marker) was enriched > 32-fold compared to homogenate, while adenylate cyclase activity (basal membrane marker) was decreased to < 0.01 of the homogenate value. MVM were loaded with a pH-sensitive fluorophore, 6-carboxyfluorescein, 0.5 mM Mg^{2+} and optionally with 0.2 mM ATP by hypotonic lysis. Transmembrane proton flux was measurements from the rate of change of fluorescence as MVM (pH = 6.0) were added to pH 7.5 buffer. Na$^+$/H$^+$ exchange activity was calculated as the difference in proton flux between measurements in the presence and absence of 50 mM Na$^+$. Prior to assay, MVM were incubated at room temperature for 15 min with Ca^{2+} or TPA (12-0-tetradecanoyl-phorbol-13-acetate), factors shown to increase protein kinase C activity. In Ca^{2+}-free control experiments (5 mM external EGTA plus Ca^{2+}-ionophore ionomycin), intravesicular ATP did not stimulate the basal Na$^+$/H$^+$ exchange rate 4.8 ± 1.5 neq/(s$_2$mg) significantly in five separate MVM preparations. Neither 0.2 mM Ca^{2+} in the presence of ionomycin nor TPA (1 μM) affected Na$^+$/H$^+$ activity in the absence of ATP. In the presence of ATP however, 0.2 mM Ca^{2+} plus ionomycin stimulated Na$^+$/H$^+$ activity by 84 ± 11 per cent (P < 0.01; n = 5). TPA stimulated Na$^+$/H$^+$ activity by 132 ± 36 per cent (P < 0.01; n = 5) in the presence of ATP. Stimulation of Na$^+$/H$^+$ exchange activity by high levels of Ca^{2+} (> 1 μM) or by TPA is consistent with activation of the transporter by protein kinase C-mediated phosphorylation. This demonstrates that Na$^+$/H$^+$ exchange may be regulated by factors which modulate protein kinase C and the regulatory receptor-coupling G proteins, $G_{s\alpha}$, $G_{I\alpha}$ and $G_{\beta/\alpha}$ are present in these membranes. In combination with these elements, the coupling between protein kinase C and the Na$^+$/H$^+$ exchanger allow us to explore the microvillous signal transduction chain using a functional, physiologic marker.

Is there dual regulation of syncytial adenylate cyclase ?

M.M. Jacobs, N.P. Illsley

Department of Obstetrics, Gynaecology and Reproductive Sciences, University of California, San Francisco, CA 94143, USA

Syncytial adenylate cyclase is highly polarized ; it is present on the basal membrane (BM) and absent on the microvillous membrane (MVM). Receptor polarization is also evident ; stimulatory β-adrenergic receptors are found only on BM and inhibitory kappa opioid receptors only on MVM. Placental homogenates have been used as a rich source for purification of the stimulatory and inhibitory adenylate cyclase regulatory components $G_{5\alpha}$, $G_{1\alpha}$ and $G_{\beta\gamma}$, although their placental location is not known. We asked if basal membrane adenylate cyclase is under both stimulatory and inhibitory control and if there is evidence for polarization of syncytial-G-protein subunits. Paired microvillus (MVM) and basal membrane (BM) particulates were prepared from term placentae obtained at elective caesarean section, prior to labour. Adenylate cyclase and alkaline phosphatase activity were found in separate fractions designated BM and MVM respectively. Studies of adenylate cyclase activation in BM showed stimulation by isoproterenol and PGE. However, inhibitory stimuli such as α_2-adrenergic (norepinephrine plus propranolol), muscarinic (carbachol) and adenosinergic (adenosine) agonists could not decrease GTP, forskolin or agonist-induced cyclase stimulation. Cyclase inhibition is sensitive to Na^+ concentration : therefore we tested possible inhibitors in the presence of 0, 5, 50 and 100 mM Na^+ but no effect was observed. To determine if G protein subunit composition could explain the absent inhibition, $G_{5\alpha}$ and $G_{1\alpha}$ were quantitated in BM and MVM with Western Blots using specific antibodies against individual subunits. There was 1.5-2 fold more $G_{5\alpha}$ and $G_{1\alpha}$ in MVM than BM but the ratio of the two was not different between the membranes. The presence in MVM of opioid receptors and $G_{1\alpha}$ allows testing of syncytial $G_{1\alpha}$-receptor coupling. Radioligand experiments were carried out in MVM examining the competition of diphenorphine for the kappa opioid agonist ^3H -U69593. There was a 25-fold reduction in diphenorphine affinity after addition of GTP, confirming functional $G_{1\alpha}$-opioid receptor inter-action. To further evaluate nucleotide binding to G proteins in BM, were examined the potency of several guanine nucleotide analogues to stimulate cyclase. The order of potency was GTPγS > GppNHp > GTP > GDPβS and was not altered by isoproterenol or PGE. Our data point to absent or ineffective inhibitory receptors in BM, since the necessary distal cyclase components are present ; both G-protein subunits are found in BM, G_5 can couple to receptors and cyclase and $G_{1\alpha}$ can couple to receptors in MVM. It is possible that $G_{1\alpha}$ cannot couple to cyclase in BM, perhaps because of compartmentalization. If pertussin toxin treatment of BM enhances cyclase stimulation, this will indicate $G_{1\alpha}$-cyclase interaction.

Neurotransmitters in human term placenta: biochemistry and immunochemistry

L. Kaplan, J.J. Lopez Costa*, S.E. Carbone**, O. Mastronardi**,
D.C. Rondina**, J. Pecci Saavedra*, J.A. Moguilevsky**

INSERM U.166, maternité Baudelocque, 75014 Paris, France
* Institut de Biologia Celular and ** Departamento de Fisiologia, Facultad de Medicina
CONICET, Buenos Aires, Argentina

In the last years a series of neurotransmitters and releasing factors have been detected in human placenta despite of the inexistence of innervation, but it is difficult to explain their physiological role.

In the present work, we determined the presence of serotonin, dopamine, noradrenaline and gamma-aminobutyric acid. Using spectrofluorometric methods we found respectively: 0.24 ± 0.04; 0.28 ± 0.03; 0.33 ± 0.03 and 27 ± 7 (n = 10 $\pm$ s.d.) µg/g of fresh placental tissues. We detected also serotonin, by immunocytochemical PAP method, in cytotrophoblastic Langhans cells.

Since several of these neurotransmitters are found in the CNS, the possibility of a mechanism that regulates hormone secretion in the placenta, similar to that of the hypothalamic-pituitary axis, can be suggested.

These neurotransmitters may act like hormones in the periphery. They are probably related to the regulation of placental flux and with fetal blood pressure, in coordination with catecholamines. Serotonin has also been related to the contraction of the uterus. The study of the neurotransmitters' action on placental hormones (hCG, hCS) and steroids is currently in progress in laboratory.

Evidence for a GABA-ergic modulation of hCG secretion by human first trimester placenta tissue

P. Licht, W.E. Merz

Department of Biochemistry II, University of Heidelberg, Im Neuenheimer Feld 328, D-6900 Heidelberg, Federal Republic of Germany

The processes which regulate biosynthesis and secretion of human choriogonadotropin (hCG) by the syncytiotrophoblast are still unknown. Some evidence suggests a stimulatory role of gonadoliberin (GnRH) produced by the cytotrophoblast. Since GnRH secretion by the hypothalamus is suppressed by the neurotransmitter γ-aminobutyric acid (GABA) (1) it is of interest whether GABA interferes also with hCG secretion in placenta tissue.

Human placenta tissue (8. and 10. week of gestation) was perifused according to (2) using serum-free culture conditions (M 199 with Hank's salts, antibiotics and 10 μg/ml transferrin). A single GABA-pulse (10 μM; 1 h) caused a marked increase in hCG secretion rate beginning 4 h after the start of infusion and lasting until the end of the observation period (15 h). Furthermore, in GABA treated cultures an episodic hCG secretion pattern seemed to occur with a peak frequency of about 9 h, whereas the control cultures showed continuous secretion. When GABA was added as a continuous infusion (2.5 μM) episodic hCG peaks were observed, lasting 6-8 h and yielding a 3-8 fold increase in hormone secretion. The $GABA_A$-receptor agonist muscimol (10 μM) increased hCG concentration in medium of stationary cultures 2.5 fold.

The $GABA_A$ antagonist bicuculline (10 μM) caused a significant decrease of hCG secretion and suppressed the episodic secretion pattern. The effect of bicuculline was completely abolished by an equimolar concentration of GABA. Our data show that hCG secretion by human first trimester placenta tissue is stimulated by GABA. The secretion pattern seems to be modulated towards a episodic hormone release. These changes are probably due to a $GABA_A$-receptor mediated process.

1. C. Masotto, A. Negro-Vilar: Endocrinology 121, 2251-2255, 1987
2. W.E. Merz, C. Erlewein: Placenta 7, 448-449, 1986

VIP receptors, positively coupled with adenylate cyclase activity on fetal vascularization of human placenta

A. Malassiné, F. Mondon*, J. Besson**, M. Vial**, G. Tanguy, W. Rostène**, F. Ferré*

Laboratoire de physiologie cellulaire UFR Sciences Poitiers, France
* *INSERM U. 166, maternité Baudelocque, 75014 Paris, France*
** *INSERM U. 55, hôpital Saint-Antoine, 75012 Paris, France*

The presence of Vasoactive Intestinal Peptide (VIP) has been reported in human placenta, a non innervated organ. However, the exact origin of this neuropeptide, the precise localization of its binding sites and the mechanism(s) of action remain to be clarified.

The ^{125}I-VIP binding to placental slices was saturable and unlabelled VIP was able to compete in a dose-dependent manner with an IC_{50} value of $5.2 \pm 1.3 \; 10^{-10}$M. On the same preparation VIP stimulated specifically the cAMP synthesis with an ED_{50} value of $2.9 \pm 1.6 \; 10^{-9}$M.

Light microscopic autoradiography showed the association of numerous grains with the stem villi arteries and arterioles (muscle layer and endothelium). Veins, veinules and capillaries were less labelled. Coincubation of ^{125}I-VIP with excess of unlabelled VIP resulted in a reduction of grains. Consequently adenylate cyclase activity was studied on isolated membranes of stem villi vessels obtained by fine mechanical dissociation and ultracentrifugations. The VIP stimulated, in a dose-dependent manner, the adenylate cyclase activity of these vascular membranes.

These results confirm the presence of VIP receptors positively coupled with adenylate cyclase on the fetal vascularisation and suggest a functional role for the peptide in placental hemodynamic.

Exogenous GnRH stimulates transcriptional and secretory rates as well as episodic secretion of human choriogonadotropin (hCG)

W.E. Merz, P. Licht, P. Harbarth

Department of Biochemistry II, University of Heidelberg, Im Neuenheimer Feld 328, D-6900 Heidelberg, Federal Republic of Germany

Gonadoliberin (GnRH) which is synthetized in the cytotrophoblast was suggested to stimulate secretion of hCG. The aim of this study was to characterize the effect of exogenous GnRH on hCG biosynthesis and secretion more closely. Placenta tissue (8.-10. week of gestation) was cultivated in our superfusion system. GnRH was added to the medium (M 199) by means of micropumps in two pulses (1 µM, 30 min) 24 h and 36 or 48 h after the beginning of the culture.
Three different effects of GnRH were observed: 1) A GnRH pulse caused a promt and transient significant ($p < 0.05$) increase in hCG secretion. The velocity of the change in hCG secretion rate suggests a direct effect of GnRH on the secretory pathway. 2) Several hours after application of the last GnRH pulse a constant increase of the hCG secretion (1.5-2 fold) was observed. HCG-α-subunit mRNA levels measured by means of ^{32}P-labeled cDNA were markedly increased in cultures treated with 10-1000 nM GnRH. This suggests a specific effect for GnRH in transcriptional control of hCG-biosynthesis. High concentrations (10 µM) as well as the application of GnRH superagonists resulted in a decrease of hCG secretion probably due to desensitization. 3) The hCG secretion pattern showed episodic peaks with a frequency between $(210-270)^{-1}$ min. The frequency as well as the amplitude of episodic hCG secretion seemed to be augmented by GnRH. Visual impression was supported when the data were inspected with different alogorithms for the detection of episodic hormone secretion (modified Santen and Bardin, Pulsar, Cycle detector). This confirms for the first time by in vitro experiments findings of others who recently had described a pulsatile hCG secretion in pregnant women (1) and in normal adults (2). Our results provide evidence for a specific modulation of the biosynthesis and episodic secretion of hCG by GnRH in vitro.

(1) Owens, O'D M., Ryan, K.J., Tulchinsky, D.: J. Clin. Endocrinol. Metab. 53, 1307-1309, 1981.
(2) Odell, W.D., Griffin, J.: N. Engl. J. Med. 27, 1688-1691, 1987

Estradiol and estradiol receptors are present in and around placental arteries in guinea-pigs

W. Moll, D. Scholl, H. Caffier, R. Götz

Institut für Physiologie, Universität Regensburg, D-8400 Regensburg, and Universitäts-Frauenklinik, Universität Würzburg, D-8700 Würzburg, Federal Republic of Germany

In order to test whether estradiol (E_2) is involved in the regulation of the adaptive changes of the uterine arteries supplying the placenta, local concentrations of E_2 and E_2 receptors (ER) in and around the placental arteries have been measured using conventional methods.

Whilst E_2 concentration in plasma was only ($\bar{x} \pm s$) 0.002 ± 0.0008 ng/ml, it was 0.5 ± 0.3 ng/g (N = 4) in the placental arteries, 0.6 ± 0.2 ng/g (5) in the myometrium, 1.2 ± 0.7 (11) ng/g in the proximal part of the decidua and near zero in the placenta in mid-pregnancy. Tissue estradiol competitively displaced the tracer from the RIA antibody. 1% of decidual and myometrial estradiol was ultrafiltrable. ER density in fmol/mg protein was 20 ± 9 in the placental arteries, 5 ± 7 in the myometrial arteries, 50 ± 15 in the adjacent myometrium, 12 ± 23 in the decidua and zero in the placenta (N = 9).

We conclude that E_2 is locally concentrated in and around the arteries supplying the placenta of the guinea pig and participates in the regulation of the placental arteries.

High-affinity specific binding for [^{125}I]ET-1 in membranes from human placenta vessels

F. Mondon, C. Robaut*, F. Ferré

INSERM U.166, maternité Baudelocque, Paris, France
* Rhone-Poulenc Santé, C.R.V., 13, quai Jules-Guesde, BP 14, 94403 Vitry-sur-Seine Cedex, France

Endothelins are a new family of potent vasoactive peptides produced by human [ET-2], porcine [ET-1] and rat [ET-3] endothelial cells. Their main pharmacological effect is a sustained contraction in several smooth muscles. The aim of this investigation was to assess whether binding sites for [^{125}I]ET-1 are present in membranes from human placenta blood vessels.

Stem villi vessels were separated from surrounding tissue by mechanical dissociation. Then, they were subjected to ultracentrifugation to prepare membranes for [^{125}I]ET-1 binding.

[^{125}I]ET-1 binding displayed high affinity and specificity. Furthermore, it was saturable, displaceable and confined to a single class of recognition sites. The values for the apparent dissociation constant (Kd) and for the maximum binding capacity were 0.027 ± 0.052 nM and 650 ± 88 fmol/mg prot. (n=5), respectively. The binding of [^{125}I]ET-1 was inhibited competitively by cold ET-1 with a Ki close to the Kd. Furthermore, [^{125}I]ET-1 binding was fully antagonized by ET-2 and ET-3. A variety of pharmacological agents including vasoactive peptides and calcium antagonists, did not exhibit affinity for [^{125}I]ET-1 binding sites. In contrast, sarafotoxin S6b, a snake cardiac depressant peptide toxin with structure homologous to endothelin, potently displaced the ET-1 radioligand with a Ki of 0.06 nM. The properties of the described binding do not differ from those found in rat aorta membranes, thus suggesting a strong similarity of rat and man vessel endothelin binding sites. In conclusion, [^{125}I]ET-1 binds to a population of high-affinity sites which may be counterparts of ET-1 receptors that can intervene in the physiological regulation of placenta vessel resistance.

Gestational profiles of the immunocytochemical localization of rat prolactin-like protein-B, rat placental lactogen II and pregnancy-specific β1-glycoprotein in rat placenta

S. Ogilvie, L.H. Larkin, M.L. Duckworth, K.T. Shiverick

Departments of Pharmacology and Therapeutics, Anatomy and Cell Biology, University of Florida, Gainesville, FL 32610, and Department of Physiology, University of Manitoba, Winnipeg, R3E OW3 Canada

The rat placenta expresses mRNAs encoding rat prolactin-like protein B (rPLP-B), rat placental lactogen II (rPLII) and pregnancy-specific ß1-glycoprotein (PSßG) during mid to late gestation. The predicted amino acid sequence of rPLP-B and rPLII show 44 and 52% homology, respectively, to pituitary prolactin, while human PSßG exhibits sequence homology to carcinoembryonic antigen. The physiological role(s) during pregnancy of none of these proteins has been established. Our studies examined the cytochemical localization of rPLP-B, rPLII and PSßG during mid to late gestation.

Anti-rPLP-B was generated against a chemically synthesized oligopeptide representing amino acids 186-200 in the deduced amino acid sequence of rPLP-B (Duckworth et al., 1988, Mol Endocrinol 2:912). Rabbit polyclonal anti-rPLII generated against an oligopeptide representing amino acids 56-70 in the deduced amino acid sequence of rPLII was provided by M.J. Soares (Deb et al., 1989, Mol Cell Endocrinol 63:45). Rabbit anti-human PSßG was from (DAKO. Paraffin sections from rat placentas day (d) 11 to d 17 of gestation were treated with each of these three antibodies followed by avidin-biotin peroxidase complex. Rat PLP-B immunostaining is seen as early as d 11 of gestation in some of the basophilic cytotrophoblast cells of the basal zone and was localized in the perinuclear area. As gestation proceeds, the perinuclear staining becomes more intense with an increasing number of the cytotrophoblast cells staining. Interestingly, fetal red blood cells (rbc) stain intensely with anti-rPLP-B from d 11 through d 17 of gestation. In contrast, rPLII immunostaining is seen in the giant trophoblast cells at d 11 and 12 of gestation. By d 13 of gestation, not only is immunostaining observed in the giant cells, but is observed in the fetal rbc and the basophilic cytotrophoblast cells as well. This staining pattern persists through d 17 of gestation. As was observed with rPLP-B, not all of the cells stain with anti-rPLII, but there is a gradual increase in the proportion as gestation proceeds. PSßG immunostaining is observed in the giant trophoblast cells at d 11, 12 and 13 of gestation. By d 14 of gestation immunostaining is observed in the basophilic cytotrophoblast cells and by d 15 nearly all of the cytotrophoblast cells stain intensely, with only a few giant trophoblast cells now showing immunoreative product. No immunoreactive PSßG was detected in fetal rbc from d 11 to d 17 of gestation.

The progressive changes in the localization of these three proteins may indicate changes in the differentiation of gene expression in the rat placenta, as well as in the functional role(s) of the proteins during mid to late gestation.

Effect of extracellular calcium and potassium depolarization on the chorionic gonadotropin and placental lactogen releases by human placental explants

B. Polliotti, S. Meuris, P. Lebrun, C. Robyn

Human Reproduction Research Unit and Laboratory of Pharmacology, Université libre de Bruxelles, Saint Pierre Hospital, 322 rue Haute, B-1000 Brussels, Belgique

Calcium ion (Ca^{++}) is crucial for exocytosis in most secretory cells. In placental cells, the participation of Ca^{++} in the secretory mechanisms as well as the modalities of Ca^{++} entry remain unclear. The aim of the study was to investigate the effects of Ca^{++} and depolarizing K^+ concentration on the release of chorionic gonadotropin (hCG) and placental lactogen (hPL).

Placentas were obtained after vaginal delivery from normal term pregnancies. Cotyledons were minced into explants. Explants (n=20) were placed into vials containing 3ml of Krebs-Ringer solution under a 95% O_2 - 5% CO_2 atmosphere at 37°C. The medium was replaced every 30 min. Preincubation lasted 150 min and incubation 30 min. Two conditions of preincubation were used: normoCa^{++} (1.5mM) and alphaCa^{++} (no Ca^{++} and 0.5mM EGTA). During incubation, several concentrations of Ca^{++} and K^+ were investigated: normoCa^{++}, alphaCa^{++}, hyperCa^{++} (10mM), normoK$^+$ (5mM) and hyperK$^+$ (50 and 80mM). The last preincubation and incubation medium were radioimmunoassayed for hCG and hPL. The amounts of hCG and hPL released (mean ± sem) during incubation were expressed as the percentages of the amounts released at 150 min.

Preincubation	Incubation	N	hCG %			hPL %		
NormoCa^{++}	normoCa^{++} (C)	36	83.97	±	3.48	85.96	±	5.33
NormoCa^{++}	alphaCa^{++}	12	93.21	±	9.37	310.08	±	79.83**
NormoCa^{++}	hyperCa^{++}	12	238.18	±	27.75**	175.84	±	23.83**
AlphaCa^{++}	alphaCa^{++}	12	80.48	±	5.81	85.72	±	6.99
AlphaCa^{++}	normoCa^{++}	12	238.21	±	16.09**	497.51	±	198.63*
AlphaCa^{++}	hyperCa^{++}	12	752.91	±	40.21**	1117.27	±	247.41**
NormoK$^+$	normoK$^+$ (C)	12	94.25	±	6.93	94.39	±	6.59
NormoK$^+$	50mM K$^+$	12	83.73	±	5.42	99.59	±	9.37
NormoK$^+$	80mM K$^+$	12	94.45	±	7.19	108.74	±	7.24

Significance of comparisons with controls (C), * p<0.05 and ** p<0.001

Releases were always significantly stimulated when concentration of Ca^{++} was higher during incubation than preincubation. The significant increase of hPL observed in alphaCa^{++} condition after normoCa^{++} preincubation was not observed when explants were washed, before incubation, 5x 1 min with the alphaCa^{++} medium (104.70 ± 15.20 %). Increasing amounts of K^+ in the incubation medium had no significant effect on hCG and hPL releases.

These data indicate that extracellular Ca^{++} stimulates hCG and hPL releases from trophoblastic tissue, as reported for numerous secretory cells. The failure of high K^+ concentrations to affect hPL and hCG releases suggest that modifications in membrane potential are not essential for these normal secretory processes.

The presence of gonadotropin receptors in human placenta, amnion, chorion and decidua

Ch.V. Rao, Z.M. Lei

Department of Obstetrics and Gynaecology, University of Louisville, School of Medicine, Louisville, KY 40292, USA

    Light microscope immunocytochemistry with monoclonal antibody to rat luteal human chorionic gonadotropin/luteinizing hormone (hCG/LH) receptors was used to investigate the possible presence of these receptors in the human fetoplacental unit. The receptor antibody cross reacted with human and bovine hCG/LH receptors and appears to be directed against the receptor rather than other proteins including HLA class I antigens. Mid pregnancy placenta, amniotic epithelium, chorionic cytotrophoblasts and decidual cells contained receptor antibody binding sites which indicates the presence of hCG/hLH receptors. At term pregnancy, while receptors in fetal membranes and decidua continue to be detected, placental tissues did not show any detectable receptors unless the tissues were pretreated with neuraminidase. This indicated that term pregnancy placenta contained hCG/hLH receptors masked by sialic acid residues. Comparison of immunostaining indicated that syncytiotrophoblasts contained more receptors than cytotrophoblasts at mid pregnancy; mesenchymal cells or blood vessels contained no detectable receptors. There were more receptors in decidua than fetal membranes at mid and term pregnancy. While the amniotic epithelial receptors decreased, the receptors in chorionic cytotrophoblasts and decidual cells increased from mid to term pregnancy. In summary, the presence of hCG/hLH receptors in human placenta, fetal membranes and decidua suggests that hCG may possibly regulate functions of human fetoplacental unit by autocrine, paracrine and endocrine mechanisms. (Support NIH HD14697).

Sterylsulfatase activity of human placental cells in monolayer culture

B. Ugele, L. Dibbelt*, E. Kuss

I. Frauenklinik der Universität, Maistrasse 11, D-8000 München 2, Federal Republic of Germany
*Institute of Biochemistry, Endocrinology, Medical University, Natzburger Allee 160, D-2400 Lübeck 1, République fédérale d'Allemagne

Recently, we purified and characterized the human placental steroid sulfatase (STS) (1). Immunocytochemical studies with a monospecific antiserum revealed nearly exclusive staining of syncytiotrophoblasts (2). Now we have focussed our attention on the ontogeny of this enzyme.

Human mononucleated cells from term placenta, presumably cytotrophoblasts, were separated according to Kliman et al. (3) and maintained in monolayer culture with medium 199 supplemented with 20 % fetal calf serum for up to 1 week. The STS specific activity of freshly isolated cells was low compared to that of placental tissue before trypsin treatment (about 30 %; substrate: DHEA-S). In cultured placental XY-cells this activity did not change, in XX-cells, however, this activity did increase about 2-fold within 90 h in parallel with syncytium formation. The assumption that STS activity is of cytotrophoblastic origin was supported by immunostaining of both short-term cultured cells and slices of immature placenta with clearly identifiable cytotrophoblasts (age 12 weeks). As expected, placental type alkaline phosphatase could not be detected by the corresponding antiserum in these preparations.

In summary, the results demonstrate a distinct STS activity in mononucleated placenta cells and suggest a XX-dependent increase during formation of the syncytium.

(1) Dibbelt and Kuss, Biol. Chem. Hoppe-Seyler 367, 1233 (1986)
(2) Dibbelt and Kuss, in: Mol. Biol. and Cell Regulation of the Placenta (1988)
    (11. Rochester Trophoblast Conference)
(3) Kliman et al., Endocrinology 118, 1567 (1986)

Extracellular adenine nucleotides in human trophoblastic purine nucleotide synthesis

K. Vettenranta, K.O. Raivio

Children's Hospital, University of Helsinki, SF-00290, Helsinki Finland

ATP and other purine nucleotides appear to be synthesized through reutilization of pre-existing purine bases and/or nucleosides rather than de novo in the human trophoblast throughout gestation. However, the origin of the reutilizable purines is unknown. We studied the ability of cultured human trophoblasts to use extracellular adenine nucleotides as precursors in their intracellular nucleotide synthesis. First and third trimester placentae were obtained from elective terminations and caesarean sections, respectively. Primary, short-term cultures of trophoblasts were established using collagenase treatment. The cultured cells were incubated in the presence of 2 µM deoxycoformycin and 14-C-ATP (1, 5 or 10 µM) or 14-C-ADP (1, 5 or 15 µM) for up to 4 hours. Parallel experiments in the presence of 250 µM alpha-beta-methylene-ADP or 10 µM dipyridamole plus 10 µM adenine were also performed. Purine compounds were separated using thin-layer chromatography. The metabolic integrity of the cells was monitored with the adenylate energy charge [mean 0.71 (0.50-0.89)]. Both the first and third trimester trophoblasts dephosphorylated extracellular adenine nucleotides to adenosine, which was then intracellularly rephosphorylated. Adenine nucleotides constituted more than 80% of the nucleotides thus formed. The contribution of specific 5'-nucleotidase to the dephosphorylation of extracellular AMP was $7 \pm 6\%$ in the first and $37 \pm 15\%$ in third trimester cells. Uptake of intact nucleotides did not take place. Incorporation of radioactivity into intracellular nucleotides increased as a function of time and precursor concentration in both the first and third trimester trophoblasts. The utilization of 14-C-ADP was significantly ($p<0.05$) more active in the third trimester cells while no change with gestation in that of -ATP was seen. In conclusion, it appears that throughout gestation the human trophoblast is able to degrade extracellular ATP and ADP to adenosine, which is further employed as a reutilizable substrate in intracellular nucleotide synthesis.

3.
Growth factors and differentiation
Facteurs de croissance et différenciation

Growth factors and trophoblast differentiation

D. Evain-Brion

Laboratoire de physiopathologie du développement, CNRS URA 1337, Ecole normale supérieure, 46, rue d'Ulm, 8ᵉ étage, 75230 Paris Cedex 05, France

Embryonic development is mainly the results of temporally and locally coordinated events, that are proliferation, adhesion, migration, differentiation and death of cells. The behavior of each cell depends largely on its degree of interaction with its neighbours and with extracellular environment. These interactions involve adhesion process and soluble factors.

The cell adhesion molecules (CAMs) generally large intrinsic cell surface glycoproteins mediate with specific subcellular structures, (cytoskeleton, tight and gap junctions, desmosomes, focal plaques) the ligation between cells. The substrate adhesion molecules (SAMs) are involved in the interactions between cells, basal laminae and extracellular matrix. They include the fibronectins, laminin and cytotactin (Thiery, 1989).

The cells interact also with their neighbours with soluble factors. Among them growth factors play a major role in the regulation of cell growth and differentiation.

In this paper we will briefly summarize the different types of growth factors before focusing on their specific role in placental development.

Growth factor families and development

Growth factors are polypeptide factors, circulating with or without specific binding proteins, in an active or inactive form. They bind to specific plasma membrane receptors on their target cells and subsequently initiate a rapid chain of events which eventually results in DNA duplication and cell division. First described for their mitogenic effect, these peptide growth factors have a spectrum of biological actions which include the regulation of tissue morphology, differentiation, movement and functional activity. The expanding number of growth factors can be grouped into superfamilies based on nucleotides and amino acid sequence homology as well as similar receptor binding activity.

The EGF family has three members which have a proven mitogenic activity and bind to a common receptor. TGF a was initially identified in the conditioned medium of transformed cells (De Larco and Todaro, 1978) but more recently normal cells in various tissue, including the placenta in the rodents, have also been found to produce TGF a . Whereas EGF is synthesized by exocrine cells and discharged in saliva, milk, or urine, locally produced TGF a may be the endogenous ligand for the EGF receptor in various tissues.

The family of fibroblast growth factor includes so far five members. The two prototypic FGFs, acidic and basic, originally purified as monomeric 16-18 KDa proteins from normal brain and pituitary respectively, were found to bind tightly to heparin and were shown to stimulate mitogenesis of a wide variety of cells of mesodermal and neuroectodermal origin.

The family of platelet derived growth factor (PDGF) is made up by dimers of two different but related polypeptide chains (A and B). PDGF is the main mitogen for connective tissue derived cells.

The insulin-like growth factor (IGFI, IGFII) are a pair of growth factors with amino acid sequence and structural similarity to human pro-insulin.

Transforming growth factor beta (TGF ß) is a dimer expressed in a wide variety of cell types. TGF ß illustrates the multiplicity of responses elicited by growth factor action. Depending on the cell type and on other growth factors acting on the cell, TGF ß may either promote or inhibit cell growth and differentiation.

The interleukins were first defined as signalling molecules controlling activities of cells in the immune system.

Tumor Necrosis factor and the various hematopoietic growth factors may play a specific role in the placenta which will be discussed further on.

The specificity of the signal mediated by polypeptide growth factors lies in their interaction with specific cell surface receptors. . The binding of growth factor to its receptor elicits a cascade of events including protein phosphorylation, (Hunter and Cooper, 1985) inositol lipid breakdown (Berridge, 1986; Nishizuka, 1986), ion fluxes (Goustin et al., 1986) and changes in gene expression.

Of great interest is the fact that these polypeptide factors, firstly named for their growth promoting effect, have multiple functions; some of them are of prime interest during early development. Some factors have a chemotactic activity, influencing therefore cell migration. PDGF is a chemotactic agent for fibroblasts (Seppä et al., 1982) and muscle cells, TGFß for fibroblasts (Postlewhaite et al., 1987). TGF ß has an extensive role in enhancing the formation of extra cellular matrix. TGFß stimulates the formation of both collagen and fibronectin in fibroblasts (Ignotz and Massagué, 1986; Roberts et al., 1986) and inhibits their degradation by decreasing the synthesis and secretion of proteases, and increasing the synthesis and secretion of protease inhibitors (Laiho et al., 1986; Chiang and Nelsen-Hamilton, 1986, Matrisian et al., 1986).

The b FGF has a strong affinity for extracellular matrix (ECM) heparan sulfate proteoglycans (HSPG) and glycosaminoglycans (GAG) (Saksela et al., 1988; Vigny et al., 1988). This matrix binding of FGF may provide a reservoir of growth factor which varies with the remodeling of ECM through the production of hyaluronidase.

Endly TGFß and FGF have been shown to be morphogens, inducing the formation of mesoderm in Xenopus egg (Rosa et al., 1988, Kimelman and Kirschner, 1987).

Growth factors and placental development

The placenta is a very rich source of growth factors and growth factors receptors. We will focus on the growth factors isolated from or modulating the functions of the individual cells composing the chorionic villi.

Two types of chorionic villi are found at the maternal-fetal interface: floating and anchoring villi. In floating villi the cytotrophoblast cells lie beneath the syncytiotrophoblastic covering and are separated from the underlying connective tissue stroma by a basement membrane. The anchoring villi contain an additionnal population of cytotrophoblasts that adhere to the uterine epithelium, migrate into the endometrium and invade the spiral arterioles.

Inside the chorionic villi, the cytotrophoblasts, epithelial cells, play a major role in placental development. The knowledge of the biology of these cells have been greatly improved by the ability of culturing in vitro purified trophoblast cells (Kliman et al., 1986, 1987). In the presence of fetal calf serum, the mononuclear cytotrophoblasts flatten out onto the culture surface, aggregate and form syncytia over a 24 h to 96 h period. There have been discussions about the fact that this morphological differentiation is linked to biochemical differentiation such as the ability to secrete chorionic gonadotropin (hCG) and placental lactogen (hPL) as shown by Kao et al., 1988.

Recent studies address evidences that growth factors may play a major role in the growth and differentiation of cytotrophoblasts.

Cytotrophoblasts from first trimester placenta are highly proliferative and have invasive properties. They produce IGF2 (Ohlsson et al., 1989; Brice et al., 1989) and PDGF (Goustin et al., 1985, Taylor et al., 1988), as well as specific metalloproteinases (Fisher et al., 1989) which are able of degradating the extracellular matrix. In addition they have an alteration in the glycosylation of their substrate adhesive molecules, including fibronectin, which could be related to their invasiveness (Moss et al., 1988). These properties are lost in term placenta cytotrophoblasts. IGF2 can stimulates weakly in vitro the growth of cytotrophoblasts and mesenchymal cells (Ohlsson et al., 1989).

The PDGF receptors have been found to be localized on the trophoblasts and the stromal villous cores suggesting an autocrine or paracrine effect of this growth factor.

The interleukin 1 which is produced by isolated cytotrophoblasts, monocytes and decidual cells stimulates hCG secretion by first trimester trophoblast but does not modify cell growth or differentiation (Yagel et al., 1989). The interleukin 2 gene is expressed in the syncytiotrophoblast (Boehm et al., 1989)

TGFß is expressed as shown by in situ hybridization in both the villus core and syncytiotrophoblast layer of the placenta (Boehm et al., 1988). TGFß does not influence hCG or progesterone release from cultured placental cells (Petraglia et al., 1989). However, due to its pleiotropic functions TGFß may play a major role in the regulation cytotrophoblast growth, possibly in synergism with IGF2 or PDGF. Its role is also possible in pregnancy maintenance as a locally produced immunosuppressant.

Endly, a 34Kd protein growth factor, firstly isolated from syncytial membranes of human placenta has been shown to be secreted in vitro by freshly isolated cytotrophoblasts and in cells that have fused in culture to form multinuclear syncytiotrophoblast, but its role remains to be tested (Roy-Choudhury et al., 1988).

EGF receptors are localized on the cytotrophoblasts, but the majority of them are located on the mitotically inactive syncytiotrophoblasts suggesting a role for the EGF receptor in the induction of differentiated function of trophoblast rather than trophoblast proliferation (Chegini and Rao, 1985; Magid et al., 1985).

Of interest is the fact that [^{125}I] EGF binding to isolated cultured cytotrophoblasts can be modulated by some polypeptide hormones such as parathyroid hormone (PTH). At physiological dose (Alsat et al., 1989) PTH increases by two to three fold the EGF receptor number on cytotrophoblast cells without modification in their affinity.

The expression of IGF2 and IGF1 receptors predominates in low proliferative cytotrophoblasts (Ohlsson et al., 1989), while the CSF receptors seem predominant on the syncytiotrophoblast as shown by in situ hybridization.

Proto-oncogenes and trophoblast differentiation

The placenta is an unusually active tissue in proto-oncogene expression, recently reviewed by Adamson (1987). Of interest the expression of the protooncogene c-sis homologous to the PDGF ß chain is limited to the early gestation with a peak of expression at the 10th week of gestation.

In conclusion, the trophoblast performs many important nutritional and immunological functions during pregnancy. In the trophoblastic villi, the cytotrophoblasts are fascinating cells able to growth, differentiate and invade neighbour tissue. It is clear that our knowledge of their behavior and interaction with their neighbours and extracellular environment is still very poor. The recent progress in cell and molecular biology have allowed to describe the production and the presence of receptors for some soluble factors, such as growth factors which by their pleiotropic effects on cell function play certainly a major role in regulation of trophoblast function. Progress in the knowledge of the role of growth factors in trophoblast functions may allow to understand some pathological aspects of pregnancy such as intrauterine growth retardation.

Résumé

Au cours du développement embryonnaire les fonctions cellulaires sont modelées par l'adhérence spécifique des cellules entre elles, leur adhérence avec la matrice extracellulaire et par un grand nombre de facteurs diffusibles incluant les facteurs de croissance.

Les facteurs de croissance sont des polypeptides se liant à un récepteur cellulaire spécifique; ils sont regroupés en famille selon des homologies de séquence nucléotique ou d'amino-acides; ou de communauté de récepteur.

Le placenta est une très riche source de facteurs de croissance ou de leur récepteur. A l'intérieur de la villosité choriale, les cytotrophoblastes jouent un rôle essentiel dans le développement placentaire. Les cytotrophoblastes du placenta du premier trimestre sont hautement prolifératifs et invasifs. Ils produisent de l'IGF2 et du PDGF. La présence de TGFß a été révélée par hybridation in situ, au niveau de l'axe mésenchymateux et de la villosité choriale, jouant probablement un rôle dans la régulation de la croissance des cytotrophoblastes.

Les récepteurs à l'EGF sont nombreux sur les cellules cytotrophoblastiques et augmentent avec la différentiation syncytiale, in vitro. Les récepteurs au CSF prédominent sur les syncytiotrophoblastes.

Ainsi au sein de la villosité choriale la régulation de la croissance, de la différentiation et de l'invasivité des cellules cytotrophoblastiques, sont sous la dépendance de nombreux facteurs de croissance dont certains viennent juste d'être connus.

REFERENCES

Adamson, E.D. (1987). Expression of proto-oncogenes in the placenta. Placenta, 8: 449-466.

Alsat, E., Mirlesse, V., and Evain-Brion, D. (1989). Parathyroid hormone increases epidermal growth factor receptors in cultured human trophoblastic cells from early and term placenta. J. Clin. Invest., Submitted.

Berridge, M.J., 1986, , (1986). Intercellular signalling through inositol triphosphate and diacylglycerol. Biol. Chem. Hoppe Seyler, 367: 447-456.

Boehm, K.D., Engelmann, G.L., Sparks, M., Ruscetti, F., and Han, J. (1988). Transforming growth factor-beta (TGF-β1) gene expression in developing human placenta (Abstract 264). J. Cell Biol., 107: 6-.

Boehm, K.D., Kelley, M.F., Ilan, J., and Ilan, J. (1989). The interleukin 2 gene is expressed in the syncytiotrophoblast of the human placenta. Proc. Natl. Acad. Sci. USA, 86: 656-660.

Brice, A.L., Cheetham, J.E., Bolton, V.N., Hill, N.C.W., and Schofield, P.N. (1989). Temporal changes in the expression of the insulin-like growth factor II gene associated with tissue maturation in the human fetus. Development, 106: 543-554.

Chegini, N., and Rao, C.V. (1985). Epidermal growth factor binding to human amnion, chorion, decidua and placenta from mid- and term pregnancy: Quantitative light microscopic autoradiographic studies. J. Clin. Endocrinol. Metab., 61: 529-535.

Chiang, C.-P., and Nelsen-Hamilton, M. (1986). Opposite and selective effects of epidermal growth factor and human platelet transforming growth factor-beta on the production of secreted proteins by murine 3T3 cells and human fibroblasts. J. Biol. Chem., 261: 10478-10481.

De Larco, J.E., and Todaro, G.J. (1978). Growth factors from murine sarcoma virus-transformed cells. Proc. Natl. Acad. Sci. USA, 75: 4001-4005.

Fisher, S.J., Cui, T.-Y., Zhang, L., Hartman, L., Grahl, K., Guo-Yang, Z., Tarpey, J. and Damsky, C.H. (1989). Adhesive and degradative properties of human placental cytotrophoblast cells in vitro. J. Cell Biol., 109: 891-902.

Goustin, A.S., Betsholtz, C., Pfeifer-Ohlsson, S., Persson, H., Rydnaert, J., Bywater, M., Holmgren, G., Heldin, C.H., Westermark, B. and Ohlsson, R. (1985). Coexpression of the sis and myc oncogenes in developing human placenta suggests autocrine control of trophoblast growth. Cell, 41: 301-312.

Goustin, A.S., Leof, F.B., Shipley, G.D., and Moses, H.L. (1986). Growth factors and cancer. Cancer Res., 46: 1015-1029.

Hunter, T., and Cooper, J.A. (1985). Protein tyrosine kinases-A. Rev. Biochem., 54: 897-930.

Ignotz, R., and Massagué, J. (1986). Transforming growth factor-beta stimulates the expression of fibronectin and collagen and their incorporation into the extracellular matrix. J. Biol. Chem., 261: 4337-4345.

Kao, L.-C., Caltabiano, S., Wu, S., Strauss III, J.F., and Kliman, H.J. (1988). The human villous cytotrophoblast: Interactions with extracellular matrix proteins, endocrine function, and cytoplasmic differentiation in the absence of syncytium formation. Develop. Biol., 130: 693-702.

Kimelman, D., and Kirschner, M. (1987). Synergistic induction of mesoderm by FGF and TGF ß and the identification of an mRNA coding for FGF in the early Xenopus embryo. Cell, 51: 869-877.

Kliman, H.J., Nestler, J.E., Sermasi, E., Sanger, J.M., and Strauss III, J.F. (1986). Purification, characterization and in vitro differentiation of cytotrophoblasts from human term placentae. Endocrinol., 118: 1567-1582.

Kliman, H.J., Feinman, M.A., and Strauss III, J.F. (1987). Differentiation of human cytotrophoblast into syncytiotrophoblast in culture. Trophoblast Res., 2: 407-421.

Magid, M., Nanney, L.B., Stoscheck, C.M., and King, Jr. L.E. (1985). Epidermal growth factor binding and receptor distribution in term human placenta. Placenta, 6: 519-526.

Matrisian, L.M., Leroy, P., Ruhlmann, C., Gesnel, M.-C., and Breathnach, R. (1986). Isolation of the oncogene and epidermal growth factor-induced transin gene: Complex control in rat fibroblasts. Mol. Cell Biol., 6: 1679-1686.

Moss, L., Fisher, S.J., and Damsky, C.H. (1988). Stage and tissue specificity in the glycosylation of human trophoblast integrins (Abstract 865). J. Cell Biol., 107: 6-.

Nishizuka, Y. (1986). Perspectives on the role of proteinkinase C in stimulus-responsive coupling. J. Natn. Cancer Inst., 76: 363-370.

Ohlsson, R., Holmgren, L., Glaser, A., Szeppecht, A., and Pfeifer-Ohlsson, S. (1989). Insulin-like growth factor 2 and short-range stimulatory loops in control of human placental growth. EMBO J., 8: 1993-1999

Petraglia, F., Vaughan, J., and Vale, W. (1989). Inhibin and activin modulate the release of gonadotropin- releasing hormone, human chorionic gonadotropin, and progesterone from cultured human placental cells. Proc. Natl. Acad. Sci. USA, 86: 5114-5117.

Postlewhaite, A.E., Keski-Oja, J., Moses, H.L., and Kang, A.H. (1987). Stimulation of the chemotactic migration of human fibroblasts by transforming growth factor ß. J. Exp. Med., 165: 251-254.

Roberts, A.B., Sporn, M.B., Assoian, R.K., Smith, J.M., Roche, N.S., Wakefield, L.M., Heine, U.I., Liotta, L.A., Falanga, V., Kehrl, J.H. and Fauci, A.S. (1986). Transforming growth factor type-beta: Rapid induction of fibrosis and angiogenesis in vivo and stimulation of collagen formation in vitro. Proc. Natl. Acad. Sci. USA, 83: 4167-4171.

Rosa, F., Roberts, A.B., Danielpour, D., Dart, L.L., Sporn, M.B., and Dawid, I.B. (1988). Mesoderm induction in amphibians: The role of TGF-ß2-like factors. Science, 239: 783-785.

Roy-Choudhury, S., Sen-Majumdar, A., Murthy, U., Mishra, V., Kliman, H.J., Nestler, J.E., Strauss III, J.F. and Das, M. (1988). Biosynthesis and turnover of a 34-kDa protein growth factor in human cytotrophoblasts. Eur. J. Biochem., 172: 777-783.

Saksela, O., Moscatelli, D., Sommer, A., and Rifkin, D.B. (1988). Endothelial cell derived heparan sulfate binds basic fibroblast growth factor and protects it from proteolytic degradation. J. Cell Biol., 107: 743-751.

Seppä, H., Grotendorst, G., Seppä, S., Schiffmann, E., and Martin, G. (1982). Platelet derived growth factor is chemotactic for fibroblasts. J. Cell Biol., 92: 584-588.

Taylor, R.N., Wilcox, J.N., Doldsmith, P.C., and Williams, L.T. (1988). In situ localization of PDGF and PDGF-receptor in the midtrimester human placenta (Abstract 2694). J. Cell Biol., 107: 6-.

Thiery, J.P. (1989). Cell Adhesion in Morphogenesis. In "Cell to Cell Signals in Mammalian Development". S.W. de Laat, J.G. Bluemink, C.L. Mummery, eds Springer Verlag. pp. 109-128.

Vigny, M., Ollier-Hartmann, M.P., Lavigne, MFayein, N., Jeanny, J.C., Laurent, M., and Courtois, Y. (1988). Specific binding of basic fibroblast growth factor to basement membrane-like structures and to purified heparan sulfate proteoglycan of the EHS tumor. J. Cell. Physiol., 137: 321-328.

Yagel, S., Lala, P.K., Powell, W.A., and Casper, R.F. (1989). Interleukin-1 stimulates human chorionic gonadotropin secretion by first trimester human trophoblast. J. Clin. Endocrinol. Metab., 68: 992-995.

Biochemical and biological characterization of a crude growth factor extract (EAP) from placental tissue

J. Tiollier, S. Uhlrich, V. Chirouze, M. Tardy, J.-L. Tayot

IMEDEX BP 38, 69630 Chaponost, France

Placenta is the richest human source of proteins, glycoproteins, hormones and lipids, In the past decade, several new placental proteins have been isolated and studied.
Also, most known growth factors had already been described in placenta : Fibroblast Growth Factor (FGF), ß-Transforming Growth Factor (TGFß), monocyte-macrophage Colony Stimulating Factor (CSF1), Insuline-like Growth Factor 1 and 2 (IGF), etc..... Therefore, placental tissue may be a useful source of human growth factors on an industrial scale.

In this paper, we report the optimization of crude growth factor extract obtained from placental tissue. Biochemical characterization of its content in proteins, hormones, lipids and growth factors and the biological characterization of its growth factor activities in cell culture assays are described.

The EAP is prepared at an industrial scale from human placental tissue by acid extraction and alcohol fractionation. The results described below were obtained on three different pilot batches of EAP which ensure the reproducibility of the results.

The major proteins (gammaglobulins (11 %), transferrin (3 %), albumin (6 %), peptides < 20 kDa (65 %)) of this extract were characterized by polyacrylamide gel electrophoresis. The concentrations of the different known proteins, particularly plasma proteins, lipoproteins, transport and adhesion proteins, were determined by immunodiffusion (Table 1).

	Concentration (mg/l)
Total Proteins	10000 - 13000
Albumin	170 - 320
Transferrin	130 - 270
IgG	450 - 850
IgA	40 - 60
Hemopexin	5 - 7
Haptoglobin	20 - 40
α-Antitrypsin	70 - 120
α-Antichymotrypsin	40 - 70
α_1-Acid glycoprotein	4 - 11
Lactoferrin	40 - 60
β_1 sp$_1$ glycoprotein	110 - 130
Antithrombin III	5 - 15
Fibronectin	100 - 120
Apolipoprotein A	400 - 650

Table 1 : Quantitative characterization of majors proteins in EAP
Titrations made by radial immunodiffusion techniques.
Average concentration in the three representative batches.

The lipids were identified by thin-layer chromatography. The extracted lipids were essentially composed of phospholipids (56 %), free fatty acids (12 %) and free cholesterol (29 %). The steroid hormones were determined by RIA titrations and only low levels (under ng) could be detected.

The EAP was mainly manufactured to get a growth factor extract. Its content in growth factors was estimated by specific techniques designed for each factor.
The lowest levels were detected for IGF-2 (15-140 ng/ml), while EGF and bFGF concentrations were in the same range, i.e. 100-250 ng/ml. The highest levels were detected for IGF-1 (350-1300 ng/ml) and especially TGFß (0.9-4.8 µg/ml)(Table 2).

Growth factors	Type of titration	Concentration in EAP (ng/ml)	Concentration in calf serum (ng/ml)
bFGF	Biochemical : Chromatography +electrophoresis	100 - 250	< 10
EGF	Radioreceptor assay	100 - 130	15 - 30
TGFß	Radioreceptor assay	900 - 4800	
IGF-1	RIA	350 - 1300	30 - 400
IGF-2	Radioreceptor assay	15 - 140	

Table 2 : Quantitative growth factor composition in EAP

The growth promoting activities of EAP were determined by cell culture assays. EAP stimulates proliferation of various cells (MRC5 and various fibroblasts, corneal endothelial cells). The ED50 (half maximal stimulation) has been found to be in the range of 100-200 µg/ml(Fig. 1)

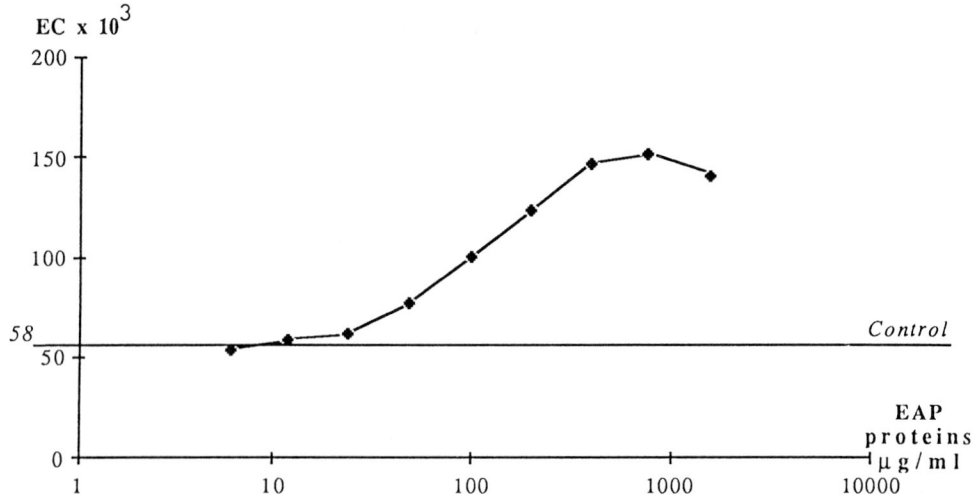

Fig. 1 : Effect of increasing concentrations of EAP on bovine corneal endothelial cells

The data presented here show that EAP exhibits growth factor activities. Thus, the main points of interest of such an extract are :

- for purification to homogeneity of isolated (known or unknown) growth factors ;

- for serum replacement in serum free or low serum protein cell culture media to promote proliferation and differentiation of many different cells ;

- as a potentially useful phamacological agent in wound healing.

Workshop on growth factors and placental differentiation

D. W. Morrish, Ch. V. Rao

7-117 Clinical Science Building, Department of Medicine, Edmonton, Alberto IGG 263, Canada

The workshop focused on the emerging role of growth factors in placental cell proliferation and differentiation. Dr. D. Morrish introduced the session with a summary of growth factors and differentiation as they relate to the placenta. A growth factor was defined as a substance that is not an ion, co-factor, or a substrate and which induces proliferation and/or differentiation in a target cell and/or differentiated function. In addition to classical growth factors, some hormones such as hCG and insulin, etc., may also qualify for this definition. Differentiation of the trophoblast was described as twofold: the morphological fusion of cytotrophoblasts to form a syncytium and the modulation of biochemical properties of syncytio as well as cytotrophoblasts with respect to steroids, peptide hormones, growth factors and eicosanoids synthesis and secretion, receptor expression and ion/substrate transport functions. It should be noted that the cytotrophoblasts are not just proliferating stem cells but have considerable endocrine properties such as synthesis and secretion of a variety of hormones and growth factors, expression of receptors for them (Belisle et al, J Receptor Res 8:391, 1988), (EGF abstracts by C-J Yeh et al, DW Morrish et al), (CSF-1 receptors EGF abstract by M. Garcia-Lloret et al), expression of hCG genes and cyclic AMP regulation of steroid and protein hormones (Strauss et al, J Clin Endocrinol Metab 64:1002, 1987). Many cell types in the placenta may interact to promote trophoblast growth by the secretion of growth factors. A list of the cells, factors secreted, and target cells is in the accompanying table. A variety of possible interactions based on this is shown in the schematic diagram (Fig. 1). The placenta is anatomically structured to be an autocrine/paracrine system involving these factors: cytotrophoblast underlies the syncytium, and macrophages and fibroblasts are located in close proximity in the interstitium of the villus. Our knowledge of how these different factors interact in the placenta clearly lags behind other fields such as the ovary (Clinics Endocrinol Metab 15:117, 1986). Discussion of the introduction yielded an important additional concept: that is, we still lack evidence of a clear endocrine regulatory system in the placenta. We do not know of endocrine factors that control hCG and hPL secretion for example, although many paracrine factors have been described. Does this mean that such factors do not exist and that placental regulation is purely autocrine/paracrine? In this regard, Dr. Ch.V. Rao alluded to his data (presented at the workshop on Receptors and Endocrine Regulatory Function; Reshef et al., J Clin Endocrinol Metab, 1990, In press) on the existence of hCG receptors in syncytio and cytotrophoblasts as well as in amniotic epithelium, chorionic cytotrophoblasts and decidual cells. This might enable the placenta to have feedback regulation of its own secretion, as well as secretion of steroids, protein hormones, growth

Figure 1

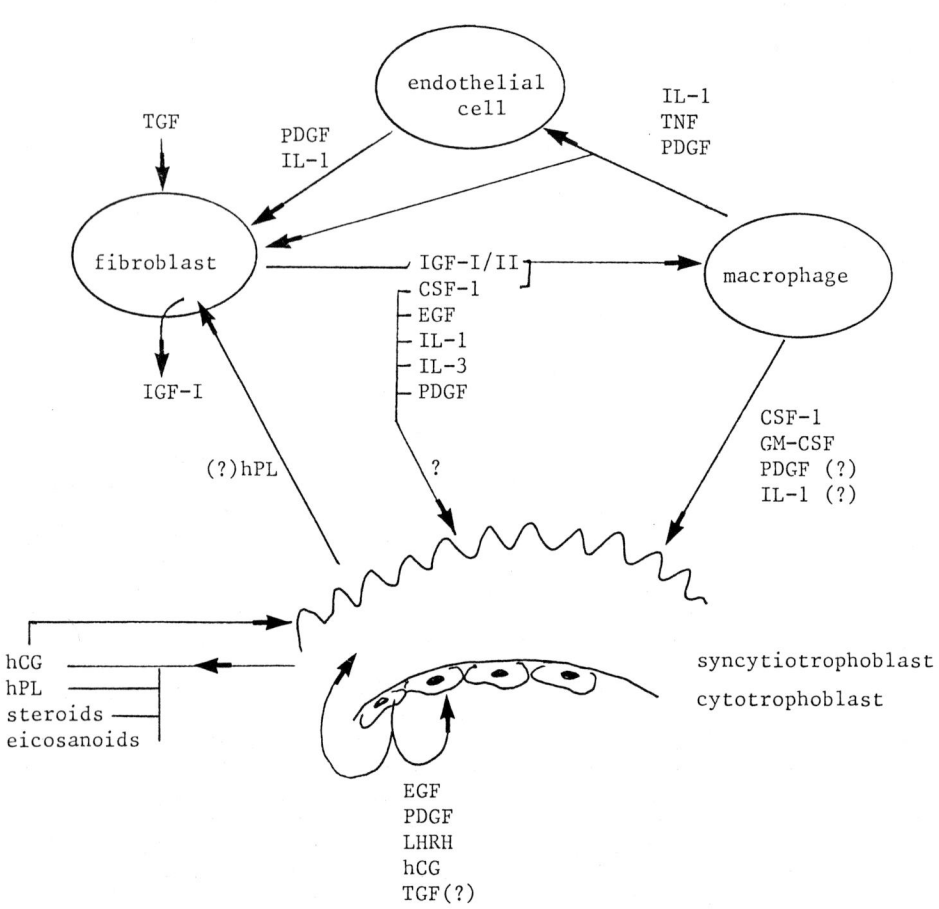

factors and eicosanoids by intracrine mechanisms. The presence of receptors in fetal membranes and decidua might allow for paracrine and/or endocrine roles for hCG in these tissues. Considerable data regarding this concept in human placenta has previously been published by Dr. Lise Cedard and her colleagues (reviewed in J Steroid Biochem, 12:17-24, 1980). hCG regulation of placenta can make this hormone as a hormone of pregnancy maintenance not only through a luteotropic effect during the first 9 weeks of pregnancy but also a placentotropic effect throughout pregnancy. The presence of hCG/LH receptors in uterine blood vessels and myometrial smooth muscle has also recently been described by Dr. Rao and his colleagues (Reshef et al., J Clin Endocrinol Metab, 1990, In press, Society for Gynecologic Investigation, 1990, In press). This could mean that hCG can also contribute to pregnancy maintenance through its relaxation effect on myometrium and increase of uterine blood flow during pregnancy. Finally, it has recently been shown that hCG is a mitogen for cytotrophoblasts (Yagel et al., Am J Obstet Gynecol, 160:938-945, 1989).

A further point of discussion centered on the role of fetal-placental control by growth factors. An example of such communication is the demonstration that hPL stimulates insulin and IGF-I production in fetal pancreas in culture (J Endocrinol 113:297, 1987). it was noted that it will be difficult, for ethical reasons, to study these interactions in the human. These important areas deserve attention in future research.

Dr. D. Morrish also noted that his diagrams omitted discussion of interactions with the decidua and fetal membranes, both of which secrete factors of possible importance, such as prolactin. For example, decidual prolactin can inhibit placental hCG synthesis (Yuen et al., Am J Obstet Gynecol, 154:336-340, 1986) and high hCG concentrations typically found in normal pregnancy can stimulate prolactin synthesis in decidua (Rosenberg and Bhatngar, Am J Obstet Gynecol, 148:461-465, 1984).

Three presentations then followed. Dr. I. Henrichs presented his data regarding the metabolic effects of IGF-I on placental carbohydrate metabolism in which he demonstrated a higher rate of glycogenolysis, glucose utilization and lactate production after IGF-I treatment of trophoblast. Discussion brought out the aspect that much attention has been given to proliferative functions of growth factors, but little to metabolic effects, which have been shown to be significant in many other cell systems. Much work is needed in this area for the placenta. Dr. D. Morrish then presented data on the role of growth factors in inducing differentiation of the trophoblast. EGF induces morphological differentiation of term cytotrophoblasts with syncytial formation and this is accompanied by hCG and hPL secretion (J Clin Endocrinol Metab 65:1282, 1987). Dr. A. Barnea noted that EGF did not induce this differentiation in early pregnancy placenta. Thus it could be that EGF and other growth factors may have different and sometimes even opposing effects depending on the developmental stage of the placenta. The EGF appears to be produced by the cytotrophoblast as shown by immunostaining and EGF RIA of culture medium. In addition, the cytokines GM-CSF and to a lesser extent CSF-I (EGF abstract by M. Garcia-Lloret, L. Guilbert, D.W. Morrish) showed strong effects on inducing morphological differentiation. Interestingly, no significant hCG/hPL secretion accompanied the GM-CSF effects suggesting it was a pure differentiating factor. Dr. E. Barnea then showed data using an explant superfusion system of the effects of EGF on first trimester placenta. His data showed that there was spontaneous pulsatile secretion of hCG from the placenta. This secretion was stimulated two-fold within 10 min by addition of 50 ng/ml EGF but decreased over 2 h with repeated stimulation. The data indicated that there is a rapid-release pool of hCG and that EGF could cause its release. It was of note that a growth factor thought of as having chronic trophic effects also had an acute secretagogue effect. Discussion of the data suggested that the decline of secretion probably represented exhaustion of tissue stores, but receptor down-regulation may have contributed. It was also clear that explant systems as well as isolated cell preparations, would continue to play an important role in elucidating placental physiology.

A general discussion then followed in which it was noted that a very large number of growth factors have significant effects on the placenta. This brought up the teleological question of why there should be so many factors on cell proliferation and differentiation. Does this simply represent redundancy of control? How does the cell know which factor to respond to at which time? A further aspect of the existence of multiple factors acting on the placenta is that it is essentially unknown whether the effects observed are physiological or pharmacological. Although many studies use concentrations equal to levels found in body fluids, we do not know the tissue concentrations prevalent in any autocrine or paracrine system. Are these lower or higher than the body fluid levels? At present there is no method to measure these. It would be of great

great importance to have such information. The profusion of factors makes careful in vivo studies also important to determine the integrated effect of growth factors. A few studies of this type have been done, for example, in perfused sheep fetuses. In conclusion, the workshop highlighted a variety of known and new effects of growth factors on placental proliferation and differentiation. There remain however, many basic unanswered questions in regard to an endocrine level of placental control, interaction of the placenta, fetus and decidua integration of growth factor effects and metabolic effects of growth factors.

Table 1

Cell	Secretion Products	Responds to, or has Receptors for
Macrophage	TNF Il-1 PDGF GM-CSF	CSF-1
Endothelial Cell	IL-1 PDGF GM-CSF	TNF
Fibroblasts	INF IL-1 *IGF-I/II *EGF *CSF-1	PDGF EGF *IGFI/II TGFβ *CSF-1
Trophoblast (Mouse)	PDGF TGF IL-3	IL-3 GM-CSF
Human: Cytotrophoblast	*PDGF *EGF LHRH hCG	*EGF hCG PDGF Insulin IGF-I
Syncytiotrophoblast	*EGF hCG hPL Steroids hCG	*EGF hCG LHRH

* = shown in placenta

Regulation of epidermal growth factor receptors in human trophoblastic cells in culture: effect of parathyroid hormone

E. Alsat, V. Mirlesse, M. Dodeur, D. Evain-Brion

Laboratoire de physiopathologie du développement. CNRS-ENS, 46, rue d'Ulm, 75230, Paris Cedex 05, France

Specific binding of ^{125}I-human Epidermal Growth Factor (EGF) to cultured trophoblastic cells from first trimester and term human placentae was measured. Cytotrophoblasts were isolated by a trypsin/DNase dispersion method, purified on Percoll gradient and cultured in DMEM supplemented with 20% fetal calf serum.

Syncytia formation was observed within 24-48 hours after plating. During this period, the EGF binding capacity of these cultured cells is increased by 2-fold and remains at the same level the following three days. Scatchard analysis from binding data obtained at 4°C, reveals a biphasic plot compatible with two classes of binding sites with respective Kd of 0.3 nM and 9.5 nM and 2×10^4 sites of high affinity per cell.

Human parathyroid hormone (hPTH 1-34) added in the culture medium produces a more than 2-fold increase of the EGF binding to the cultured trophoblastic cells, as compared to the control. This effect is concentration dependent with a maximum obtained at 1nM. PTH acts by increasing the number of high affinity sites without modification of their affinity. Similar results are obtained with placental cells from both gestational stages studied and must be related to our previous studies showing an increased secretion of hCG by cells cultured in presence of PTH and EGF.

Trophoblastic cells in culture offer a nice model to study the interaction between growth factors and peptide hormones in the regulation of cellular endocrine functions.

Epidermal growth factor receptors in rat placenta, amnion and yolk sac : characteristics of specific binding are dependent on gestational age

S.W. D'Souza, P. Ali, J.L. Smart

University of Manchester, St Mary's Hospital, Manchester M13 OJH, UK

Studies of ^{125}I-EGF binding to the rat placenta, amnion, and yolk sac were carried out on days 14, 17, and 20 of gestation. In the placenta EGF binding was detectable on all 3 days; in the amnion EGF binding was undetectably low on day 14 but was present on days 17 and 20; while in the yolk sac EGF binding was undetectably low on all 3 days. Although Scatchard analysis of EGF binding to placental tissue raised the possibility of high and low affinity receptors, a statistical analysis of the ligand binding data was consistent with the presence of only one type of EGF receptor. The overall affinity of the receptors did not change with stage of gestation. However, the concentration of EGF receptors was lower in placental tissue on day 17 than on days 14 or 20 of gestation; the receptor concentrations were similar on days 14 and 20. It is suggested that EGF binding to the placenta, amnion, and yolk sac may reflect the levels of cell proliferation in those tissues in the latter part of gestation.

Acknowledgment: Medical Research Council, and Dr. H. Gregory.

Anti-TGF β antibodies as tools for placental TGF β purification

J.J. Feige, M. Keramidas, L. Multigner, J.J., Bourgarit, J.M. Saez*, S. Uhlrich**, E.M. Chambaz

INSERM U.244, DEF/BRCE, CENG 85X, 38041 Grenoble Cedex, France
* INSERM U.307, hôpital Debrousse, Lyon, France
** IMEDEX, ZI des Troques, 69630 Chaponost, France

Transforming growth factor ß is a multifunctional molecule initially characterized for its ability to induce anchorage-independent growth of NRK fibroblasts. It has been characterized as a growth inhibitor or a growth stimulator or a regulator of differentiated functions depending on the target cell type. In our laboratory, we reported that TGF ß is a very potent inhibitor of adrenocortical cell steroidogenesis, acting mainly at the level of the 17 alpha-hydroxylase gene expression (J.J. Feige et al., 1987, J. Biol. Chem., 262, 13491-13495).

In the work presented here, we characterized the presence of TGF ß in an industrial fraction prepared by acid/ethanol extraction of human placental tissue. Both a specific radioreceptor assay and a biological assay based on the inhibition of adrenocortical cell cortisol production indicated the presence of 75 ug TGF ß per kg of tissue. In order to get a rapid and efficient purification of placental TGF ß(s), we decided to use immunoaffinity chromatography techniques and attempted to raise rabbit polyclonal antibodies against several synthetic peptide fragments corresponding to the N-terminal or the C-terminal regions of TGF ß1 and/or TGF ß2. These peptides were chosen after a computer-assisted analysis of the immunogenic domains of both molecules. The best immune response was obtained with TGF ß1 (91-102). The antiserum against this peptide presented a 1:5000 titer in ELISA and appeared to be specific for TGF ß1 (no cross-reaction with TGF ß2). It was able to inhibit the binding of TGF ß1 to its receptors on adrenocortical cells and to block the biological activity of TGF ß1 on these cells (inhibition of steroid production).

Therefore antibodies to this peptide appear as useful tools for further purification of placental TGF ß(s).

Expression of nuclear proto-oncogenes c-fos and c-myc messenger ribonucleic acids in human placenta

S. Hauguel de Mouzon, D. Evain-Brion, M. Forestier, E. Alsat

Centre de nutrition. CNRS, Meudon, et laboratoire de physiopathologie et du développement, 46 rue d'Ulm, 75005 Paris, France

Quantification of c-fos and c-myc mRNAs was performed in first trimester and term placentas. RNA extraction was carried out using the single step procedure either in whole tissue homogenate or after isolation and purification of trophoblast cells. Northern analysis was followed by hybridation with ^{32}P cDNA radiolabeled probes coding for c-myc and c-fos. Very low signals for myc and fos mRNAs were obtained in whole tissue extracts either in early or term placentas. By contrast, trophoblast cells isolated after trypsin digestion and purified in the presence of 10 % deactivated fetal calf serum displayed very high signals i.e. 20 fold augmentation as compared to levels obtained in crude homogenates from the same placentas. When trophoblast cells were analyzed for c-myc and c-fos mRNAs expressions as a function of gestational age, they were detectable as early as 7 weeks of gestation. c-fos mRNA levels did not vary throughout pregnancy while c-myc levels were enhanced by 3 fold in term placentas.

These data demonstrate that in human isolated trophoblast cells, the nuclear protooncogenes c-myc and c-fos are expressed at very early stages of pregnancy. A possible up regulation of myc and fos expression by several growth factors present in the serum during the purification procedure was ruled out since identical levels of expression were observed in the absence of serum. Thus, we believe that the high constitutive signals observed in isolated trophoblast cells as compared to crude placental homogenate are related to the high degree of purification obtained under our experimental conditions. Whether or not this high constitutive expression of myc and fos mRNAs is a necessary event for normal placental growth and proliferation remains to be tested.

Variable epidermal growth factor receptor regulation among normal trophoblast, fibroblasts and tumor cells

D.W. Morrish, R. Sasi, C.C. Lin, D. Bhardwaj, S. Shiferaw

Departments of Medicine and Pathology, University of Alberta, Edmonton, Alberta, Canada T6G 2G3

Epidermal growth factor (EGF) induces proliferation of fibroblasts, some tumor cells, and also induces differentiation of normal human trophoblast (Morrish et al, J Clin Endocrinol Metab 65:1282, 1987). Because both up- and down-regulation of EGF receptors (EGF-R) have been reported, and because we wished to determine the molecular events during proliferation and differentiation of cells, we studied EGF effects on its receptor in two breast cancer cell lines (AA, LW), A431 cells (epidermoid carcinoma), normal human term placental fibroblasts, and normal isolated term human trophoblast. AA and LW cells were grown from biopsies. Trophoblast was prepared as described using a trypsin-DNAse I procedure (above reference), and fibroblasts were grown from placental preparations. All cell types were cultured for 3 days prior to testing in serum-free medium (DMEM or Ham's F10), then exposed for 3 and 24 h to 10 ng/ml EGF. Controls received BSA. Cells were removed after EGF exposure and RNA extracted with guanidinium isothiocyanate and CsCl gradient centrifugation. RNA was slot blotted to nylon membranes and hybridized with the human EGF receptor cDNA probe. Results showed no change after 3 h EGF exposure in any cell type except a slight increase in AA cells. After 24 h of EGF exposure, EGF-R mRNA expression did not change in AA, LW, A431 or fibroblast cells. However, trophoblast demonstrated a 20-50% decrease in EGF-R mRNA compared to controls. We thus conclude that normal trophoblast down-regulates its EGF-R mRNA, but that normal placental fibroblasts and three tumor cell lines do not. Trophoblast in this culture system is non-proliferative compared to other cells used. Therefore, differences in EGF-R response to EGF among trophoblasts, fibroblasts, and tumor cells may be useful in determining controlling mechanisms of normal and neoplastic cell growth. Supported by the Medical Research Council of Canada.

Expression of c-erbB-2 protein product in human placental villi as compared to EGF-receptor

J. Mühlhauser, C.A. Schroeter*, P. Kaufmann, M. Castellucci

Department of Anatomy, Aachen, Federal Republic of Germany
* Department of Surgery, Maastricht, The Netherlands

The protein production of c-erbB-2 oncogene is a truncated version of the EGF-R molecule. Because of this close relationship it was interesting to compare the expression of both in first trimester and full term placentas. A monoclonal antibody to the synthetic peptide of EGF-R and a polyclonal antibody for c-erbB-2 protein product were used by fluorescence and immunoperoxidase techniques.

In first trimester placentas in addition to a diffuse stromal reaction, the protein product of c-erbB-2 was localized in the cytoplasm of the villous cytotrophoblast and syncytiotrophoblast. In the full term placental villi some focal immunopositive reaction products were at the level of syncytiotrophoblast.

In first trimester placentas EGF-R was found to be localized along the cell membranes of the villous cytotrophoblast. The plasma membranes of the syncytiotrophoblast were only weakly stained. In full term placentas fluorescence and avidin/biotin immunoperoxidase reaction were found principally along the apical membrane of the syncytiotrophoblast, and to a lesser extent along the basal side of the syncytiotrophoblast. The majority of the mature villous cytotrophoblast stains negatively.

C-erbB-2 oncogene protein product is expressed in different types of adenocarcinomas. EGF-R has been described in numerous growing tissues. In spite of the structural similarities of both proteins, in human placental villi they differ in three aspects: timing of expression, localization in the various tissues, and subcellular localization.

Growth factors as stimulators of placental cell growth

V. Toder, E. Kochavi, H. Altaratz, N. Gleicher

Department of Embryology and Teratology, Tel Aviv University Medical School, Israel, and Dept. of Obstetrics & Gynaecology, Mount Sinai Hospital, Chicago, USA

According to recent developments it is more and more clear that maternal T lymphocytes act upon fetally derived placental cells to improve their proliferative potential. The lymphocytes perform these actions through soluble products, the lymphokines, augmenting cell proliferation and controlling several cell functions. We present here data showing the influence of various recombinant growth factors on the placental cells proliferation. Colony-stimulating factors were found to be the most potent stimulators of trophoblast, decidua and placental lymphoid cells. Interleukin-1, but not Interleukin-2 could cause placental trophoblast proliferation. In parallell, Epidermal growth factor, Fibroblast growth factor and Platellet-derived growth factor were found to be capable of triggerring placental cell proliferation. Small doses of progesterone and estrogen synergized with growth factors influence on trophoblast functioning although higher doses were found to be inhibitory. Possible applications of these findings seem to be of great importance.

Biochemical and biological characterization of bFGF extracted from human placenta

S. Uhlrich, J. Tiollier, M. Tardy, J.L. Tayot

IMEDEX, BP 38, 69630 Chaponost, France

Basic Fibroblast Growth Factor is a multifunctional polypeptide which controls the proliferation and differentiation of mesoderm and neuro-ectoderm cells. We report here the quantitative purification of bFGF from human placental tissue. A crude placental extract was obtained by acid extraction of the placental tissue and further alcohol fractionation. The bFGF fraction was then purified to homogeneity by ion exchange chromatography on CM-Trisacryl followed by affinity chromatography on Heparin-Sepharose. The yield of the purification was 10 mcg bFGF per kg of placenta.

The human placental bFGF fraction consists of two proteins with apparent molecular weights 16kDa and 18kDa. Both proteins were separated by polyacryl-amide gel electrophoresis followed by reverse phase HPLC. Both proteins have the same N-terminal sequence "Pro-Ala-Leu-Pro-Glu-Asp-Gly-Gly-Ser-Gly-Ala-Phe" which is identical to that of human bFGF. Furthermore, the polyacrylamide gel electrophoresis of the placental bFGF fraction, run under reducing conditions, shows a single protein with an apparent molecular weight of 16kDa. Thus, the 16kDa and 18kDa proteins are only two molecular forms of the same bFGF protein which differ one from each other by their oxydative state.

The biological characterization of growth factor activity of these molecules was performed by serum free cell culture assays with bovine corneal endothelial cells (EC) and CCL39 fibroblasts. The ED50 (half maximal stimulation) for stimulation of proliferation of CCL39 or EC was found to be in the range of 0.5 to 1 ng/ml for bFGF. Cell growth depends on the concentration of FGF present in the medium and reaches a plateau at 20 ng/ml.

The main properties of bFGF, observed in vitro, on EC include acceleration of cellular proliferation in serum free media on collagen coating, prolongation of life span of primary cultures and stimulation of the proliferation of cell monolayers.

Growth factors of the human placenta

M. Vučković-Tomanović, D.C. West, O. Genbačev

Institute of Endocrinology, Immunology and Nutrition, INEP, Zemun, Yugoslavia
* Christie Hospital and Holt Radium Institute, Manchester, UK

While it is known that the term placenta contains several growth factors, namely FGF, TGFα and TGFβ, little is known with regard to their distribution in this tissue. The present study examines 1st and 3rd trimester placenta for their content and distribution of bFGF, aFGF and TGFβ. Immunohistochemical staining for bFGF, aFGF and TGFβ showed the presence of these growth factors, but indicated that distribution differed between the placentas of different gestational age. Using salt precipitation, DEAE chromatography and heparin affinity chromatography we confirmed the presence of these growth factors. Molecular weight determination and identification was carried out using SDS gel electrophoresis and Western blotting, the proteins being detected using the same antibodies which were used for staining. Immunological and biochemical techniques confirmed the presence of TGFβ and bFGF in placenta of both gestational ages. However, in 1st trimester placenta, much of TGFβ was in a latent form not present at term. Also, the distribution of bFGF differed with age.

4.

Trophoblast antigens and immunological aspects of feto-maternal relationship

Antigènes trophoblastiques et aspects immunologiques des relations fœto-maternelles

Interferon as a fetoplacental signal in human pregnancy

T. Chard

Department of Obstetrics, Gynaecology and Reproductive Physiology, St Bartholomew's Hospital Medical College, and the London Hospital Medical College, London EC1A 7BE, UK

Until recently it has been assumed that the functions of the interferons (IFNs), and in particular IFN-alpha, were concerned solely with molecular signalling between lymphoid cells. However, it is now apparent that production of IFN-alpha is characteristic of the fetus and/or placenta in a number of species including man (Chard, 1989). In at least two species, the sheep and cow, IFN-alpha may play a key endocrine role in the maintenance of early pregnancy.

In the human IFN-alpha cannot be detected in maternal blood and tissues, but is widely distributed in the fetus and its immediate surroundings (fetal blood, fetal organs, placenta, membranes, amniotic fluid and decidua) (Chard et al, 1986). The levels of IFN in human amniotic fluid are the same in early and late pregnancy. There is some evidence that the levels are higher in first than subsequent pregnancies (Chard, 1989).

The sheep blastocyst before implantation secretes a protein with anti-luteolytic properties; this is referrred to as ovine trophoblast protein 1 (oTP-1) (reviewed by Flint et al, 1988). Gene sequencing has shown clear homology between oTP-1 and human IFN-alpha (Imakawa et al, 1987). Amino acid analysis has yielded similar results, with considerable homology in the 40 N-terminal amino acids of the two proteins (Stewart et al, 1987). In particular, there is a highly conserved sequence (Cys-Ala-Trp-Glu) in all IFNs and in oTP-1. The early bovine conceptus also secretes a protein (bovine trophoblast protein 1 [bTP-1]) from day 15 onwards. Sequencing of cDNAs corresponding to the mRNAs for bTP-1 has shown considerable homology with IFN-alpha (Imakawa et al, 1989). Other lymphokines have been identified in the human placenta and membranes including interleukin 2 (Soubiran et al, 1987), granulocyte-macrophage colony stimulating factor (Cukrova and Hrkal, 1987), and tumour necrosis factor (Jaatela et al, 1988).

Although interferon-alpha appears to be a characteristic pregnancy-associated protein in a variety of species the stimulus to its production is not yet known. It is well-recognised that the contact between fetal and maternal tissues yields the potential for a classic graft-versus-host reaction (maternal v fetal, or vice versa). Thus it is interesting that in the mouse, both acute and chronic graft-versus-host disease are associated with an increase in IFN-alpha production. However, because the actions of the IFNs are complex, and the nature of the fetomaternal immune relationship is ill-understood, it is not yet possible to draw any simple conclusions as to how IFNs and the immune system might interact in pregnancy.

The evidence for the endocrine effects of IFN-alpha, at least in the sheep and the cow, now seems to be much more clearcut than the evidence for immunological effects. Indeed, a very strong case can be made that IFN-alpha is the main embryonic signal leading to the maintenance of the corpus lutem in these species. In man the main luteotrophic factor in early pregnancy is presumed to be chorionic gonadotrophin (hCG). Recent studies have shown that IFN-alpha may have a specific relationship to production of hCG (Iles and Chard, 1989). The experimental system involved three hCG-producing bladder carcinoma cell lines growth in vitro; the rate of hCG production was increased 2-3 fold in the presence of IFN-alpha while IFN-gamma had no effect.

There are other examples of interactions between lymphokines and hormones. Most notable is interleukin 1 (IL-1) which in tissue culture experiments has been shown to stimulate pituitary hormone release (Bernton et al, 1987) and inhibit progesterone release by granulosa cells (Fukuoka et al, 1988). Elevated IL-1 levels are also found in the blood during the luteal phase of the menstrual cycle (Cannon and Dinarello, 1985).

REFERENCES

BERNTON EW, Beach JE, Holaday JW, Smallridge RC, Fein HG (1987) Release Release of multiple hormones by a direct action of interleukin-1 on pituitary cells. Science 238, 519-526.
CHARD T, (1989) Interferon in pregnancy. J Developmental Physiol (in press.
CHARD T, Craig PH, Menabawey M, Lee C (1986) Alpha interferon in human pregnancy. Br J Obstet Gynaecol 93, 1145-1149.
CUKROVA V, Hrkal Z (1987) Purification and characterization of granulocyte-macrophage colony stimulating factor from human placenta J Chromatog 413, 242-246.
FLINT APF, Lamming EG, Stewart HJ (1988) A role for interferons in the maternal recognition of pregnancy. Mol Cell Endocr 58, 109-111.
FUKUOKA, M, Mori T, Taii S, Yasuda K (1988) Interleukin-1 inhibits luteinization of porcine granulosa cells in culture. Endocrinology 122, 367-369.
ILES R, Chard T (1989) Enhancement of ectopic beta-human chorionic gonadotrophin expression by interferon alpha. J Endocr (in press).
IMAKAWA K, Anthony RV, Kazemi M, Marotti KR, Polites HG, Roberts RM (1987) Interferon-like sequence of ovine trophoblast protein secreted by embryonic trophectoderm. Nature 330, 377-379.

IMAKAWA K, Hansen TR, Malathy PV, Anthony RV, Polites HG, Marotti KR, Roberts RM (1989) Molecular cloning and characterization of complementary deoxyribonucleic acids corresponding to bovine trophoblast protein-1: A comparison with ovine trophoblast protein-1 and bovine interferon-alpha II. Mol Endocr 3, 127-139.

JAATELA M, Kiuusela P, Saksela E (1988) Demonstration of tumour necrosis factor in human amniotic fluids and supernatants of placental and decidual tissues. Lab Invest 58, 48-52.

SOUBIRAN P, Zapitelli JP, Schaffar L (1987) IL2-like material is present in human placenta and amnion. J Reprod Immunol 12, 225-234.

STEWART HG, McCann SHE, Barker RJ, Lee KE, Lamming GE, Flint APF (1987) Interferon sequence homology and receptor binding activity of ovine trophoblast antiluteolytic protein. J Endocr 115, R13-R15.

Résumé

INTERFERON EST UN SIGNAL FETO-PLACENTAIRE PENDANT LA GROSSESSE.

L'interféron alpha est sécrété par le trophoblaste et est un facteur anti-lutéolytique puissant pendant la grossesse de la brebis. L'interféron alpha est aussi produit en quantité significative pendant la grossesse humaine. Chez l'homme, néammoins, l'hormone gonadotrophine chorionique est le signal embryonnaire le plus important pour permettre la survie du corps jaune. Recemment nous avons montré une corrélation étroite entre les taux interféron alpha et d'hCG dans les tumeurs qui sécrètent l'hCG. Il semble donc qu'un certain nombre de molécules reliées puissent être responsable des effets lutéotrophiques et éventuellement aussi anti-lutéolytique, et que ces molécules peuvent comprendre aussi bien l'hCG que l'interféron alpha.

Modèle murin de résorption fœtale à médiation NK

R. Kinsky, G. Delage*, N. Rosin, M. Nguy Thang*, M. Hoffmann, G. Chaouat

INSERM, U. 262, clinique universitaire Baudelocque, 123 bd de Port-Royal, 75674 Paris Cedex 14,
*INSERM, U. 245, hôpital Saint-Antoine, 75571 Paris Cedex 12, France

RESUME - Il est généralement admis que certains modèles murins d'avortement présentent un caractère immunologique, impliquent la participation prépondérante d'effecteurs lytiques cellulaires non spécifiques. Nous avons utilisé dans le présent travail de l'ARN synthétique à double brin, connu pour son pouvoir inducteur puissant de l'interféron et de cellules NK, le Poly(I). Poly (C) et son dérivé moins toxique le Poly (I). Poly(C12U). Ces deux substances augmentent la fréquence de résorption embryonnaire dans la combinaison CBA/J x DBA2, présentant des avortements spontanés ainsi que d'autres combinaisons de lignées n'en présentant pas en absence de traitement. Nous signalons les cinétiques optimales du phénomène qui indiquent un effet précoce agissant sur l'implantation. L'effet abortif peut être transféré adoptivement par des cellules spléniques de souris vierges traitées par le ARN injectées à des femelles gestantes au stade de préimplantation. L'effet peut être abrogé par l'élimination sélective de cellules NK. Nous discutons l'implication de ces résultats dans la survie fetale ainsi que l'utilisation de ce système dans d'autres domaines de recherche.

INTRODUCTION - Le modèles d'avortement spontané par croisement des lignées CBA/J x DBA2 (1) a été suivi rapidement par le système B10 x B10A. ce qui a permis d'analyser en détail les mécanismes impliqués dans les avortements intra-espèces d'étiologie immune probable. Plusieurs faits plaident en faveur du concept que des mécanismes non-spécifiques sont à l'origine de l'arrêt de la gestation. Premièrement, le trophoblaste est résistant à la lyse par les cellules NK classiques ou par des CTLs activées en milieu de culture conventionnel (2) bien que cette destruction puisse être obtenue par des CTLs activées en milieu Opti--MEM. Ces dernières présentent une activité proche de cellules LAK, les cellules LAK classiques étant elles-mêmes actives. Certains résultats in vivo sont en accord avec ceux obtenus in vitro. Les "avortements réels" à médiation immunologique qui présentent certaines particularités par rapport aux "résorptions" peuvent être induits par certaines tumeurs syngéniques. Les schémas utilisés semblent induire des effecteurs non-spécifiques, plutôt que spécifiques, bien que l'on puisse argumenter que des celles CTL activées par des antigènes associés à la tumeur agiraient de manière croisée avec un antigène placentaire. Deuxièmement, dans le système CBA/J x DBA2 lui-même, il a été démontré que l'infiltration par des cellules à caractère NK précède celle des CTLs. Cette observation a été renforcée par les résultats de BAINES et coll. (3), qui ont montré une corrélation entre infiltration cellulaire Asialo-Gm1+ des sites d'implantation et les taux de résorption dans le même système. Les résultats obtenus par le traitement au Poly (I).Poly (C) paraissent aller dans le même sens. Nous avons confirmé ces résultats en utilisant soit le Poly (I).Poly(C) ou son dérivé moins toxique, le Poly (I). Poly (C12U). Parallèlement, nous avons aussi démontré qu'un avortement pouvait être provoqué chez la souris par un recombinant TNF alpha en accord avec le système LPS/TNF de PARAND (4).De fortes doses d'interféron gamma recombinant ainsi que de fortes doses de rIL-2, seul ou en association avec de l'indométhacine présentent également un effet abortif. Dans la présente communication nous étendons les résultats obtenus dans le système CBA/J x DBA2 à plusieurs combinaisons normalement non abortives et nous présentons les conditions optimales de l'action du Poly (I). Poly (C12U) en apportant une première caractérisation des cellules effectrices elles-mêmes.

MATERIEL ET METHODES - Souris - Toutes les souris (à l'exception de CBA/J) proviennent du Centre d'élevage IRSC (CNRS, Villejuif) et sont utilisées à l'âge de 8 semaines. Les souris CBA/J également âgées de 8 semaines, proviennent d'IFFA-CREDO, l'Arbresle, France. Les accouplements CBA/J x DBA2 ont été réalisés à l'hôpital Cochin, U262, INSERM, animalerie commune. Les autres accouplements ont été faits au service animalier de l'INSERM à l'hôpital Saint-Antoine, Paris dans des conditions gnotobiotiques strictes. Le jour de l'accouplement effectif est vérifié chaque matin par la présence de bouchons vaginaux. Les femelles positives sont séparées au hasard dans les groupes expérimentaux et témoins. Les souris Nu/nu ne sont conservées que moins d'une semaine ddans notre animalerie sous iso-capes stériles suivant un usage établi.

Traitement par ARN à double brin : Le Poly(I). Poly (C12U) est préparé suivant la méthode décrite par GREENE et TSO (5). Une seule dose de 20ug est administrée par voie i.p. en 0,2ml d'eau physiologique contenant du MgC12. Les souris témoins ne reçoivent que le diluant. Le même traitement est appliqué dans le système CBA/J x DBA2 réalisé à l'hôpital Cochin.

Récolte d'embryons - Les femelles expérimentales et témoins sont sacrifiées le jour 13 de gestation, les embryons viables ou résorbés comptés et les utérus inspectés sous une loupe pour déceler des modifications morphologiques éventuelles (épaississement utérin ou microrésorptions) témoignant d'une perte embryonnaire précoce.

Evaluation de l'activité NK - Nous avons utilisé le test classique de libération de 31Cr de 4 heures (CRT) à partir de cellules cibles radiomarquées Yac-1 en faisant appel à des spénocytes effecteurs de souris traitées ou témoins envue de l'évaluation de l'activité NK stimulée par le Poly (I). Poly(C12U) et de la recherche d'une corrélation éventuelle avec la fréquence de résorptions fetales.

Sérum de lapin anti-NK - Des lapins sont immunisés à deux semaines d'intervalle par 3 injections s.c. et i.M. de splénocytes Balb/c Nu/nu, et une rate par lapin receveur était utilisée par chaque injection en substance adjuvante de Freund complète. Les lapins sont saignés à blanc par ponction cardiaque 10 jours après le dernier rappel, le sérum absorbé 5x sur des thymocytes Balb/c, C3H, CBA/J et C57Bl/6, centrifugé et conservé à -20°C. Après vérification de l'inocuité sur thymocytes de souris, le sérum est utilisé à la dilution de 1:10 en présence de C de lapin et cobaye en vue d'une déplétion de cellules NK et la viabilité cellulaire vérifiée par l'exclusion du bleu trypan. A la dilution indiquée, le sérum supprime totalement la lyse par cellules NK spléniques de Balb/c Nu/nu, CBA/J Nu/nu ou C57Bl/6, Nu/nu et la cible classique Yac-1. Un résultat semblable est obtenu dans le test CR I' de 4 heures en utilisant des splénocytes C3H. Par contre, la lyse médiée par CTL (ex. les cellules C3H anti-Balb/c MLR/CTL lysant Yac-1 n'est que très modérément affectée. De plus, la lyse de L-1210 et SP2O, lignées cellulaires réputées CTL sensibles et NK resistantes n'est pas modifiée significativement dans le test CRT de 4 heures. Tous les tests étaient réalisés aux rapports Effecteurs : Cibles de 25:1 et 100:1 présentant la sensiblité maximum.

Analyse statistique - Les pourcentages d'embryons vivants ou résorbés ainsi que ceux des femelles gestantes au jour du sacrifice sont comparés au moyen du test X^2.

RESULTATS - Augmentation de résorptions dans la combinaison CBA/J x DBA2 : Trois expériences représentatives concernant l'administration de Poly (I). Poly(C) et de Poly (I). Poly(C12U) montrent qu'au jour 7.5 après l'accouplement CBA/J x DBA2 il apparaît que lesdeux préparations d'ARN présentent un effet abortif en accord avec BAINES et coll. .

Il est à noter que l'activateur NK non toxique, le Poly (I). Poly (C12U) était moins abortif. Les fréquences élevées de résorption sont observées dans cette combinaison (CBA/J x DBA2) abortive, ainsi que dans celles de la combinaison non-abortive (CBA/J x Balb/c).

-Cinétique de l'effet du Poly (I). Poly (C12U) : Dans une première série d'expériences portant sur la combinaison non abortive (Balb/c Balb/c) il a été observé qu'une seule injection de Poly (I). Poly (C12U) , le jour 3 après l'accouplement, augmentait la fréquence de résorptions dans cette lignée. En vue de préciser le temps optimal de l'injection du produit pour obtenir le maximum de résorptions, des séries de femelles Balb/c présentant des bouchons vaginaux positifs sont injectées à différents temps de gestation. La fréquence maximale de résorptions était observée lorsque l'injection de Poly (I). Poly (C12U) était faite le jour 3.

- Le Poly (I). Poly (C12U) est abortif dans certaines combinaisons de lignées non abortives. Le même traitement administré en une seule injection le jour 3 à des femelles d'accouplements Balb/c x Balb/c, CBA/J x Balb/c et C57Bl/6 xBalb/c donne des résultats semblables. Il est à souligner que le produit s'est montré abortif dans toutes les 3 combinaisons testées. Cependant, le nombre des femelles Balb/c gestantes au jour 13 est significativement réduit ($p < 0.05$) dans le groupe traité. Les utérus épaissis chez les femelles "non gestantes" indique un échec précoce d'implantation. De plus, la disparité allo-antigénique entre la mère et le fetus (allo-antigènes paternels) est inversement proportionnelle à l'effet du Poly (I). Poly (C12U) suggérant l'intervention de l'ancien concept de la " vigueur hybride".

- Transfert d'effets abortifs par cellules NK actives. En une première évaluation d'un effet abortif transférable par cellules, 50 x10^6 de splénocytes de femelles Balb/c vierges, traitées par Poly (I). Poly (C12U) deux jours avant le transfert, sont injectées par voie i.v. à des femelles gestantes depuis 6 jours. Le traitement, proche des conditions d'administration directe du Poly(I). Poly(C12U), donne une augmentation des résorptions de 14% (chez les receveurs de cellules normales non activiées) à 47% ($p < 0.01$) chez les receveurs expérimentaux, le traitement des cellules activées expérimentales par le sérum de lapin anti-NK normalise les taux de résorption à 14% ($p < 0.05$).

DISCUSSION - Quelques soient les interprétations de ces études nos résultats confirment que le déroulement favorable de la gestation peut être modulé au niveau de lignées de souris non-abortives par l'activation de cellules NK. Le modèle se prête parfaitement à l'étude détaillée du ou des phénotypes des cellules abortives dans différents systèmes. Il serait ainsi possible d'élucider l'équilibre délicat entre les évènements locaux et généraux. Des tentatives de contrecarrer les effets du Poly

(I). Poly (C12U) par des facteurs libérés par des lymphocytes de femmes gravides saines, traités par la progestérone, à la différence de ceux provenant de sujets non gravides, se sont révélées positives (SZEKERES-BARTHO et coll. (6). Il apparaît donc, qu'en dehors de son intérêt fondamental, le modèle présenté offre en plus la possibilité d'étudier de nouvelles méthodes permettant de prévenir des avortements à répétition d'étiologie immune.

SUMMARY

There is now ample evidence that some models of immunologically mediated murine abortion can involve non specific cellular lytic effectors. In this paper, we use 2 double stranded synthetic RNAs, known as potent interferon inducers and NK activators, the Poly (I). Poly (C) and the less toxic Poly(I). Poly (C12U). They both enhance significantly resorption rates in abortion prone and non abortion prone strains of mice. The optimal kinetics of the phenomenon is described showing an anti-implantation like effect following injection at an early pregnancy stage. The abortifacient effect can be transferred to naive recipients by spleen cells from ds RNA injected donors. Such an effect is abrogated if the cells are treated by an anti-NK antiserum. The relevance of these findings to the survival of the conceptus, and the potential use of this system are suggested.

REFERENCES

1) CHAOUAT,G., KOLB, J.P., KIGER, N. STANISLAWSKI, M. and WEGMAN, T. G. - 1985. Immunological concommitants of vaccination against abortion in mice. J; Immunol., 134, 1594-1602.
2) CHAOUAT, G. and J.P. KOLB. Immunoactive products of placenta. IV.) Impairment by placental cells and their products of CTL function at effector stage. J. Immunol. 1985. 135. 215-221.
3) GENDRON, R. and BAINES,M. Infiltrating decidual Natural Killer cells are associated with spontaneous abortion in mice. 1988. Cell. Immunol., 113-261.
4) PARAND,M. Role of TNF in non specific stimulation of mouse resistance to infection. Special issue (International conference on Tumor Necrosis Factor and related cytokines, Heidelberg, 1987.), 175, 1/2, 26.
5) ZARLING, J., SCHLAIS, J.ESKRA, L., GREENE, J.J., TS'O P.O.P. and CARTER, W.A. Augmentation of human Natural Killer activity by polyisonic acid polycytidilic acid and its non toxic mismatched analogs. J. Immunol., 1980, 124, 1852-1857.
6) CHAOUAT, G., SZEKERES- BARTHO, J., MENU, E.,KINSKY, R., THANG, M.N., DY, M. and MINKOWSKI, M., 1989, In : Experimental models in Obstetrics and gynaecology, An International workshop. C. Romanini, Ed., Parthenon Publishing, Ltd., In Press, 1989.

Trophoblastin oTP, embryonic interferons

J. Martal, N. Chene, M. Charlier, M. Guillomot, P. Reinaud, J. Bertin, G. Danet, K. Zouari, G. Charpigny

INRA, Unité d'Endocrinologie de l'embryon, Station de physiologie animale, 78350 Jouy-en-Josas, France

In the ewe, Moor and Rowson (1966) demonstrated that the conceptus (embryo and its envelopes), using a local mechanism, inhibits the cyclic luteolytic action of uterine origin. Daily uterine infusions of ovine trophoblast homogenates from days 14-16 of gestation to recipient cyclic ewes, in progesterone phase, maintain luteal function for several months (Rowson and Moor, 1967; Martal and Lacroix, 1978; Martal et al., 1979). Surgical removal of ovine 21-23-day old embryos, by flushing, leads to the extended luteum corpus lifespan for several months. On the other hand, the injection of older trophoblasts (days 21-23) is not effective in the maintenance of luteal structures. This antiluteolytic embryonic signal is a protein since it is inactivated by pronase and synthesized from day 12 to 22. It has been designated as trophoblastin (Martal et al., 1979), then protein X (Godkin et al., 1982), oTP-1 (ovine Trophoblastic Protein-1) (Godkin et al., 1984a) and also oTPB (ovine Trophoblastic Protein B) (Martal et al., 1984a) and oTP (Trophoblastin) (Charlier et al., 1989). This protein (oTP-1) injected to cyclic ewes by the intrauterine route partially prevents corpus luteum regression (Godkin et al., 1984b).

The physiochemical properties of trophoblastin have been well established. In the ewe, trophoblastin is represented by a protein of $M_r \simeq 20,000$ with a pI of 5.3. The N-terminal amino acid sequence has been evidenced. It reveals a 55 per cent identity with class-I interferon α and a 64 per cent homology with class-II bovine interferon α (Charpigny et al., 1988a). The entire amino acid sequence of trophoblastin has been deduced after cDNA cloning. It represents a single chain polypeptide with 172 amino acids and a 23-residue signal sequence (Imakawa et al., 1987; Charlier et al., 1989; Stewart et al., 1989b). The whole oTP molecule shares about 45 to 55 per cent sequence homology with human, bovine, porcine and murine α-interferons. The homology between oTP cDNA and interferons αs cDNA is 80 and 70 per cent, respectively, with bovine and human IFN-αII and 63 per cent with boIFN-αI. The higher degree of nucleotide conservation between oTP and boIFN-αII has been observed in the untranslated 5'-sequence (88 per cent), in the region coding for the peptide signal (97 per cent) and in mature protein (83 per cent) (Charlier et al., 1989).

The purified trophoblastin shows an antiviral activity which is characteristic for interferons (Pontzer et al., 1988; Charpigny et al., 1988b; Martal et al., 1988). A very high antiviral activity can be detected in the culture medium of simian cells after trophoblastin cDNA transfection (Charlier et al., 1989).

Purified oTP is able to bind to endometrial receptors and can be displaced by recombinant interferons αI (Stewart et al., 1987). Intrauterine injections of recombinant class-I interferons α partially inhibit luteolysis in the ewe (Stewart et al., 1989a) and in recipient cows (Plante et al., 1988). The presence of trophoblastin in epithelial or fibroblastic endometrial cells inhibits prostaglandin F2α secretion (Charpigny et al., unpublished results) known to be the uterine luteolytic hormone. It can be assumed that trophoblastin acts as an antiluteolytic and embryonic interferon.

After HPLC purification, five isoforms of trophoblastin have been determined by analysis of their N-terminal (Charpigny et al., 1988; Martal et al., 1988). All these isoforms are endowed with a high antiviral activity ($\simeq 10^8$ IU/mg), characteristic for IFNs (Charpigny et al., unpublished results). The N-terminal sequences of these isoforms slightly differ from that of oTP-1 described by Imakawa et al. (1987). Moreover, let us emphasize that these embryonic IFNs are naturally expressed without viral induction contrary to others IFNs α, ß, γ described in the literature. According to the authors, a conceptus can express one or more embryonic IFN isoforms (unpublished data). Thus, several genes can be transcribed simulteanously in early pregnancy. The short period of trophoblastin expression has been shown in various ways, i.e. homogenates on 25 (Rowson and Moor, 1967) or 21-23-day-old trophoblasts (Martal et al., 1979), injected into the uterus of recipient cyclic ewes, do not inhibit luteolysis. The transfer of ovine trophoblastic vesicles ($\simeq$ 12-day-old) into recipient cyclic animals prolongs corpus luteum lifespan and demonstrates the trophoblastic origin of oTP (Heyman et al., 1984; Martal et al., 1984a). In this case, the embryonic disc is not involved in oTP production. Its localization only into trophectoderm has been achieved by immunocytology (Godkin et al., 1984a), immunofluorescence and in situ hybridization (Farin et al., 1989; Guillomot et al., unpublished data). The implantation of the trophoblast on maternal caruncles might inhibit oTP synthesis (Guillomot et al., unpublished data). The analysis of the oTP mRNA expression by Northern blot (from 11 to 21 days) (Hansen et al., 1988; Charlier et al., 1989) is in agreement with previous results. In the cow, the presence of trophoblastin has been revealed by injection of bovine embryonic extracts (Northey and French, 1980; Dalla-Porta and Humblot, 1983) or of bovine trophoblastic vesicles (Heyman et al., 1984). The latter transferred to cyclic ewes (J12) lead also to the maintenance of corpus luteum activity and likewise ovine trophoblastic vesicles (J12) injected into cyclic heifers (J12) lead to the same results (Martal et al., 1984b), suggesting a close relationship between ovine and bovine trophoblastin. There is a crossed immunological reaction between ovine and bovine trophoblastin (Helmer et al., 1987). The bTP cDNA has been recently cloned (Imakawa et al., 1989) and reveals a nucleotide sequence homology which is greater between oTP and bTP (85 per cent) than between bTP and bIFN-αII (79 per cent). In the same way, there is a higher amino acid sequence homology between bTP and oTP (80 per cent) than between bTP and bIFNαII (70 per cent). Unlike oTP, bTP is glycosylated as most IFNαI, its glycosylation site is located at position 78. oTP and bTP have two disulfide bonds (1-99, 29-139) specific of IFNs and of their antiviral activity, as well as the common sequence 139-146 (Cys-Ala-Trp-Glu-Ile-Val-Arg-Val). Trophoblastin has been evidenced in another domestic ruminant: the goat (Godkin et al., 1987). Homologous mRNA to oTP have been detected by Northern blot in 17-day-old caprine embryos (Charlier et al., 1989). In culture, these embryos develop significant antiviral activity (Zouari et al., unpublished results).

The ruminant trophoblastin seems to be correlated to the embryonic IFN family, eliciting antiluteolytic and consistent antiviral properties, sharing a very high structural homology with IFNs, specially IFN-αII. Trophoblastin most likely acts by regressing luteolysis via alteration of the endometrial $PGF_{2\alpha}$ secretion, in ruminants.

Antiviral activities have been demonstrated in embryos of species different from ruminants, in the pig (Godkin et al., 1987; Cross et al., 1989; Labonnardière et al., unpublished results) and in the rabbit (Martal et al., unpublished results). These IFN-like activities are characterized by their restricted period of expression and their lower production than in ruminants. Their involvment in the luteal regulation has not yet been evidenced in these species.

RESUME

La trophoblastine, dénommée également oTP (Protéine trophoblastique ovine), représente un signal embryonnaire antilutéolytique émis par le conceptus ovin pendant une courte durée (12-22 jours chez la brebis). En inhibant la sécrétion de $PGF_{2\alpha}$ par un mécanisme local, la trophoblastine permet ainsi le maintien du corps jaune et sa sécrétion de progestérone.

La détermination de sa séquence en acides animés (N-terminale) et nucléotidique révèle une homologie de séquence de 45 à 55 % avec les interférons α de classe I humain, bovin, porcin et murin. La trophoblastine s'apparente plus particulièrement à la sous-famille des interférons α de classe II humain et bovin (respectivement 80 et 70 % d'homologie). Le conceptus bovin sécrète une glycoprotéine (ou bTP) présentant une homologie nucléotidique de 85 % avec la trophoblastine ovine. Deux ponts disulfures, ainsi qu'une séquence commune (139-146) spécifiques des interférons, se retrouvent à la fois dans l'oTP et la bTP. Chez la chèvre, un ARN messager homologue à celui de l'oTP est mis en évidence dans l'embryon de 17 jours.

De plus, ces interférons embryonnaires présentent une forte activité antivirale. L'administration d'interférons α-I recombinants à des ruminants en cycle inhibent partiellement la lutéolyse. Ces résultats montrent que l'on peut assimiler la trophoblastine, en particulier celle des ruminants, à une famille d'interférons embryonnaires.

REFERENCES

CHARLIER M., HUE D., MARTAL J., GAYE P., 1989. Cloning and expression of cDNA encoding ovine trophoblastin: its identity with a class-II alpha interferon. Gene, 77, 341-348.

CHARPIGNY G., REINAUD P., HUET J.C., GUILLOMOT M., CHARLIER M., PERNOLLET J.C., MARTAL J., 1988a. High homology between a trophoblastic protein (trophoblastin) isolated from ovine embryo and α-interferons. FEBS Letters, 228, 12-16.

CHARPIGNY G., REINAUD P., LA BONNARDIERE C., GUILLOMOT M., HUET J.C., PERNOLLET J.C., MARTAL J., 1988b. Evidence for antiviral properties of three purified isoforms of oTPB. In Proc. int. workshop on maternal recognition of pregnancy and maintenance of the corpus luteum. Jerusalem (Israël) (abstr.), p. 72.

CROSS J.C., ROBERTS R.M., 1989. Porcine conceptuses secrete an interferon during the preattachment period of early pregnancy. Biol. Reprod., 40, 1109-1118.

DALLA PORTA M.A., HUMBLOT P., 1983. Effect of embryo removal and embryonic extracts or PGE_2 infusions on luteal function in the bovine. Theriogenology, 19, 122-131.

FARIN C.E., IMAKAWA K., ROBERTS R.M., 1989. In situ localization of mRNA for the interferon, ovine trophoblast protein-1, during early embryonic development of the sheep. Mol. Endocr., 3, 1099-1107.

GODKIN J.D., BAZER F.W., MOFFATT J., SESSIONS F., ROBERTS R.M., 1982. Purification and properties of a major, low component weight protein released by the trophoblast of sheep blastocysts at day 13-21. J. Reprod. Fert., 65, 141-150.

GODKIN J.D., BAZER F.W., ROBERTS R.M., 1984a. Ovine trophoblast protein 1, an early secreted blastocyst protein binds specifically to uterine endometrium and affects protein synthesis. Endocrinology, 114, 120-130.

GODKIN J.D., BAZER F.W., THATCHER W.W., ROBERTS M., 1984b. Proteins released by cultured day 15-16 conceptuses prolong luteal maintenance when introduced into uterine lumen of cyclic ewes. J. Reprod. Fert., 71, 57-64.

GODKIN J.D., SMITH L., LIFSEY B., BAUMBACH G., GNATEK G., DUBY R.T., 1987. Comparison of in vitro protein production by caprine, bovine, ovine and porcine conceptuses during early pregnancy. J. anim. Sci., 65, suppl. 1, abstr. 517.

HANSEN T.R., IMAKAWA K., POLITES H.G., MAROTTI K.R., ANTHONY R.V., ROBERTS R.M., 1988. Interferon RNA of embryonic origin is expressed transiently during early pregnancy in the ewe. J. Biol. Chem., 263, 12801-12804.

HELMER S.D., HANSEN P.J., ANTHONY R.V., THATCHER W.W., BAZER F.W., ROBERTS R.M., 1987. Identification of bovine trophoblast protein-1, a secretory protein immunologically related to ovine trophoblast protein-1. J. Reprod. Fert., 79, 83-91.

HEYMAN Y., CAMOUS S., FEVRE J., MEZIOU W., MARTAL J., 1984. Maintenance of corpus luteum after uterine transfer of trophoblastic vesicles in cyclic cows and ewes. J. Reprod. Fert., 70, 533-540.

IMAKAWA K., ANTHONY R.V., KAZEMI M., MAROTTI K.R., POLITES H.G., ROBERTS R.M., 1987. Interferon-like sequence of ovine trophoblast protein secreted by embryonic trophectoderm. Nature, 330, 377-379.

IMAKAWA K., HANSEN T.R., MALATHY P.V., ANTHONY R.V., POLITES H.G., MAROTTI K.R., ROBERTS R.M.,. 1989. Molecular cloning and characterization of complementary deoxyribonucleic acids corresponding to bovine trophoblast protein-1: a comparison with ovine trophoblastic protein-1 and bovine interferon-α_{II}. Mol. Endocrinol., 3, 127-139.

MARTAL J., CAMOUS S., FEVRE J., CHARLIER M., HEYMAN Y., 1984a. Specificity of embryonic signals maintaining corpus luteum in early pregnancy in ruminants. Proc. 10th int. Congress on animal reproduction and artificial insemination, Urbana-Champaign, USA, III, No. 510, 3 p.

MARTAL J., CHARLIER M., CAMOUS S., FEVRE J., HEYMAN Y., 1984b. Origin of embryonic signals allowing the establishment of pregnancy corpus luteum in ruminants. 10th int. Congress on animal reproduction and artificial insemination, Urbana-Champaign, USA, III, No. 509, 3 p.

MARTAL J., CHARPIGNY G., REINAUD P., HUET J.C., GUILLOMOT M., ZOUARI K., PERNOLLET J.C., LA BONNARDIERE C., 1988. Embryonic signals and corpus luteum: why three isoforms of trophoblastin can be considered as interferons α of class II? J. Reprod. Fert., abstr. II, 2.

MARTAL J., LACROIX M.C., 1978. Importance d'une trophoblastine dans le contrôle endocrinien du corps jaune gestatif chez la brebis au moment de l'implantation. In "Implantation de l'oeuf", du MESNIL du BUISSON F., PSYCHOYOS A., THOMAS K. Eds, 193-208, Masson, Paris.

MARTAL J., LACROIX M.C., LOUDES C., SAUNIER M., WINTENBERGER-TORRES S., 1979. Trophoblastin, an antiluteolytic protein present in early pregnancy in sheep. J. Reprod. Fert., 56, 63-73.

MOOR R.M., ROWSON L.E.A., 1966. The corpus luteum of the sheep: functional relationship between the embryo and the corpus luteum. J. Endocr., 34, 233-239.

NORTHEY D.L., FRENCH L.R., 1980. Effect of embryo removal and intra-uterine infusion of embryonic homogenates on the lifespan of the bovine corpus luteum. J. anim. Sci., 50, 298-302.

PLANTE C., HANSEN P.J., THATCHER W.W., 1988. Prolongation of luteal lifespan in cows by intrauterine infusion of recombinant bovine alpha-interferon. Endocrinology, 122, 2342-2344.

PONTZER C.H., TORRES R.A., VALLET J.L., BAZER F.W., JOHNSON H.M., 1988. Antiviral activity of the pregnancy recognition hormone ovine trophoblast protein-1. Bioch. Biophys. Res. Commun., 152, 801-807.

ROWSON L.E.A., MOOR R.M., 1967. The influence of embryonic tissue homogenate infused into the uterus of life-span of the corpus luteum in the sheep. J. Reprod. Fert., 13, 511-516.

STEWART H.J., FLINT A.P.F., LAMMING G.E., M.E., Mc CANN S.H.E., PARKINSON T.J., 1989a. Antiluteolytic effects of blastocyst-secreted interferon investigated in vitro and in vivo in the sheep. J. Reprod. Fert., suppl. 37, 127-138.

STEWART H.J., Mc CANN S.H.E., BARKER P.J., LEE K.E., LAMMING G.E., FLINT A.P.F., 1987. Interferon sequence homology and receptor binding activity of ovine trophoblast antiluteolytic protein. J. Endocr., 115, R13-R15.

STEWART H.J., Mc CANN S.H.E., NORTHROP A.J., LAMMING G.E., FLINT A.P.F., 1989b. Sheep antiluteolytic interferon: cDNA sequence and analysis of mRNA levels. Mol. Endocrinology, 2, 65-70.

Effect of anti CG antibodies on baboon placental derived trophoblast *in vitro*

C.S. Bambra

Institute of Primate Research, P.O. Box 24481, Karen, Nairobi, Kenya

Effect of anti CG antibodies was studied on placental derived monolayers. The resultant monolayers were characterized using Mabs and found to be cellular trophoblast. Cellular trophoblast secreted CG which could be both localized on cells and measured in the daily spent medium.

Addition of antibodies caused cell lysis and this phenomenon was reproducible in all 6 placental cultures studied. The progress of lysis could be followed in a time course study. Statistical evaluation demonstrated a significant difference between control (normal baboon serum) and treated (CG antiserum) cultures and the effect was evident after 4 hours of incubation with the antibody. These studies suggest that the lysis was mediated by an ADCC mechanism.

Placental communications : biochemical, morphological and cellular aspects. Eds L. Cedard, E. Alsat, J.-C. Challier, G. Chaouat, A. Malassiné.
Colloque INSERM/John Libbey Eurotext Ltd. © 1990, Vol. 199, p. 133

Lymphokines at the feto-maternal interface affect fetal size and survival

G. Chaouat, E. Menu, M. Hoffmann, M. Dy[*], M. Minkowski, D.A. Clark[**], T.G. Wegmann[***]

INSERM U.262, clinique universitaire Baudelocque, 123 bd de Port-Royal, 75674 Paris Cedex 14, France
[*] *INSERM U.25, hôpital Necker-Enfants-Malades, 75015 Paris, France*
[**] *Mac Master University, Hamilton, Ontario, Canada L8N 3Z1*
[***] *University of Alberta, Edmonton, Alberta, Canada T6G 2H7*

It has been previously shown in vitro that members of the CSFs family of lympokines were growth factors for placental cells, whereas TNF was cytostatic for murine trophoblast. High doses of R-IL-2 induce LAKCs in the Asialo GM1+ lineage. These cells cannot kill murine trophoblast (nor can NKs), when endowed with a "low" lytic activity, whereas they do when optimally activated ("high" lytic capacity). Gamma interferon is cytostatic at high doses for trophoblast, and enhances MHC class I expression on spongiotrophoblat.

We describe in vivo correlates of the above experiments in the CBA x DBA/2 system : Gamma interferon and TNF alpha are abortifacient (in the Balb/c x Balb/c and C3H x C3H strains as well), and high doses of R-IL2 induce abortion too. Conversely, IL-3 and GM-CSF reduce resorbtion rates in the CBA x DBA/2 system. The doses, however, of GM CSF injected are too low to act directly on placental growth, and we will show that in GM CSF treated animals, decidual TNF content is lowered when compared to controls. The latter effect could be direct (low doses of CSFs preventing LAKCs differentiation or function) or indirect (CSF acting, via a CSF1 mediated cascade, on placental growth, resulting in turn in "TGF Beta like" placental factor release, which inactivates LAKCs and NK, hencer affecting TNF. Spongiotrophoblasts weights and size will be given in control and treated animals, and the relevance of these data for clinics will be discussed.

Do human trophoblast, leukocyte and sperm share a common antigen recognized by GB24 ?

P. Fenichel, C. Grivaux, G. Dohr*, M. Samson, C. Milesi-Fluet, C.-J. G. Yeh, B.-L. Hsi

INSERM U.210, faculté de médecine, 06034 Nice Cedex, France
* Institut fur Histologie und Embryologie der Universität Graz, 8010 Graz, Austria

GB24 is a murine monoclonal antibody (IgGI) raised against human placental microvilli. By using immunofluorescence, GB24 reacted with syncytiotrophoblasts of first trimester (5 weeks of gestation), second and third trimester placenta as well as peripheral blood leukocytes. On human germ cells, GB24 did not react with non-fertilized oocytes. However, it reacted with the acrosomal region of intra-testicular, epididymal and fresh ejaculated sperm after the spermatozoa have been fixed and permeabilized by acetone. The membrane immunofluorescence showed that fresh ejaculated sperm did not react with GB24, but 40 per cent of live motile spermatozoa were positive after the induction of acrosome reaction by calcium ionophore A23187 (10 µM in B2 medium). These results suggested that the antigen recognized by GB24 is localized on the inner acrosomal sperm membrane. This localization was confirmed by using immunogold electron microscopy. By SDS-polyacrylamide gel electrophoresis, GB24 immunoprecipitated two proteins of 62 and 75 kDa from placental microvilli. However, a wide variability of the intensity of each band could be observed from one placenta to another. The biochemical analysis made on umbilical cord blood lymphocytes showed the same results as that of the corresponding placenta. These data indicate that the antigen recognized by GB24 on sperms is inducible by the biological process of fertilization or by artificial activation of calcium ionophore, the biochemical properties of the antigen on placenta is heterogeneous, and the characteristics of this antigen on each placenta is identical to the lymphocytes of the same donor. Thus trophoblast and lymphocytes share a common antigen recognized by GB24. Biochemical analysis of the acrosomal sperm antigen recognized by GB24 is currently under investigation.

Functional expression of CSF-1 receptors on normal human trophoblast

M. Garcia-Lloret, L. Guilbert, D.W. Morrish

Departments of Medicine and Immunology, University of Alberta, Edmonton, Alberta, Canada T6G 2G3

CSF-1 is a hematopoietic growth factor originally identified by its ability to stimulate the proliferation, differentiation, and survival of the monocyte-macrophage lineage, effects mediated by binding to a single class of high affinity receptors, whose structure is closely related to the c-fms oncogene. Recent reports of Bartocci et al (JExp Med 164:956, 1986) suggest an important role of CSF-1 in normal pregnancy. CSF-1 increases 1000 fold in uterine tissues during murine pregnancy, resulting in increase in hCG production. Expression of c-fms RNA has been demonstrated in choriocarcinoma cell lines (J Clin Invest 77:1740, 1986) and human placenta (Muller et al, Nature 299:640, 1982). GM-CSF is an important macrophage growth factor, but it has not been studied in the placenta. Previous work from this laboratory has demonstrated that choriocarcinoma cell lines (BeWo, JAR, JEG-3) have CSF-1 receptors, respond to this factor, and secrete it. We therefore wished to determine the effects of CSF-1 and GM-CSF on normal human trophoblast. Purified term villus trophoblast was prepared as previously described (Morrish et al, J Clin Endocrinol Metab 65:1282, 1987). The isolated cells were over 98% pure cytotrophoblast as demonstrated by negative immunostaining to vimentin (fibroblasts), OKM1 (macrophages), and GZ121 (syncytiotrophoblast), but positive for H315 (syncytio- and cytotrophoblast), cytokeratin (epithelial cells), and oncomodulin (cytotrophoblast). CSF-1 and GM-CSF used were recombinant human preparations. Trophoblasts bound 125-I-CSF-1 in a specific and saturable manner and expressed c-fms RNA on northern analysis. In situ hybridization of first trimester placenta indicated the syncytium and some cytotrophoblast contained c-fms. Addition of 5,000 U/ml CSF-1 induced a significant increase in integrated hCG secretion (299.2 $\pm$ 99.2 ; control = 137.9 $\pm$ 46.4; $p<0.05$) and hPL secretion (277.7 $\pm$ 86.1; control = 213.5 $\pm$ 65.1; $p<0.05$) in six experiments. Addition of 100 U/ml GM-CSF induced slight hPL secretion, but no significant hCG secretion. Morphological studies showed slightly increased syncytial formation in CSF-1 treated cells, and dramatically increased syncytial formation after GM-CSF exposure. We thus conclude that CSF-1 receptors are present in trophoblast and that these cells respond with an increase in hCG and hPL secretion, and possibly with morphologic differentiation. In contrast, GM-CSF has small effects on hPL secretion, little or no effect on hCG secretion, but induces dramatic morphologic differentiation. The results thus show that these two cytokines appear to be significant factors in trophoblast function and differentiation.

Comparison of HLA class-I antigen expression in cultured human extravillous trophoblast, fetal skin and JEG-3 cells

A. Grabowska, G. Chumbley, Y.W. Loke

Department of Pathology, University of Cambridge, UK

The culture method developed in our laboratory provides us with trophoblast cells that have characteristics of the extravillous population which invade the decidua, including positive staining with W6/32. We have examined the nature of this Class-I-like molecule and have observed that it has a heavy chain of molecular weight lower than that associated with B_2 microglobulin in classical HLA, and lacks the classical polymorphic determinants. The molecule is only weakly expressed on the surface of the cell. Surface expression is increased in the presence of γ-interferon, but only to a small extent. Northern blotting shows that mRNA from these cells readily hybridises to an HLA-A,B,C cDNA probe. These characteristics of the trophoblast HLA molecule have been compared with those of the choriocarcinoma-derived cell-line, JEG-3, and fetal skin cells of a similar gestational age.

Effect of 1,25 dihydroxycholecalciferol on IL-2 dependent CILL-2 proliferation and cAMP cell content. Role at the feto-placental interface

B. Hamelin, J. Demignon, E. Menu, G. Chaouat, C. Rebut-Bonneton

INSERM U.262, clinique Baudelocque, 123 bd de Port-Royal, 75010 Paris, France

A potent metabolite of cholecalciferol, the 1,25 dihydroxyvitamin D3 (1,25-(OH)2D3) is known to exert immunoregulatory functions in man. 1,25-(OH)2D3 plasma concentration was shown to increase during pregnancy owing to a high 25-hydroxyvitamin D3 hydroxylase level in kidney and mainly in placenta. Therefore we thought that 1,25-(OH)2D3 should be a likely candidate for the suppression of the maternal immunological attack against the fetus probably in conjonction with other placental products
The purpose of the present work was to investigate via the 3'5' cyclic adenosine monophosphate (cAMP) cell content, the mechanism of the inhibitory effect exerted by 1,25-(OH)2D3 on a non-allogenic cytotoxicity, an event which is likely to contribute to the fetal acceptance by the mother. The proliferative stimuli provided by the interaction of IL-2 with its specific receptor is far less clear.
Prostaglandine E2 increased within the first seconds the cAMP concentration of 60% and reduced IL-2 driven lymphocyte proliferation. However, the intracellular cAMP concentration was related to the IL-2 concentrations in the incubation medium $0.49 \pm 0.02; 3.5 \pm 0.8;$ and 1.58 ± 0.04, pmoles/10^6 cells without IL2, with normal IL2 and half concentration of IL2 respectively.

Moreover, 10^{-10} M 1,25-(OH)2D3, wich did not reduce IL2 dependent -CILL2 proliferation nor change signifcally the cAMP cellular concentration, was shown to reduce the CILL2 cytotoxicity.
Taken together, these data put in question a membrane signalling mechanism for IL2 stimulation of murine T lymphocyte.

Monoclonal antibodies GZ 100 and GZ 116 recognize different trophoblast antigens

M. Hartmann, G. Dohr, G. Pilz, D. Ribitsch, G. Siwetz, G. Desoye*

Department of Histology and Embryology and
* Department of Obstetrics and Gynaecology, University of Graz Austria

GZ 100 and GZ 116 are monoclonal mouse IgGs. SDS-PAGE under reducing conditions and western blot analysis identified two bands at 51 kD and 48 kD for the GZ 100 antigen, GZ 116 recognizes an antigen with molecular weight of 57 kD as was assessed by SDS-PAGE under non-reducing conditions and western blotting.
On first trimester placentae (week 8 and 9 of gestation) the villous syncytiotrophoblast reacted with GZ 100, whereas the villous cytotrophoblast and trophoblast cell columns were not stained. The antigen recognized by GZ 116 was not present on these very early placentae, but on second trimester placenta (week 23 of gestation) GZ 116 stained the villous syncytiotrophoblast.
On term placenta both antibodies react with villous syncytiotrophoblast and cytotrophoblast of the amniochorion. Cytotrophoblastic cells in the basal plate are stained by both antibodies as well, but GZ 116 recognizes only a small subpopulation.
The GZ 100 and GZ 116 antigen is present on the human choriocarcinoma cell lines, BeWo and Jar.
GZ 116 did not react with other human tissues tested so far (lymph node, mucosae of stomach, small intestine and colon, ovary, uterus, thyreoidea and skin), the GZ 100 specific antigen was expressed by mucosa of stomach and epithelia of sweat glands.

The expert technical assistance of A.Blaschitz, S.Richter and R.Schmied is gratefully acknowledged.

Fcγ-receptors in sera from pregnant women

T.S. Jensen, E. Ulvestad, R. Matre

Broegelmann Research Laboratory for Microbiology, University of Bergen, Norway

A competitive ELISA-technique was used to quantify soluble FcR in sera from pregnant women. Costar ELISA-plates were coated with amniotic extract containing FcR. A monoclonal antibody, B1D6, isotyped against a 40kD placenta FcRII was added together with undiluted serum. Peroxidase-conjugated rabbit anti-mouse IgG was added, and the reaction visualized using o-phenyldiamine/H_2O_2.

Blood samples were collected at 8 - 12, 16 - 20, 24 - 28, 32 - 36, and 40 - 42 weeks of gestation, and six weeks postpartum. Cord sera were sampled at term. Sera from age-matched females were used as controls.

The level of FcR measured at week 8 - 12 was not different from that of the controlgroup. We found a significant decrease in soluble FcR from weeks 8 - 12 to 24 - 28 ($p<0.01$). At week 40 the levels had increased significantly ($p<0.05$), and in cord sera levels were highly significant compared with week 24 - 28 ($p<0.001$). Levels in cord sera were also significantly increased compared with week 8 - 12 ($p<0.05$).

Soluble FcRs are thought to have immunoregulatory functions. In vitro studies have shown a suppressive effect on antibody production in a time and doserelated manner. The significance of soluble FcR in pregnancy remains to be shown.

Interaction of decidual NK and LAK cells with human trophoblast *in vitro*

A. King, Y.W. Loke

Division of Cellular and Genetic Pathology, Department of Pathology, University of Cambridge, UK

We have previously shown that trophoblast cells are resistant to lysis by NK cells from both peripheral blood and decidua. However, decidual cells do exhibit NK activity against the NK-sensitive cell line K562. We have now examined whether trophoblast cells have NK targets on their surface or whether trophoblast can actively inhibit NK activity. We have also sought to identify the cells responsible for NK activity in the decidual cell populations used. By using a single cell conjugate cell assay, it appears that decidual cells form conjugates with K562, but not with cultured trophoblast cells. In addition, immunostaining of the K562-decidua conjugates shows Leu19+ LGL are invariable the effector cells. These LGLs have been shown to be in large numbers in early decidua and have an unusual phenotype. Cold target inhibition experiments have shown that although cold K562 cells can inhibit decidual NK cells from lysing K562, trophoblast cells do not. These findings indicate that first trimester trophoblast does not have the decidual Leu19+ NK target on its cell surface.

In addition, we have examined the effect of rIL-2 on proliferation and cytolytic function of decidual LGLs against both normal and malignant trophoblast. Decidual LAK cells can be generated by culturing LGSs with rIL-2. These LAK cells are cytolytic to both normal trophoblast and JE3-choriocarcinoma cells.

Human placental supernatant and IL - 2 and IL - 4 dependent proliferation

E. Menu, D. Jankovic*, J. Thèze*, G. Chaouat

INSERM U.262, clinique Baudelocque, 123 bd de Port-Royal, 75674 Paris Cedex 14, France
* Institut Pasteur, 28 rue du Docteur-Roux, 75015 Paris, France

Human placental supernatants, obtained after 48 hours culture explants, block CTL generation in MLR, inhibit at effector stage the lytic function of NK cells and the non MHC restricted cytotoxicity of alpha-beta T cell clones.
We have investigated the effects of such supernatants on lymphokine dependent T cell proliferation using the CTLL-2 reference murine cell line. A profound, dose dependent, inhibition of both IL-2 and IL-4 dependent cell proliferation was observed. Purification of the material by HPLC yielded a 68-70 Kd peak at pH 7.4, but the peak activity was at a smaller molecular weight (one fraction before Cytochrome c) if pH 2.9 acetic acid, KCl buffer was used. CTLL-2 cells incubated with such a material did not reveal any variations in IL-2 receptor expression when monitored by FACS analysis with an anti-IL-2 receptor monoclonal antibody and there is no direct competition of the placental factor with IL-2 binding on its receptor. The same fractions were inhibitor for IL-2 and IL-4 driven proliferation ; We conclude that the material acts on CTLL-2 cells as a general growth inhibitor, independently of stimulation.

The synthesis of tumour necrosis factor by human placental and decidual tissue

P.M. Starkey, G.S. Vince, S.C. Shorter, I.L. Sargent, C.W.G. Redman

Nuffield Department of Obstetrics and Gynaecology, John Radcliffe Hospital, Oxford, OX3 9 DU, UK

Tumour necrosis factor (TNF) has the potential to play a role in early pregnancy either by regulating the maternal immune response to the developing fetus, or by directly inhibiting trophoblast growth. Its ability to stimulate fibroblasts to release prostaglandins (Dayer et al., 1985) may also be relevant, later in pregnancy, to the control of parturition.

We have measured TNF synthesis both with a bioassay which measures cytotoxicity against the TNF-sensitive WeHi cell line (Espevik & Nissen-Mayer, (1987), and with an ELISA specific for TNFα. Human decidual or chorionic villous tissue samples, taken from first trimester terminations or at term after delivery, were incubated in medium (1g of chopped tissue per 10 ml) for 24 hours. With all types of tissue, the conditioned medium was found to contain significant amounts of TNF. Addition of lipopolysaccharide, which stimulates TNF synthesis by macrophages (Beutler et al., 1985), increased TNF release from most but not all decidual and chorionic villous tissue samples.

To identify which cell types are responsible for TNF synthesis in decidual and chorionic villous tissue, we have used flow cytometry to isolate pure cell populations. First trimester decidua contains large numbers of bone marrow-derived cells, the most abundant cell types being large granular lymphocytes or LGL (45%) and macrophages (19%), with small numbers of T cells (8%) (Starkey et al., 1988). The same cell types are present in term decidua, though the proportion of LGL (4%) is much smaller. Macrophages are also found in the stroma of chorionic villi, with cytotrophoblast and syncytiotrophoblast being the other major cell types.

Decidual cell dispersions were prepared by enzymatic digestion, labelled with an antibody to HLA class II, and separated by flow cytometry into class II-positive macrophages (95% pure) and class II-negative cells (Vince et al., 1988). TNF was detectable only in medium conditioned by incubation with the purified macrophages; no TNF was secreted by the non-macrophage populations. Similar results were found with cells from either first trimester or term decidua. Cytotrophoblast, flow cytometrically purified from dispersions of amniochorion/decidua, secreted no TNF during overnight incubation.

TNF synthesis was also measured at the level of mRNA. RNA isolated from tissue samples or from purified cell preparations was transferred to nitrocellulose in a slot blot apparatus and TNF mRNA detected with a radiolabelled cDNA probe specific for TNFα (800 bp EcoR1 fragment, Genentech Inc., California, USA). Confirming the bioassay results, TNFα mRNA was detected only in RNA of macrophages purified from either term or first trimester decidua. Term cytotrophoblast contained no TNF mRNA.

Our results demonstrate that TNF is synthesised in decidual and chorionic villous tissue in the first trimester and at term. In decidua, TNF synthesis is confined to the macrophages. Term cytotrophoblast synthesise no detectable TNF, and it seems probable that macrophages are also responsible for TNF synthesis in chorionic villous tissue.

REFERENCES

Beutler, B., Mahoney, J., Trang, N.L., Pekala, P. & Cerami, A. (1985): Purification of cachectin, a lipoprotein lipase-suppressing hormone secreted by endotoxin-induced RAW 264.7 cells. J. Exp. Med. 161: 984-995.

Dayer, J.M., Beutler, B. & Cerami, A. (1985): Cachectin/tumour necrosis factor stimulates collagenase and prostaglandin E_2 production by human synovial cells and dermal fibroblasts. J. Exp. Med. 162: 2163-2168.

Espevik, T. & Nissen-Meyer, J. (1987): A highly sensitive cell line WeHi 164 clone 13, for measuring cytotoxic factor/tumour necrosis factor from human monocytes. J. Immunol. Meth. 95: 99-105.

Starkey, P.M., Sargent, I.L. & Redman, C.W.G. (1988): Cell populations in human early pregnancy decidua: characterisation and isolation of large granular lymphocytes by flow cytometry. Immunology 65: 129-134.

Vince G., Jackson, M., Redman, C.W.G., Sargent, I.L. & Starkey, P.M. (1988): Isolation and characterisation of human decidual macrophages. Proc. Roy. Microscope. Soc. 23: 374.

Résumé

SYNTHESE DE FACTEURS DE NECROSE DE TUMEUR ALPHA PAR LE PLACENTA HUMAIN.

Le facteur de nécrose de tumeurs ,TNF ,joue sûrement un rôle dans la grossesse précoce, soit en régulant la réponse maternelle anti-foetale ou en bloquant directement la croissance des cellules trophoblastiques. Nous avons mesuré la synthèse du TNF, soit en utilisant la lignée cellulaire WeHI, c'est une lignée sensible à l'effet du TNF soit par un ELISA spécifique pour le TNFalpha. Les milieux provenant de surnageants de culture de cellules déciduales ou de tissu de villosité choriale, que ce soit dans le premier trimestre ou à terme, contenait des quantités significatives de TNF. Nous avons aussi mesuré la quantité de TNF par les taux de RNA messagers spécifiques. Après extraction de RNA total de tissus déciduaux ou de villosité choriale ou même mieux, de lignées cellulaires purifiées, nous avons procédé à des analyses par des techniques de Northern et de slot blot avec des sondes spécifiques pour le TNF alpha. Des macrophages ont été isolés, par cytométrie de flux après marquage avec des anticorps appropriés, de populations cellulaires obtenus par dispersion de déciduale à terme ou du 1er trimestre et de la même façon nous avons obtenu des populations pure de cytotrophoblaste à partir d'amniochorion à terme. Nous avons détecté des mRNA spécifiques pour le TNF alpha dans la déciduale et les villosités choriales. Dans la déciduale les messagers spécifiques du TNF étaient confinés dans la population macrophagique. Le cytotrophoblaste à terme n'avait pas de mRNA spécifique pour le TNF et ne sécrétait pas de cytokine, c'est à dire de TNF théoriquement actif.

The identification of mRNA for the IL3-related cytokines in placental and decidual tissue

S.C. Shorter, G.S. Vince, P.M. Starkey

Nuffield Department of Obstetrics and Gynaecology, John Radcliffe Hospital, Headington, Oxford, OX3 9DU, UK

The IL3-related cytokines or CSFs; interleukin 3 (IL3), macrophage-colony stimulating factor (M-CSF), granulocyte-colony stimulating factor (G-CSF) and granulocyte/macrophage-colony stimulating factor (GM-CSF), direct the differentiation of pluripotent bone marrow cells and modulate the behaviour of differentiated macrophages. Studies in mice have indicated a possible role for these factors in the development of foetal trophoblast cells. We are investigating the synthesis of CSFs in placental and maternal decidual tissue in human pregnancy.

Production of granulocyte-colony stimulating factor has been measured in tissue-conditioned medium, using a murine myeloid cell line which is dependent on G-CSF and a human bladder carcinoma cell line which constitutively produces G-CSF as the positive control.

Both first trimester decidual and chorionic villous tissue were found to synthesize G-CSF. Levels in villi conditioned media were constant with gestational age, whereas, with decidual conditioned media, as gestational age increased, the presence of an inhibitory factor became evident.

G-CSF mRNA has been detected by slot blot analysis of RNA isolated from decidual cells. We have isolated macrophages (95% pure) from first trimester decidua, by flow cytometry and shown that G-CSF mRNA is confined to decidual macrophages. Further evidence on the distribution of mRNA for G-CSF will be presented, including slot blots of term cytotrophoblast isolated from amniochorion.

Similar investigations are being carried out on the distribution of mRNA for IL3, GM-CSF and M-CSF in first trimester and term decidual and placental cells, by Northern and slot blot analysis.

This work was supported by the Cancer Research Campaign.

Human syncytiotrophoblast plasma membranes (STPM) inhibit production of IL2 and expression of IL2 receptor in actived Jurkat cells

G. Thibault, D. Degenne, J.-M. Guillaumin, A.-C. Girard, P. Bardos

Laboratoire d'Immunologie, Faculté de médecine, 2 bis, boulevard Tonnelé, 37000 Tours, France

Syncytiotrophoblast membranes of full term human placentae inhibit the induced proliferation of human lymphocytes (Amer. J. Reprod. Immunol. Microbiol. 1985, 8 : 20-26).

In order to clarify the mechanism of action of STPM, the early biochemical events including interleukin 2 production (on CTLL2 proliferation) and IL2 receptors expression (by flow cytometry) on Jurkat cells was studied in response to PHA, OKT3, anti CD2 mAbs, 1, 2-O-tetradecanoyl phorbol 13-acetate (TPA) and calcium ionophore A23187 alone or in association.

The production of IL2 and the expression of IL2 receptors were inhibited during stimulation by calcium ionophore and PHA but not by TPA and OKT3.

These findings favor the hypothesis according to which STPM acts on Ca++ influx and protein kinase C.

This work was supported by INSERM and Fondation Langlois.

Immunocytochemical investigations of two «new» fetal antigens (FA-1 and FA-2) isolated from amniotic fluid

D. Tornehave, B. Teisner*, J. Chemnitz, J.G. Westergaard**, H. Boye***, J.G. Grudzinskas****

*Institute of Anatomy and Cytology, *Medical Microbiology, **Obstetrics and Gynaecology, ***Pathology, University of Odense, Denmark, and **** Acad. Unit of Obstetrics and Gynaecology, The London Hospital Medical College, London, UK*

Using fractions of second trimester amniotic fluid as immunogens, we recently described the findings of two hitherto unknown antigens. Based on the compartmental distribution: high concentration in second trimester amniotic fluid and in the fetal blood circulation, the primary source of origin was suggested to be the fetus, and the antigens are referred to as fetal antigen 1 and 2 (FA-1, FA-2). Monospecific antisera against FA-1 and FA-2 were produced and the IgG fractions were used in tissue localization studies of a 7 week fetus from an ectopic pregnancy. FA-1 was found exclusively within the cytoplasm of fetal hepatocytes whereas FA-2 seemed to be associated to the fetal connective tissue, in particular the basement membrane zone and the sclerotome. Results obtained during systematic localization studies of FA-1 and FA-2 of a normal 11 weeks old fetus further established the findings of FA-1 being observed intracellularly in fetal hepatocytes. Moreover, FA-2 was found intracellularly in the osteoblasts during ossification of bones, independently of whether this occurs in membrane or in cartilage. These findings will thoroughly be discussed.

Cytokine production by human decidual macrophages and cytotrophoblast cells

G.S. Vince, S.C. Shorter, P.M. Starkey, I.L. Sargent, C.W.G. Redman

Nuffield Department of Obstetrics and Gynaecology, John Radcliffe Hospital, Headington, Oxford, OX3 9DU, UK

Macrophages comprise around 20% of the cells in human decidual tissue both at term and in the first trimester. Macrophages have the potential to secrete a variety of cytokines which may modulate placental development. PDGF stimulates trophoblast growth whilst TGFß has immunosuppressive activity. To investigate the role of these cytokines at the materno-fetal interface, pure populations of macrophages have been isolated from human maternal decidua at term and in the first trimester by flow cytometric sorting of antibody-labelled cell dispersions. Using similar techniques, pure populations of cytotrophoblast have been obtained from placental membranes at term.

PDGF has been assayed by its ability to stimulate proliferation of quiescent 3T3 cells. Preliminary results using conditioned media from first trimester decidua and villi samples suggest that mitogenic activity is present in some first trimester samples.

Total cellular RNA has been extracted from the decidua and chorionic villi from term and first trimester samples and from pure populations of macrophages and cytotrophoblast. The level of mRNA production for specific cytokines was assessed using the relevant cDNA probes with Northern and slot blot analysis as appropriate

The possible role of local cytokine production by macrophages within the decidua will be discussed.

5.
Placental circulation
Circulation placentaire

Plasminogen activator inhibitors of the placenta and placental bed in normotensive and hypertensive pregnancy

B.L. Sheppard, C. Boyle, N. Gleeson, M. Jordan, L. Daly, J. Bonnar

Trinity College Department of Obstetrics and Gynaecology, Rotunda, and St James's Hospitals, Dublin, Ireland

Fibrin deposition occurs in uterine spiral arteries of the placental bed as the vessels undergo physiological adaptations, required to facilitate increased blood flow to the fetoplacental unit, following the invasion of endovascular trophoblast in early pregnancy (Brosens et al 1967; Sheppard and Bonnar 1974). Morphological studies have shown that increased fibrin formation and deposition within the uteroplacental circulation is intimately concerned with the pathology of uteroplacental spiral arteries which occurs in pregnancies complicated by maternal hypertension, particularly pre-eclampsia (PET) and intrauterine fetal growth retardation (IUGR) with or without the presence of hypertension (Sheppard and Bonnar 1981, 1988). These pathological changes within the uteroplacental vessels in PET and IUGR are associated with a reduced blood flow through the placenta and myometrium (Browne and Veall 1953; Kaeaer et al 1980).

In normal pregnancy, compared to endothelial cells, trophoblast cells lining decidual spiral arteries and syncytiotrophoblast of placental chorionic villi have a reduced capacity to lyse fibrin (Sheppard and Bonnar 1978). Fibrinolysis is activated by the release of plasminogen activators from vascular endothelial cells, the activity of which is mainly controlled by highly specific plasminogen activator inhibitors. At least two such inhibitors, known to be important in the fibrinolytic enzyme system during pregnancy, have been identified: plasminogen activator inhibitor 1 (PAI-1), previously termed endothelial cell PA-inhibitor or fast acting PA-inhibitor (Dosne et al 1977; Kruithof et al 1984) and plasminogen activator inhibitor 2 (PAI-2) or placental type PA-inhibitor (Astedt et al 1987).

We are investigating the mechanism of fibrinolytic activation and inhibition within the placental bed in pregnancy and report here observations of PAI-1 and PAI-2 at Caesarean section in normal pregnancy and pregnancy complicated by PET and IUGR.

PATIENTS AND METHODS

Biopsies, of the placenta and placental bed, and blood samples were obtained at Caesarean section from three groups of patients:

Group 1. 12 normal pregnancies.
Group 2. 10 pregnancies complicated by pre-eclampsia (presence of hypertension, $140/90$ mm Hg or above, and proteinuria, 1 gm or more per 24 hours) resulting in the delivery of an infant of appropriate birthweight for gestational age.
Group 3. 12 pregnancies complicated by intrauterine fetal growth retardation (less than the 10th centile) of which 6 were normotensive and 6 were hypertensive.

Full and informed consent was given for each placental bed biopsy to be taken, and the study was approved by the Hospital Ethics Committee.

Peripheral vein, uterine vein and umbilical cord blood samples (5 ml) were taken into plastic syringes following delivery of the infant but before delivery of the placenta. The blood samples were immediately transferred to tubes containing sodium citrate as an anticoagulant: the plasma aliquots obtained by centrifugation were snap frozen in liquid nitrogen and stored at $-80^{\circ}C$. Plasminogen activator inhibitor content of the plasma was estimated by TintElize PAI-1, enzyme immunoassay for the determination of human PAI-1 antigen (plasminogen activator inhibitor 1, endothelial type) (Biopool, Sweden) and TintElize PAI-2, enzyme immunoassay for the determination of human PAI-2 antigen (plasminogen activator inhibitor 2, placental type) (Biopool, Sweden).

A wedge-shaped biopsy of the placental bed was obtained under direct vision following removal of the placenta and a biopsy of villous tissue, which appeared macroscopically normal, was also immediately taken from the delivered placenta. Each biopsy was divided into three parts. One part was fixed in formal saline for examination by light microscopy. A second part was snap frozen and stored for immunohistochemical localisation of plasminogen activator inhibitors. Immunostaining of inhibitors was carried out on 6-8 micron cryostat sections of the biopsies, air-dried and fixed in acetone, using the Avidin-Biotin technique and PAI-1 and PAI-2 antihuman mouse monoclonal antibodies of IgG type (Biopool, Sweden). Negative and positive controls were included with each assay. The third part of each biopsy was frozen, following further dissection, as placenta, placental bed decidua or myometrium for tissue extraction and subsequent quantitative estimation of inhibitors. Tissue extraction involved homogenisation in 2M sodium thiocyanate (pH 7.75) and dialysis of homogenate supernatants with 0.14M phosphate buffered saline (pH 7.4) Inhibitor content of these tissue extracts was assayed with the same TintElize PAI-1 and PAI-2 enzyme immunoassays used for the plasma samples. Following protein estimations of the tissue extracts, PAI-1 and PAI-2 levels were estimated in relation to protein content of extracts of the placenta, placental bed decidua and myometrium.

RESULTS

In the placenta immunoreactive material was seen in the endothelial cells of blood vessels of chorionic villi, when using PAI-1 antibodies, and in the syncytiotrophoblast lining the chorionic villi when using PAI-2 antibodies (Fig 1a and b). No staining was seen in the stroma or stromal cells of the villi. PAI-2 was also localised in trophoblast cells of decidual spiral arteries which had undergone the physiological changes of normal pregnancy (Fig 2).

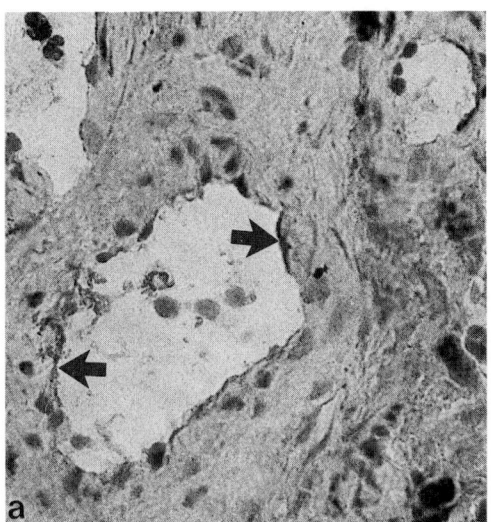 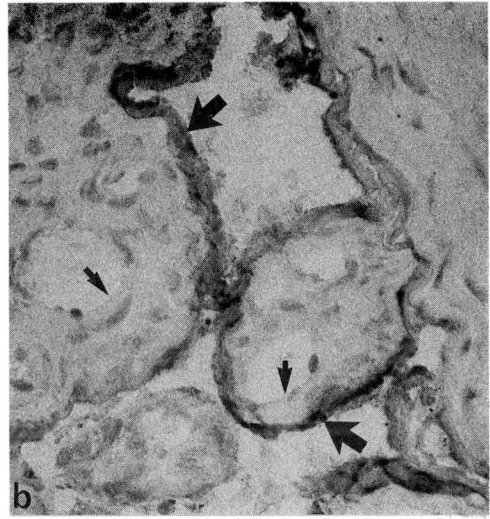

Fig. 1. Immunocytochemical staining of (a) PAI-1 in vascular endothelium (arrows) within placental villi and (b) PAI-2 in syncytiotrophoblast (large arrows) but not in vascular endothelium (small arrows) of placental villi.

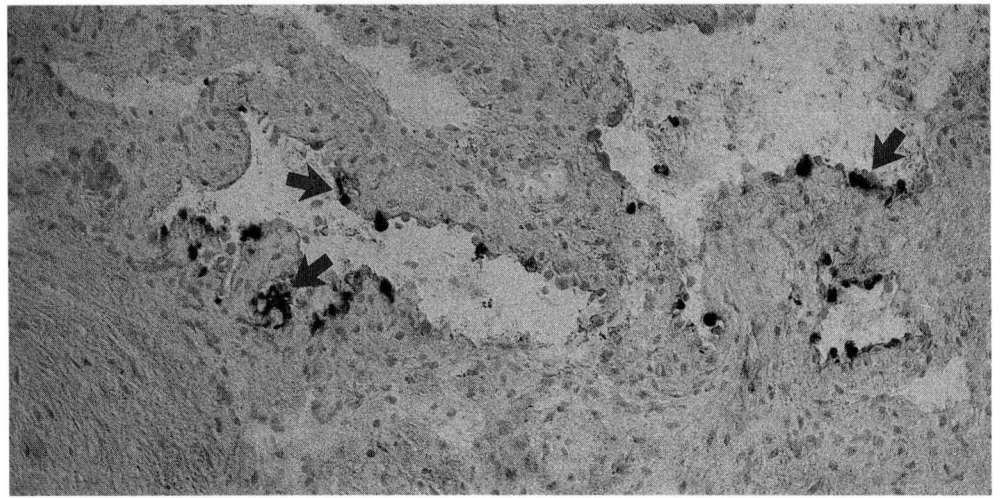

Fig. 2. PAI-2 immunostaining of trophoblast (arrows) within the wall of a decidual spiral artery in normal pregnancy.

In normal pregnancy, PET and IUGR the placenta was found to contain significantly higher levels of PAI-2 than PAI-1 ($p < 0.001$). The levels of placental PAI-1 were significantly higher in PET and IUGR than in normal pregnancy ($p < 0.02$) and levels of PAI-2 were lower, particularly in IUGR - although not significantly (Fig 3). Higher levels of PAI-1 than PAI-2 were seen in the extracts of placental bed decidua in normal pregnancy, PET and IUGR ($p < 0.01$). Although the levels of decidual PAI-1 and PAI-2 appeared higher in PET and IUGR than in normal pregnancy the differences were not significant (Fig 4). Similar levels of both PAI-1 and PAI-2 were found in the placental bed myometrium in normal pregnancy (Fig 5). Although the level of PAI-2 was lower in PET and the levels of PAI-1 and PAI-2 were lower in IUGR the differences from normal were not significant.

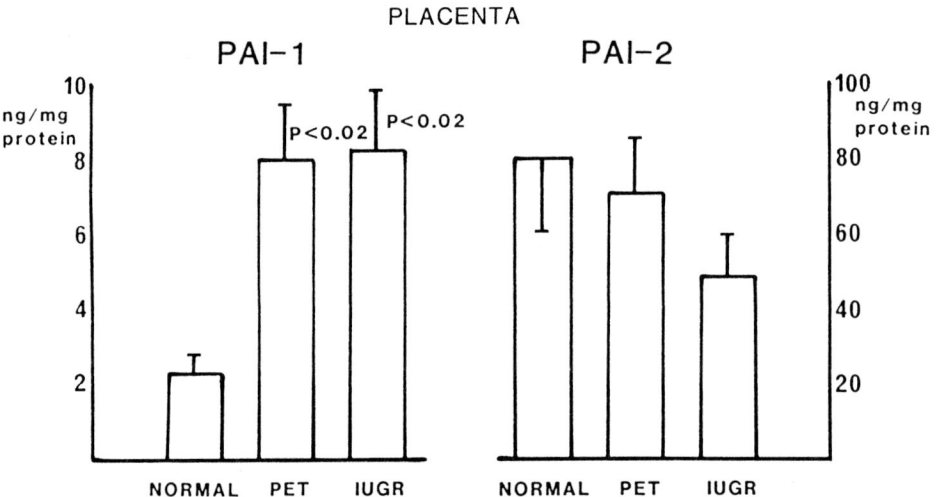

Fig. 3. Levels (Mean ± SEM) of PAI-1 and PAI-2 in extracts of placenta from normal pregnancy, pre-eclampsia (PET) and intrauterine fetal growth retardation (IUGR).

In normal pregnancy peripheral vein plasma at Caesarean section contained significantly higher levels of PAI-2 than PAI-1 ($p < 0.001$). PAI-1 levels were significantly higher in peripheral vein plasma in PET and IUGR ($p < 0.02$) and PAI-2 levels significantly lower in IUGR ($p < 0.001$), although not in PET, than in normal pregnancy. Similar levels of PAI-1 were detected in umbilical cord plasma from normal pregnancy, PET and IUGR. No PAI-2 was detected in umbilical cord plasma. Uterine vein plasma reflected similar, but higher levels of inhibitors to that observed in peripheral vein plasma. Significantly higher levels of PAI-1 ($p=0.03$) and lower levels of PAI-2 ($p < 0.001$) were seen in IUGR than in normal pregnancy uterine vein plasma (Fig 6). No significant differences were seen in PAI-1 or PAI-2 levels between normal pregnancy and PET in uterine vein plasma. Whereas a positive correlation was found

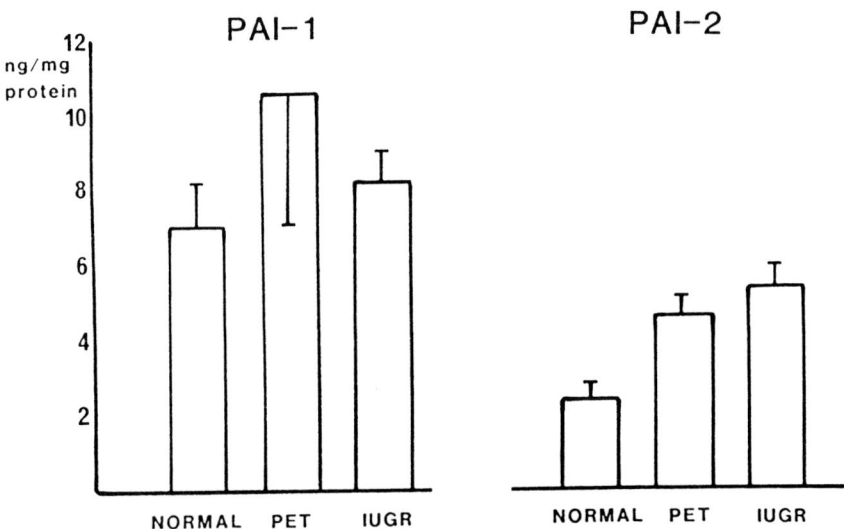

Fig. 4. Levels (Mean ± SEM) of PAI-1 and PAI-2 in extracts of placental bed decidua from normal pregnancy, pre-eclampsia (PET) and intrauterine fetal growth retardation (IUGR).

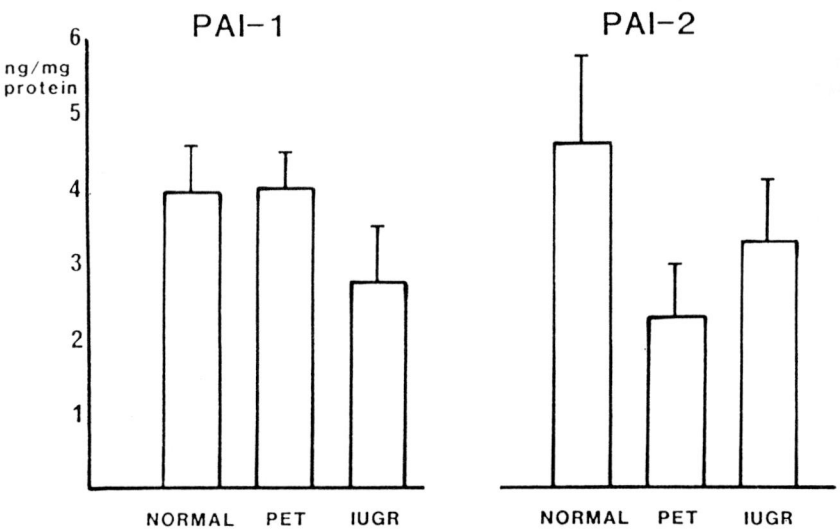

Fig. 5. Levels (Mean ± SEM) of PAI-1 and PAI-2 in extracts of placental bed myometrium from normal pregnancy, pre-eclampsia (PET) and intrauterine fetal growth retardation (IUGR).

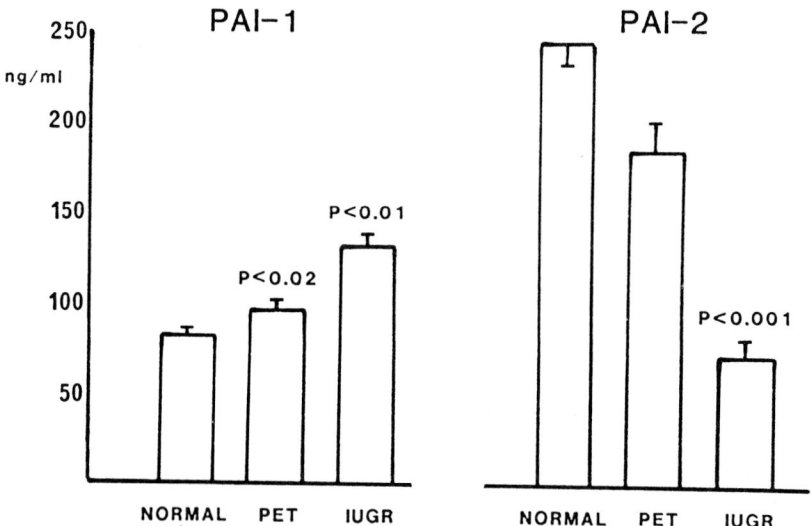

Fig. 6. Levels (Mean ± SEM) of PAI-1 and PAI-2 in uterine vein plasma at Caesarean section from normal pregnancy, pre-eclampsia (PET) and intrauterine fetal growth retardation (IUGR).

between PAI-2 levels in uterine vein plasma (r=0.69) and birthweights, a negative correlation was found between the levels of PAI-1 in uterine vein plasma (r=-0.58) and birthweights of the infants from all pregnancies included in the study (Fig 7).

DISCUSSION

The immunohistochemical results confirm previous studies which have localised PAI-1 to vascular endothelium and PAI-2 to trophoblastic epithelium of the placenta (Astedt et al 1986). It has been suggested that replacement of endothelial cells by cytotrophoblast in the decidua during normal pregnancy has the effect of reducing the fibrinolytic activity of uteroplacental arteries (Sheppard and Bonnar 1978) and the present study has shown that this is probably due to the presence of PAI-2 in the cytoplasm of these endovascular cytotrophoblast cells.

The levels of PAI-1 and PAI-2 in the placental extracts in normal pregnancy, PET and IUGR were reflected in the plasma levels of the uterine, and to a lesser extent, the peripheral vein. It was noticeable that lower levels of PAI-2, but significantly higher levels of PAI-1 were found in the placenta and uterine vein plasma of the low birth weight infants. Occlusive lesions in maternal uterine vasculature may be the major cause of the infarction and impairment of placental function in pregnancies complicated by PET and IUGR. The lower levels of PAI-2 and higher levels of PAI-1 found in the placenta in IUGR may reflect irreversible hypoxic

UTEROPLACENTAL INHIBITORS OF FIBRINOLYSIS
CORRELATION of BIRTHWEIGHT with UTERINE PAI-1

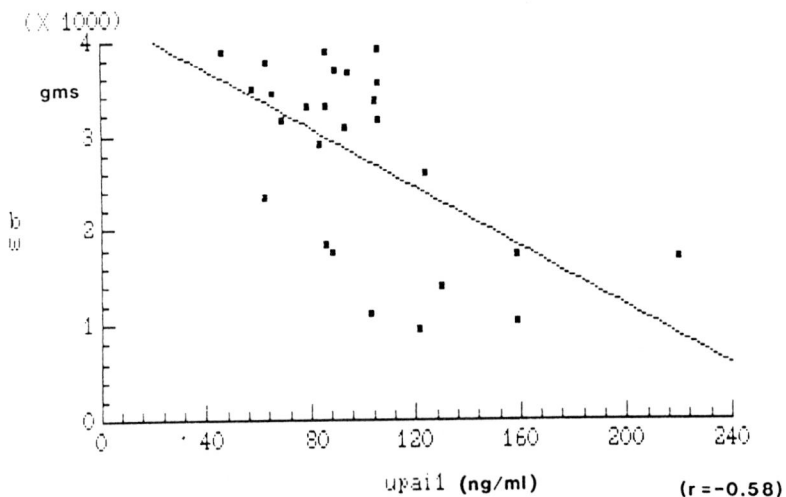

Fig. 7. Correlation of uterine vein plasma PAI-1 levels and birth weights of infants.

effects on the placenta associated with a reduced uteroplacental blood flow. Degenerative changes of placental villi have been shown to be related to the degree of luminal occlusion of decidual spiral arteries (Sheppard and Bonnar 1980). Whereas complete occlusion of decidual spiral arteries is associated with total necrosis of adjacent placental villi, partial occlusion may be coupled with degenerative changes in the villous syncytiotrophoblast with a normal underlying circulation in fetal blood vessels. This would suggest that in such pregnancies the degenerating placental villi would have a reduced capacity to produce PAI-2, although PAI-1 would be released from the normal endothelium of the villous blood vessels.

Kawano et al (1968) were the first to identify a specific inhibitor of fibrinolysis in the placenta, and it has been suggested that the placenta in IUGR contains increased quantities of fibrinolytic inhibitors (Elder and Myatt 1976). The findings of the present study suggest that a disordered inhibition of uteroplacental fibrinolysis occurs in PET, which is more exaggerated in IUGR: placental trophoblast appears less able to produce PAI-2 and the increased fibrin deposition, and persistance of microthrombi, in the uteroplacental circulation in IUGR is probably due to increased levels of PAI-1.

ACKNOWLEDGEMENTS

We are grateful for the co-operation of the theatre and labour ward staff of the Rotunda Hospital. The study was supported by a grant from the Wellcome Trust.

REFERENCES

Astedt, B., Haegerstrand, I., and Lecander, I. (1986): Cellular localisation of placental type plasminogen activator inhibitor. Thromb. Haemostas. 56: 63-65.

Astedt, B., Lecander, I., and Ny, T. (1987): The placental type plasminogen activator inhibitor PAI-2. Fibrinolysis 1: 203-208.

Browne, J.C.M., and Veall, N. (1953): The maternal placental blood flow in normotensive and hypertensive women. J. Obstet. Gynaecol. Br. Emp. 60: 241-247.

Brosens, I., Robertson, W.B., and Dixon, H.G. (1967): The physiological response of the vessels of the placental bed to normal pregnancy. J. Pathol. Bact. 93: 569-579.

Dosne, A.M., Dupuy, E., and Bodevin, E. (1978): Production of a fibrinolytic inhibitor by cultured endothelial cells from human umbilical vein. Thromb. Res. 12: 377-387.

Elder, M.G., and Myett, L. (1976): Coagulation and fibrinolysis in pregnancies complicated by fetal growth retardation. Br. J. Obstet. Gynaecol. 83: 355-364.

Kawano, T., Morimoto, K., and Uemura, Y. (1968): Urokinase inhibitor in human placenta. Nature 217: 253.

Kaeaer, K., Jouppila, P., Kuikka, J., Luotola, H., Toivanen, J., and Rekonen, A. (1980): Intervillous blood flow in normal and complicated late pregnancy measured by intravenous ^{133}Xe method. Acta Obstet. Gynecol. Scand. 59: 7-10.

Kruithof, E., Tran-Thang, C., Ransiju, A., and Bachmann, F. (1984): Demonstration of a fast-acting inhibitor of plasminogen activators in human plasma. Blood 64: 907-913.

Sheppard, B.L., and Bonnar, J (1974): The ultrastructure of the arterial supply of the human placenta in early and late pregnancy. J. Obstet. Gynaecol. Brit. Commonw. 81: 497-511.

Sheppard, B.L., and Bonnar, J. (1978): Fibrinolysis in decidual spiral arteries in late pregnancy. Thromb. Haemostas 39: 751-758.

Sheppard, B.L., and Bonnar, J. (1980): Ultrastructural abnormalities of placental villi in placentae from pregnancies complicated by intrauterine fetal growth retardation: their relationship to decidual spiral arterial lesions. Placenta 1: 145-156.

Sheppard, B.L., and Bonnar, J. (1981): An ultrastructural study of uteroplacental spiral arteries in hypertensive and normotensive pregnancy and fetal growth retardation. Br. J. Obstet. Gynaecol. 88: 695-705.

Sheppard, B.L., and Bonnar, J. (1988): The maternal blood supply to the placenta in pregnancy complicated by intrauterine fetal growth retardation. Trophoblast. Res. 3: 69-81.

Résumé

Les inhibiteurs de l'activité plasminogène du placenta de grossesse normale et en hypertension.

B.L. Sheppard, C. Boyle, N. Gleeson, M. Jordan, L. Daly & J. Bonnar

Hôpitaux St. James et Rotunda
Département d'Obstétrique et de Gynécologie
Dublin, Irlande.

Les inhibiteurs de l'activité du plasminogène (facteurs PA I et PA II) ont pour rôle d'empêcher la lyse des dépôts vasculaires de fibrine. Ces inhibiteurs ont été étudiés dans les placentas de patientes prééclamptiques et d'enfants hypotrophes qui présentent généralement des troubles circulatoires utéro-placentaires.

Le PA I est localisé dans l'endothélium des villosités placentaires; le PA II dans le syncytiotrophoblaste des villosités et dans le cytotrophoblaste endovasculaire. L'activité PA II est normalement bien supérieure à l'activité PA I dans les tissus et dans le sang veineux. Les activités PA I des tissus placentaires et du sang rétroplacentaire utérin sont très augmentées chez les patientes prééclamptiques et à enfant hypotrophe. Dans ces deux groupes de patientes, on note aussi une diminution de l'activité PA II.

Les modifications du rapport d'activité PA I/PA II caractériseraient les troubles vasculaires utéro-placentaires en rapport avec la fibrinolyse (caillot, thrombi, etc...) rencontrés dans ces pathologies, et peut-être leurs conséquences telles que la dégénérescence du trophoblaste et les réactions vasculaires foetales.

Circulation placentaire : résumé de l'atelier

A.M. Carter, W. Moll*

Department of Physiology, Odense University, Campusvej 55, DK-5230 Odense M, Denmark
*Department of Physiology, University of Regensburg, Postfach 377, D-8400 Regensburg 2, Federal Republic of Germany

Cet atelier consacré à la circulation placentaire a rassemblé 40 participants environ. La conférence d'introduction présentée par P. Kaufmann a été suivie de six autres présentations qui ont donné lieu à des échanges très intéressants entre participants.

Les idées essentielles qui se dégagent de cet atelier sont les suivantes :

- L'hypoxie serait à l'origine d'une hypercapillarisation due à une subdivision anormale des vaisseaux. Cette situation permettrait de réduire la résistance des tissus, en augmentant le volume vasculaire.

- Un regain d'intérêt pour la compréhension du rôle physiologique de la vésicule ombilicale a été observé. Le débit circulatoire dans la vésicule ombilicale a été déterminé chez le Cobaye en recourant aux microsphères radioactives. Sa valeur est étroitement et directement liée à la saturation en oxygène du sang artériel.

- Les facteurs rhéologiques (déformabilité et agglomération des hématies) de même que la nature du lit vasculaire peuvent intervenir dans la circulation du sang utéro-placentaire. Si les facteurs rhéologiques ont peu d'influence, la relation inverse observée entre la différence artério-veineuse de pression hydrostatique et le débit circulatoire suggère que le lit vasculaire se comporte comme une structure poreuse et non comme une structure tubulaire rigide. Cette observation ne tient cependant pas compte du rôle régulateur du système sympathique qui devrait être évalué afin de vérifier cette hypothèse.

- Trois présentations ont été consacrées à la vélocimétrie en Doppler qui constitue une méthode non traumatisante d'investigation de la circulation foetale. Des simulations dans un modèle digital montrent que la vélocité diastolique et la forme du signal de vélocité dépendent de la résistance périphérique, mais également de la fonction cardiaque, de la compliance et de la résistance vasculaires. Les facteurs peptidiques natriurétiques de l'oreillette influencent le signal de vélocité. Des corrélations entre ce signal et les conditions pathologiques ont été établies. La signification physiologique des changements de vélocité observés dans ces pathologies reste cependant à éclaircir.

The workshop on placental circulation attracted about 40 participants. Peter Kaufmann (Aachen) introduced the subject by giving an overview of the fetal and maternal circuits in the placenta. There were six further presentations, each of which engendered a lively discussion, from which the following points emerged.

It has now been established that a moderate rise of capillary density occurs in the placenta of guinea pigs during hypoxia. This is evidently achieved by enhanced branching, ensuring low resistance to blood flow.

The yolk sac placenta seems to be attracting increasing interest. Yolk sac blood flow has now been measured and shown to be strongly dependent upon arterial O_2 saturation.

Rheological factors such as the porous structure of the placenta, the deformability of red cells and red cell agglomeration, may affect placental blood flow. However, further work will be needed to establish whether placental structure has any major influence on the overall placental resistance to blood flow.

Three papers were devoted to Doppler flow velocimetry, which offers to provide insight into the undisturbed fetal circulation in man. However, observations in a digital model show that, besides peripheral resistance, myocardial function and the distribution of compliance and resistance determine diastolic flow and the shape of the pulse wave. Interesting interrelationships between atrial natriuretic peptide (ANP) and changes in the flow pulse wave were shown. Observations of correlated abnormal pulse wave forms and fetal distress were also reported. The discussion revealed that there is still no general consensus about the physiological and pathophysiological interpretation of Doppler measurements of fetal placental blood flow.

An abstract of the individual presentations is given below.

MATERNAL AND FETAL CIRCUITS IN THE PLACENTA: AN OVERVIEW.

Peter Kaufmann, Department of Anatomy, RWTH, Aachen, F.R.G.

Fetal Vessel System. In the human placenta at term, 60-70 villous stems arise from the chorionic plate, each branching into one villous tree, called a fetal cotyledon.

Fetal arteries and veins are restricted to the stem villi, whereas arterioles are mainly located in the intermediate villi. Capillary loops are present as a paravascular network in the stem villi as well as in the intermediate villi, and in the highest frequency as terminal capillary loops in the terminal villi. In contrast to the famous drawings by Arts, the venous limbs of the capillary loops are not positioned in the center of the terminal villi, located central to the arterial limbs, but both limbs are arranged symmetrically on opposing sides of the villus and equally near to the villous surface.

Especially in the terminal villi, long segments of the capillary loops are dilated sinusoidally. It has been incorrectly assumed that the sinusoids are dilated venous parts of the capillary loops. However, closer examination of the fetal vessel arrangement in a terminal villus reveals that the undilated capillaries are located mostly near the base, in the so-called neck region. The sinusoids prevail nearer to the villous tips, where they cause the bulbous enlargement of the terminal villi. The sinusoidal dilatations of the long terminal capillary loops (about 4000 µm in length) may serve to reduce blood flow resistance. Accordingly, the shorter paravascular capillaries (1000 to 2000µm in length) do not show sinusoids. These are also absent in the web-like capillary nets of all immature, proliferating villous types (Bøe, 1952). Labyrinthine placentas with short (about 1000µm) fetal capillaries, like that of the guinea-pig, do not need such sinusoids. In contrast the longer capillaries (2000 to 3000µm) of the otherwise identical placenta of the capybara are sinusoidally dilated.

Maternal Circulation. The human placenta belongs to the haemochorial, villous type: after leaving the spiral arteries, the maternal blood circulates through the diffuse intervillous space, flowing directly around the villi. The arterial inlets are normally situated close to the centres of the villous trees. In contrast to the arteries, the venous openings in the basal plate are arranged around the periphery of the villous trees. Each fetomaternal circulatory unit, consisting of a villous tree with a corresponding centrifugally perfused part of the intervillous space, is called a placentone. Normally several placentones belong to one maternal placental lobule, as seen from the basal surface of the placenta.

The centers of typical placentones exhibit loosely arranged villi, most of them on the immature intermediate type, providing ample intervillous space for the maternal arterial inflow. In the periphery of the placentones, most of the villi are of the mature intermediate or terminal type, located very close to each other. They are separated only by narrow capillary-like intervillous clefts. Here the maternal blood stream comes into intimate contact with the terminal villous surfaces. Near the chorionic plate and in the outer, lateral periphery of the placentones, the arrangement of the villi becomes looser again, thus providing space for the venous backflow. Such placentones are easily identified at the periphery of the placenta, where one villous tree fits into one basally visible placental lobule. However, in the central parts of the organ, large villous trees may be situated so close together that they partly overlap, 3 - 4 being present in one lobule, and the typical placentone architecture is lost.

The radioangiographic studies of Ramsey in the rhesus monkey placenta, and of Borell in the human placenta, are consistent with this placentone concept. They have revealed a rather rapid arterial filling of the centers of the maternal circulatory units, originally called a 'jet' or 'spurt', and subsequently a rather slow centrifugal spreading of the blood toward the subchorial and the peripheral venous outflow areas of the placentones.

The interrelations of fetal and maternal blood flows are rather complicated. Considering the structural arrangement and the effectiveness of the system, nowadays there is general agreement that the human placenta belongs to the multivillous type of exchanger.

Reference

Boe, F. (1952) Studies on the vascularization of the human placenta. Acta obstet gynecol scand 32, suppl. 5, 1-92.

THE MECHANICAL INFLUENCE OF HAEMORHEOLOGY AND DRIVING PRESSURE ON UTEROPLACENTAL PERFUSION.

Carla M. Verkeste and Louis L.H. Peeters, Department of Obstetrics and Gynaecology, State University Limburg, Maastricht, The Netherlands.

The placenta contains neither capillaries nor precapillary sphincters, and various studies have provided evidence that it lacks autoregulation. Theoretically this might imply that uteroplacental blood flow (UBF) is related to driving pressure and/or the intrinsic rheological properties of blood within the placental microvasculature. The effect on UBF of blood viscosity, which was varied by changing the haematocrit (Peeters, Verkeste, Saxena and Wallenburg, 1987) and red cell deformability (Verkeste, Boekkooi, Saxena and Peeters, submitted), have been investigated in our laboratory in awake 57 days pregnant guinea-pigs. The results of these studies indicate that variations in blood viscosity have little effect on UBF. In a separate study we attempted to elucidate the relationship between blood pressure (BP), i.e. driving pressure, and UBF. An increase in blood pressure is generally thought to lead to an increase in UBF. This is based on the assumption that the placental microvasculature behaves like a rigid tubular system. However, the placenta differs from almost all other vascular beds in its porous rather than tubular microvascular architecture. According to physical laws a porous structure may respond to a rise in perfusion pressure with an increase rather than a decrease in resistance to flow in the intervillous space (Dullien, 1979). This response may be potentiated by the effect of a rise in sympathetic activity- often associated with hypertension - on the radial arteries proximal to the placenta. We compared BP with UBF (percent of cardiac output), determined by the microsphere method, in 37 awake late-pregnant guinea-pigs, which had been studied in other protocols in our laboratory. Data obtained in baseline conditions were compared with those 24 hours after different manipulations. In 27 animals BP and UBF changed in opposite directions ($p<0.02$ sign test). These results support the concept that an increase in BP is paralleled by a rise in perfusion resistance of the entire uteroplacental vascular bed. However, no information is given on the separate contributions of the physical properties of the uteroplacental porous structure and sympathetic activity. This data can only be obtained either by studying the relationship between the pressure gradient between spiral artery and uterine vein, on the other hand, and intervillous flow on the other; or by evaluating the across pressure gradient uterine arterial vessels proximal to the spiral arteries in relation to flow at various arterial blood pressures.

References

Dullien, F.A.L. (1979) Porous media, Fluid transport and pore structure, p. . New York, Academic Press.

Peeters, L.L.H., Verkeste, C.M., Saxena, P.R. and Wallenburg, H.C.S. (1987) Relationship between maternal hemodynamics and hematocrit and hemodynamic effects of isovolemic hemodilution and hemoconcentration in the awake late-pregnant guinea pig. Pediatr. Res. 21, 584-589.

ATRIAL NATRIURETIC PEPTIDE IN THE HUMAN FETUS AND ITS EFFECT ON THE VASOREACTIVITY OF THE UMBILICAL CIRCULATION

John C. P. Kingdom, The Queen Mother's Hospital, Glasgow, Scotland.

The application of Doppler velocimetry to assess the umbilical circulation in pregnancy has suggested a common mechanism by which an adverse perinatal outcome may occur in both pre-eclampsia and growth-retardation (IUGR): this is that reduced, or even absent diastolic flow indicates increased downstream impedance in the fetal placental circuit. However, the significance of such waveforms remains uncertain, in part due to our poor understanding of the mechanisms by which a high-flow low-resistance circuit is maintained in normal pregnancy.

One possible factor is Atrial Natriuretic Peptide (ANP), a 28 amino-acid peptide which is released from atrial myocytes in response to atrial distension. In man, ANP is a vasodilator, promotes salt and water excretion in the kidney, and inhibits aldosterone production. In the fetus, ANP has been detected in cardiac tissue from as early as 12 weeks gestation. In later fetal life, ANP is present at higher concentrations than the corresponding maternal values (1). Intravascular transfusion of the fetus for rhesus isoimmunisation results in a rise in circulating ANP consistent with atrial distension (1). As in the adult, ANP in fetal life comes from the heart, but both the atria and ventricles manufacture and release ANP prior to birth.

Laboratory experiments utilising the sheep preparation have defined some of the actions of ANP in the fetus. ANP induces a fetal diuresis (2), inhibits adrenal steroid production and reduces lung fluid production. As such ANP might play a role in the regulation of amniotic fluid volume, the measurement of which is now a key element in biophysical fetal assessment.

The demonstration of receptors for ANP in the placenta (3) indicates that this may be an important target organ for ANP. We have identified and characterised ANP receptors in a fetal placental artery membrane fraction. We have been able to demonstrate the presence of both high- and low-affinity receptors in normal term placenta. Interestingly we have found evidence of down-regulation of the high-affinity receptor in unexplained IUGR.

The binding characteristics of the high-affinity receptor suggest that it is likely to be functional within the known range of circulating ANP in fetal blood. Utilising a simple perfusion technique, we have been able to demonstrate that ANP infused in the physiological range will significantly reduce the vasoconstictor effect of angiotensin II, supporting a physiological role for ANP in the fetal placental circulation.

These preliminary findings suggest that ANP may be contributing to the maintenance of a low-resistance circuit on the fetal side of the placenta. The dynamic ANP response in the fetus to intravascular transfusion may allow the fetus to cope with volume expansion by diverting blood to the placenta. In addition, down-regulation of fetal placental ANP receptors in unexplained IUGR may indicate a mechanism by which absent umbilical diastolic flow occurs.

The author would like to thank the following for their support and contributions to the work presented: Dr. M.Whittle, Queen Mother's Hospital, Dr's A. Jardine, J. McQueen & J. Connell, MRC Blood Pressure Unit, Western Infirmary, and Dr. A. Templeton, Institute of Physiology, Glasgow.

References

1. Kingdom, J.C.P., Jardine, A.G., Doyle, J., et al. (1989) Atrial natriuretic peptide in the fetus. Br Med J, 298, 1221-1222.

2. Brace, R.A., Cheung, C.Y. (1987) Cardiovascular and fluid responses to atrial natriuretic factor in the sheep fetus. Am J Physiol, 253, R561-R567.

3. Hatjis, C.J., Grogan, D.M. (1988) Atrial natriuretic peptide receptors in normal human placentae. Am J Obstet Gynecol, 159, 587-591.

MATERNAL OXYGEN SUPPLY AS A REGULATOR OF FETAL PLACENTAL CAPILLARISATION.

Iris Scheffen and Peter Kaufmann, Aachen, F.R.G.

It is well known that hypoxia causes fetal villous hypercapillarisation. Two different types of hypercapillarisation have been described: (a) abnormally long, unbranched, highly dilated capillary loops, and (b) dense, highly branched networks of short, narrow capillaries. So far it has proved to be impossible to determine which of these morphological reaction patterns is due to hypoxia, as in the human this usually occurs together with other pathogenetic factors and often coincides with placental prematurity. The aim of the present study was to examine this problem under experimental conditions.

Fourty-four pregnant guinea pigs were kept under uninterrupted isobaric hypoxia (12% oxygen) from day 15 to day 60 of gestation. Physiological parameters including haematocrit, P_{50}, pO_2, and reoxygenation time were studied at intervals of 20 days. Around day 60 the morphology of the fetal capillaries was examined by scanning electron microscopy of vessel casts, semithin histology, and transmission electron microscopy. The vessel casts showed a considerably higher degree of branching and coiling of the capillary bed as compared to the control placentas. Transmission electron microscopy revealed a remarkably reduced diameter and an increased number of capillary cross sections following hypoxia. The materno-fetal diffusion distance was shortened by thinning of the trophoblast. The morphological effects were verified morphometrically.

These findings demonstrate that oxygen is an important regulator of fetal capillarisation. Chronic hypoxia stimulates capillary branching and is only responsible for type (b) hypercapillarisation, characterized by dense networks of narrow and short capillaries, reducing fetal blood flow resistance. In agreement with several pathohistological indications in the human, type (a) hypercapillarisation, expressed by long, unbranched, dilated capillaries, probably is a result of placental prematurity

BLOOD FLOW TO THE YOLK SAC PLACENTA AS A FUNCTION OF ARTERIAL OXYGEN CONTENT.

Anthony M. Carter, Department of Physiology, University of Odense, Denmark.

The yolk sac shrivels and disappears early in human development, but in other mammals it is retained until term and participates in fetomaternal exchange. The vitelline vessels which supply the inverted yolk sac placenta of the guinea pig pursue a different course from the blood vessels that supply the chorioallantoic placenta. The vitelline artery can therefore be catheterized without interrupting the umbilical blood flow, and can be used to obtain a reference sample of blood when determining blood flow to the lower body and placenta by the microsphere technique (Carter, 1984). Obviously, tying a catheter into the vitelline artery precludes determination of blood flow to the yolk sac placenta. An alternative approach, which we have used when measuring cerebral blood flow (Carter & Gu, 1988), is to catheterize the axillary artery. However, because of preferential streaming of blood and microspheres in the fetal circulation, a reference sample from a branch of the ascending aorta cannot be used to measure blood flow in organs supplied from the descending aorta.

Blood flow to the yolk sac placenta of the guinea pig was therefore estimated by an indirect approach. Microspheres were injected in the saphenous vein of 7 fetuses, and the ratio found between the radioactivity in the yolk sac and the radioactivity in the gut. This ratio, which is the same as that between yolk sac blood flow and gut blood flow, was then multiplied by a reference value for blood flow to the gut.

$$Q_{yolk\ sac} = CPM_{yolk\ sac}/CPM_{gut} \times Q_{gut}$$

The reference value for gut blood flow (2.50 ml/min) was taken from a separate microsphere study in which the vitelline artery had been catheterized to obtain a reference blood sample (Carter, 1984).

Arterial oxygen content and pH was measured in blood taken from an axillary artery. A positive linear correlation was found between estimated yolk sac blood flow and arterial O_2 content ($r = 0.90$, $P<0.01$). Blood flow to the gastrointestinal tract, which was used as a reference for these estimates, shows little variation with arterial O_2 content in the 2 mM to 6 mM range (Peeters et al., 1979). It may therefore be concluded that yolk sac blood flow decreases when the arterial O_2 content falls.

In normoxaemic guinea pig fetuses near term, yolk sac blood flow was 100-300 µl per min, which is equivalent to 0.3-1.0% of the biventricular cardiac output. This must be regarded as a minimum estimate, since two incisions were made in the yolk sac, to obtain access to a fetal forelimb and a fetal hindlimb. Although these incisions were far from the main vitelline vessels, there was some loss of fluid, and this may have affected yolk sac function.

References

Carter, A.M. (1984) The blood supply to the abdominal organs of the fetal guinea-pig. J Develop Physiol 6, 407-416.

Carter, A.M. & Gu, W. (1988) Cerebral blood flow in the fetal guinea-pig. J Develop Physiol 10, 123-129.

Peeters, L.L.H., Sheldon, R.E., Jones, M.D., Makowski, E.L. & Meschia, G. (1979) Blood flow to fetal organs as a function of arterial oxygen content. Am J Obstet Gynecol 135, 637-646.

EVALUATION OF PLACENTAL RISK USING PULSED DOPPLER ULTRASOUND TO ASSESS FLOW VELOCITY IN THE UMBILICAL ARTERIES: CORRELATIONS WITH HISTOMETRIC DATA AND CLINICAL FINDINGS.

T. Hitschold, E. Weiss, H. Müntefering and P. Berle, Wiesbaden and Mainz, F.R.G.

We performed a clinical study with 310 pregnant women to assess the diagnostic value of umbilical artery flow velocities determined by pulsed Doppler ultrasound. We divided the subjects into 2 groups, one with pregnancies at risk (n=118) (i.e. PIH, SGA, smoking >10 cig./day) and another with pregnancies not at risk (n=192). Doppler examination was performed within the last week before delivery and the result was not included in clinical decisions. 39 of the pregnancies at risk had pathological flow patterns in the umbilical arteries, in 27 of these a cesarean section owing to asphyxia was later necessary, and more than 25% of the newborns had postpartum acidosis or low Apgar scores. In cases with normal Doppler flow, no cesarean section for fetal distress occurred. In the non-risk group we found no correlation between the Doppler flow results and the fetal outcome. Only 4 of 11 cesarean sections for fetal asphyxia were detected antepartum by this method in the non-risk group.

We tested whether the flow velocity waveforms in the umbilical arteries are related to the intravillous blood volume, which is thought to reflect placental vascular resistance. We made a histometric study of muscle-free villous vessels in the periphery of placentones and calculated the fetal intravillous blood volume from the cross sectional area of the vessels, number of villi per mm² and placental volume.

In a total of 160 patients we compared the fetal intravillous blood volume with the Resistance Index (RI) measured in an umbilical artery. We found an a correlation coefficient of 0.70: the larger the vascular tree of the fetal placenta, the lower the RI in the umbilical artery. This correlation is in agreement with the findings of Giles, Trudinger and Baird (1985), who observed that the wave form is related to the number of small arteries in the placenta.

An estimated intravillous blood volume of 85 ml separated normal and pathological cases. Taking this value as a limit, we have listed the fetal risks in a preliminary study of 90 cases. In cases with an intravillous blood volume below 85 ml, 85% cesarean sections for fetal asphyxia, 25% reduced Apgar scores, 30% acidosis and 70% SGA babies occurred. In pregnancies with a small intravillous blood volume the RI was above the 90th percentile in 96% and a diastolic zero or reverse flow (DZRF) was found in 35%, whereas in cases with a large intravillous blood volume there were only 4% abnormal Doppler findings and not a single case of DZRF.

In summary we found that the flow patterns of the umbilical arteries are related to intravillous blood volume and have clinical significance.

Reference

Giles, W.B., Trudinger, B.J. and Baird, P.J. (1985) Fetal umbilical flow velocity waveforms and placental resistance. pathological correlation. Br. J. Obstet. Gynaecol. 92, 31-38

THE BIOPHYSICAL BASIS OF THE FLOW PULSE IN THE FETAL CIRCULATION.

Waldemar Moll, Regensburg, F.R.G.

In order to interpret the flow pulse measured by Doppler velocimetry in the fetal aorta and in the umbilical artery, the arterial blood flow velocity was derived from biophysical parameters and simulated in a digital model. In this model the arterial system is represented by 33 elements. For each element the flow rate was derived for minute time intervals (0.1 ms) from pressures and effective mass; pressures were determined from flow rates, peripheral resistance, compliance and the viscosity of blood and the tissue. The simulations demonstrated that the systolic flow velocity wave is determined by the ejection time of the heart, the compliance of the downstream arterial system and the resistance of the downstream vascular bed. In diastole, longitudinal flow occurs in the fetal aorta and especially in the central portions of the umbilical arteries. This longitudinal flow is responsible for the diastolic Doppler signal. It is related to the cranio-caudal gradient of the resistance-compliance product (RC) along the artery. Thus cardiac parameters, arterial parameters and peripheral vascular resistance are pertinent to the flow pulse wave form. If cardiac factors and arterial compliance are unaltered, the velocity wave form will give information on placental resistance, otherwise not.

Fetal placental blood flow after uterine artery ligation in the guinea-pig

A.M. Carter, A. Detmer

Department of Physiology & Biomedical Laboratory, University of Odense, Campusvej 55, DK-5230 Odense, Denmark

Ligation of the uterine artery in the pregnant guinea pig causes a reduction in placental mass and maternal placental blood flow and leads to retardation of fetal growth (IUGR). We asked how fetal placental blood flow behaved in this situation, since it is an important determinant of oxygen and substrate delivery to the fetus. The uterine artery to one horn was ligated on day 30-33 of gestation. On day 60-64 we studied a fetus with IUGR and asymmetric growth (brain:liver weight ratio > 0.9) or a control fetus from the contralateral horn. To measure placental blood flow, tracer microspheres were injected into the right atrium of the fetus and a reference sample of blood was withdrawn from the vitelline artery. IUGR fetuses (n=5) weighed 56.6 ± 2.9 g (mean $\pm$ SEM) and the controls (n=6) weighed 83.0 ± 3.5 g ($P < 0.001$). The corresponding placental weights were 3.2 ± 0.2 and 5.2 ± 0.4 g ($P < 0.001$). Heart rate was lower in IUGR fetuses than in controls, 206 ± 19 against 281 ± 7 beats/min ($P < 0.01$). Haemoglobin concentration, measured in arterial blood, was similar in the two groups. Fetal placental blood flow was 4.9 ± 1.2 ml/min in IUGR fetuses compared to 9.9 ± 1.5 ml/min in controls ($P < 0.05$). There was no significant difference between the placental perfusion rates, which were 1.5 ± 0.3 and 1.9 ± 0.3 ml/min per g tissue. It is known that uterine artery ligation restricts placental growth before affecting fetal weight (Jansson et al. Biol. Neonate 49:172-180, 1986). Our results show that the decrease in placental size is accompanied by a reduction in fetal placental blood flow. It is therefore suggested that the consequent limitation upon oxygen delivery to the fetus is a causative factor in growth retardation. This interpretation is supported by experiments on the fetal lamb, where the long-term effects of a reduction in umbilical blood flow and oxygen delivery are a fall in fetal oxygen consumption and a decrease in the rate of fetal growth (Anderson et al. Am. J. Physiol. 250: H1037-H1042, 1986).

Trans-vaginal sonography of first trimester pregnancy : investigation on the nutritional value of the secondary yolk sac and on the maturation of chorionic vessels

A. Funk, H. Fendel

Department of Gynaecology and Obstetrics of the RWTH, Aachen, Federal Republic of Germany

Diameters of extra-embryonic space, amniotic cavity, yolk sac, CRL and the blood-flow development in the umbilical cord were examined in first trimester pregnancies by means of trans-vaginal ultrasonographic techniques. The secondary yolk sac slowly grows during the embryonic period from 3,9 mm ⌀ in the 6th to 6,0 mm ⌀ in the 11th post-menstrual week.

During the fusion of amniotic and chorionic menbranes which is reflected by the quotient of the diameters of the chorionic and amniotic cavities - tending to 1 from the 6th to the 12th week - the secondary yolk sac disappears.

In normal early pregnancy the yolk sac shows a better correlation to the CRL ($N_1=0,749$, $p=0,001$) than to the gestational age ($r_2=0,70$, $p=0,001$).

It is to presume that the yolk sac transfers proteins and other substances from the chorionic fluid, identified by a higher echogenity, to the embryo.

The arterial blood-flow, showing only a short, flat systole in the 6th week, increases steadily until the occurence of prediastoles, and finally continuous diastoles.

After the 14th week the blood-flow resembles the curve known from late pregnancy, after gradual formation of continuous flow in the venous system.

The pulsatility index drops from 5,5 in the 7th to 2,0 in the 15th week. This shows the reduced resistance of the embryonic and placental vessels, caused by the maturation of the chorionic villi.

The sonographic findings of the anatomic structures and the development of blood flow implies that the secondary yolk sac is of nutritional value to the embryo until maturation of chorionic vessels.

Dopplersonographic findings in umbilical chord anomalies : an *in vivo* model to show resistance parameters in fetplacental circulation

T. Hitschold, E. Weiss, H. Müntefering*, P. Berle

Department of Gynecology and Obstetrics, Dr. H.-Schmidt Hospital, Wiesbaden, and
* Department of Pathology, University of Mainz, Federal Republic of Germany

In cases with different umbilical chord anomalies the influence of fetoplacental resistance on the flow velocity waveforms is discussed. If there is no anastomosis between the 2 umbilical arteries, the enddiastolic flow velocities are depending on the volume of the fetoplacental vessel-tree, e.g. in the arteria with the smaller placenta part higher Resistance-Indices (RI) are found, although the villous vacularization in both parts is not different. If there is only 1 umbilical artery, the peripheral resistance is relatively reduced to 50%, because the placenta part belonging to this solitary umbilical artery is two fold bigger compared to cases with 2 umbilical arteries. The result is a higher perfusion pressure gradient with higher flow velocities, lower RI-values and an underestimation of fetoplacental resistance. In cases with thombosis of chorion membrane vessels ore stem villi vessels, placenta regions are excluded of the perfusion, followed by an increasing resistance in the same manner as it happens in slowly developing failures of villous maturation and vascularization. That means that the presented cases of umbilical chord anomalies and function disorders of allantois vessels can be interpreted as an in-vivo-model to show pressure- and resistance parameters in fetoplacental circulation. The results are in good accordance with our histometric placental data, indicating a high correlation between Doppler flow patterns in the umbilical arteries and the fetal-placenta-blood-volume. An antepartum estimation of downstreem impedance using pulsed Doppler ultrasound is possible if two normal umbilical arteries are present.

Evaluation of placental risk using pulsed Doppler ultrasound of the umbilical arteries : a histometric-clinical investigation

T. Hitschold, E. Weiss, H. Müntefering*, P. Berle

*Department of Gynaecology and Obstetrics, Dr. H.-Schmidt Hospital, Wiesbaden, and * Department of Pathology, University of Mainz, Federal Republic of Germany*

To estimate the downstreem impedance of the fetoplacental vessel tree, pulsed Doppler ultrasound of the umbilical arteries was included in the management of high risk pregnancies. We compared histometric data of 90 placentas with the umbilical waveforms. We found a high correlation (k=0,765) between the feto-placental-blood-volume (calculated as a product of cross sectional area of villous vessels and placenta volume) and the flow velocity waveforms in the umbilical arteries, which were determined max. 4 days anepartum with pulsed Doppler ultrasound (Fig. 1). In cases of chronic placenta insufficiency (high ratio of placenta weight / birth weight = PKI) we found a very high correlation (k=0,819) between the flow velocity waveforms and the villous vascularization, indicating the possibility of non-invasive antepartum estimation of total fetal villous vessels volume in high risk pregnancies. In cases with low PKI the umbilical flow depended on the placenta volume allone (k=0,810) and not on the villous vascularization. The product of both components (= calculated fetal-placenta-blood-volume) does not depend on the PKI and therefore is useful for antepartum estimation of fetoplacental vessel tree's volume.

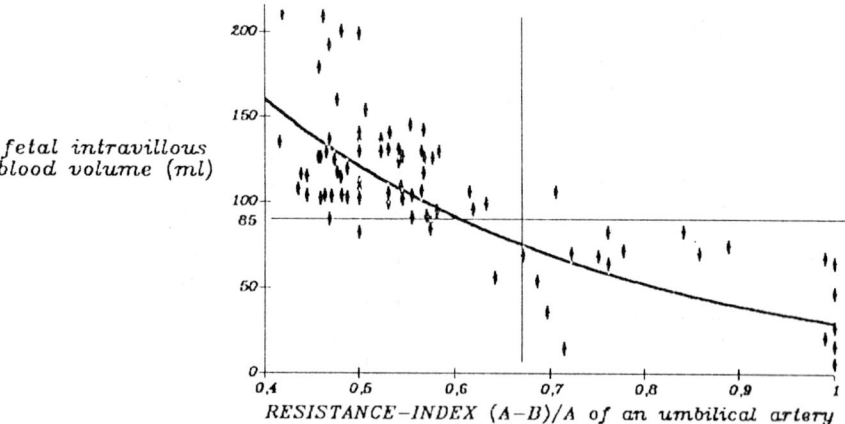

Fig 1: Resistance-Index in an umbilical artery and fetal-placenta-blood-volume - a very high correlation (k=0,765).

In summary, an antepartum registration of fetoplacental vascularization failures with pulsed Doppler ultrasound of the umbilical arteries is possible. The worse case is the loss of enddiastolic blood flow in the umbilical arteries, indicating the final point of placental lesion with reduction of all vascularization- an diffusion-parameters.

Alterations to vascular receptors for angiotensin II and atrial natriuretic peptide in placentae from pregnancies complicated by asymetrical growth-retardation

J.C.P. Kingdom, J. McQueen*, A.G. Jardine*, J.M.C. Connell*, M.J. Whittle

*Queen Mother's Hospital and *MRC Blood Pressure Unit Western Infirmary, Glasgow, UK*

Fresh placenta were collected from 8 pregnancies complicated by unexplained assymetrical growth retardation (IUGR) defined as serial estimated fetal weight measurements crossing to below 10th centile with eventual birthweight less than 5th centile for gestational age. The pregnancies were not complicated by hypertension or material systemic disease, and the women were on no drug therapy. The gestation was determined in each case by ultrasound before 16 weeks, and delivery was between 37-40 weeks, except for 1 case at 34 weeks. Fresh placentae were also collected from 9 normal deliveries between 37-40 weeks' gestation.

Fetal arterial vessels to isolated cotyledons were dissected free of connective tissue and homogenised in 0.25M sucrose. The membrane fraction was prepared by differential centrifugation and re-suspended in 50mM TRIS HCL prior to storage at -70C. Binding of Atrial Natriuretic Peptide (ANP) and Angiotensin II (Ang II) was determined by incubation of the membrane fraction with varying amounts of unlabelled peptide and a fixed amount of I^{125}- labelled peptide. Non-specific binding and ligand degradation were determined and the data presented as Scatchard transformation of specific ligand binding.

In normal placentae, the number of high affinity Ang II receptors (K_D-8.764 log.mol/L) increased with the Calcium (Ca^{2+}) concentration in the assay buffer. A similar effect was seen in IUGR samples, and though the K_D was similar (-8.754) log mol/L) the receptor number (Bmax) was significantly lower (p 0.05 in 1.2 mmol/L and 12.5mmol/L Ca^{2+}). Low affinity receptors were identified in both groups with a smaller receptor number in the IUGR group. The concentration of these receptors fell with increasing Ca^{2+}- an effect seen in both groups.

High (K_D-9.678 log mol/L) and low (K_D-7.256 log mol/L) affinity binding sites for ANP were identified in both normal and IUGR placentae. The concentration of these receptors was markedly lowered in IUGR tissue, and in both groups the numbers of high-affinity receptors were unaffected by assay Ca^{2+}.

These results suggest a pattern of chronic down-regulation of the Ang II receptor in IUGR, consistent with reports of raised circulating fetal renin in IUGR. As fetal ANP is not reported to be increased in IUGR, the alterations to ANP receptors may represent heterologous regulation by Ang II, and indicate a possible mechanism whereby fetoplacental vascular resistance may be increased in pregnancies complicated by IUGR.

Atrial natriuretic peptide opposes the vasoconstrictor effect of angiotensin II in a simple placental perfusion model

J.C.P. Kingdom, A.G. Jardine*, J.M.C. Connell, A. Templeton**, M.J. Whittle

*Queen Mother's Hospital, *MRC Blood Pressure Unit Western Infirmary and **Institute of Physiology, Glasgow, UK*

Fresh term placentae from normal pregnancies were studied. Immediately after delivery, an umbilical artery was cannulated at the base of the cord and flushed freely with heparinised 0.9% saline to re-establish flow through the umbilical vein. The placenta was connected to the perfusion system and immersed in 0.9% saline at 37°C. It was perfused with an electrolyte solution containing HEPES and FICOL and oxygenated with 95% O_2, 5% CO_2 at 37°C to achieve a pH of 7.45. The flow rate was adjusted to give a basal pressure of about 20mmHg (usually between 4-8 mls/min). Under these conditions the preparation remained stable with no evidence of tachyphylaxis to bolus injections of Angiotensin II given at 10 minute intervals over a 2 hour period.

Incremental bolus doses of Angiotensin II (10^{-14} to 10^{-8} M) produced dose-dependent transient rises in perfusion pressure. Bolus injections of Atrial Natriuretic Peptide (10^{-14} M) given 30 seconds prior to Angiotensin II (10^{-10} M) diminished the contraction response by 60%. Adding Atrial Natriuretic Peptide to the perfusion fluid at the approximate concentration to that in fetal blood ie 100pg/ml [1], produced a significant reduction in the dose response curve to Angiotensin II (P<0.001 for all values).

We suggest that Atrial Natriuretic Peptide may be an important factor in the maintenance of a low-resistance circuit on the fetal side of the placenta.

(1) KINGDOM J.C.P., JARDINE A.G., DOYLE J. et al Br.Med.J. 1989; 298:1221-2.

Maternal oxygen supply as a regulator of fetal placental capillarisation

I. Scheffen, L. Philippens, P. Kaufmann, R. Leiser*, V. Mironov**

Department of Anatomy, 5100 Aachen, RWTH Federal Republic of Germany
* Department of Veterinary Anatomy, Bern, Switzerland
** Ivanovo, USSR

It is well-known from many studies that hypoxia causes fetal villous hypercapillarisation. Two different types of the latter have been described: A. abnormally long, unbranched, highly dilated capillary loops, and B. dense, highly branched networks of short, narrow capillaries. So far it proved to be impossible to figure out which of both morphological reaction patterns is due to hypoxia, since the latter in the human is mostly combined with other pathogenetic factors and often coincides with placental prematurity. The present study aims to examine this problem under experimental conditions.

44 pregnant guinea pigs were kept under uninterrupted isobar hypoxia (12% O_2) from day 15 to day 60 of gestation. Physiological parameters like hematocrit, P50, pO_2, and reoxygenation time were studied in intervals of 20 days. Around day 60 the morphology of the fetal capillaries was examined by means of scanning electron microscopy of vessel casts, semithin histology, and transmission electron microscopy. The vessel casts showed a considerably higher degree of branching and coiling of the capillary bed as compared to the control placentas. By transmission electron microscopy in hypoxia a remarkably reduced diameter and an increased number of capillary cross-sections is evident. The materno-fetal diffusion distance is shortened by thinning of the trophoblast. The morphological effects were verified morphometrically.

The findings demonstrate that oxygen is an important regulator of fetal placental capillarisation. Chronic hypoxia stimulates capillary branching, and is only responsible for type B of hypercapillarisation, characterized by dense networks of narrow and short capillaries, thus reducing fetal blood flow resistance. In agreement with several pathohistological hints in the human, type A of hypercapillarisation expressed by long, unbranched, dilated capillaries probably is a result of placental prematurity.

Anti-platelet activity of human trophoblast

M.H.F. Sullivan, L. Patel, M.G. Elder

Institute of Obstetrics Gynaecology, Royal Postgraduate Medical School, Hammersmith Hospital, Du Cane Road, London W12 OHS, UK

Normal development of the human fetus is totally dependent on the provision of nutrients from the maternal circulation. This in turn is dependent on a good blood flow, so maintenance of vascular patency is of great importance. We have investigated whether trophoblast has features which would tend to decrease platelet aggregation, and maintain blood flow.

Keratin is a unique component of human trophoblast, in that the vascular endothelium (which is keratin-free) normally separates the blood from keratin-containing cells. We have found that human keratin at a concentration of 1ug/ml significantly decreased human platelet aggregation (stimulated by ADP or collagen) by $37.0 \pm 14.3\%$ (means ± S.D. n=15). As well as containing an anti-aggregatory compound, trophoblast cells also have the capacity to produce anti-aggregatory arachidonic acid metabolites. The major metabolite produced by cultured trophoblast has a retention time on HPLC which is consistent with a tentative identification of this metaboilite as 14,15-epoxyeicosatrienoic acid (14,15-EET). <u>In vitro</u> studies have shown that this is a potent anti-aggregatory agent, which suggests that it may also have this function <u>in vivo</u>.

These experiments suggest that at least two different compounds, namely keratin and 14,15-EET may be involved in the maintenance of placental vascular patency during pregnancy.

The effects of leukotrienes and their interaction with angiotensin II in the perfused human placenta

P.R. Tranter, M.H.F. Sullivan, M.G. Elder

Royal Postgraduate Medical School, Institute of Obstetrics and Gynaecology, Hammersmith Hospital, Du Cane Road, London W12 OHS, UK

The role of prostaglandins and thromboxanes in the fetal-placental vasculature has been well established. However, less is known about the involvement of the lipoxygenase products of arachidonic acid metabolism, in particular the leukotrienes.

The aim of this study is to investigate the effects of leukotrienes on the fetal-placental vasculature and to determine how leukotrienes affect the vasoconstrictor response produced by angiotensin II. The experiments were performed on a human placental cotyledon that was perfused in vitro on both the fetal and maternal sides.

Leukotriene B4 (LTB4) and angiotensin II (AII) were introduced separately and together into the fetal circulation and the pressure levels were monitored. 500 ng LTB4 or 5 ug AII increased the pressure on the fetal side but not on the maternal side. Initial results indicate that when the LTB4 and AII are added simultaneously the response is less than that seen with each agonist alone.

There appears to be an interaction between LTB4 and AII in the perfused placental system and experiments are being undertaken to investigate this further.

The effect of reduced red cell deformability on placental perfusion in the awake late-pregnant guinea-pig

C.M. Verkeste, P. F. Boekkooi, L.L.H. Peeters

State University Limburg, Department of Obstetrics/Gynaecology, Maastricht, The Netherlands

Introduction.
Complicated pregnancies such as preeclampsia are often associated with an increased maternal whole blood viscosity which is partly caused by reduced red cell deformability. Reduced red cell deformability may hinder the passage of these cells across the villous tree inducing enhanced impedance to flow. The effect of diminshed maternal red cell deformability on placental perfusion was studied in 57 days pregnant guinea pigs (n=15). A group of non-pregnant guinea pigs (n=6) was subjected to the same protocol and served as controls.

Experiments.
The experiments were performed 6 days after catheter implantation. Under steady-state condition a 10 ml cocktail containing equal amounts of artificially rigidified- (0.025% glutaraldehyd) and control red blood cells, obtained from a non-pregnant donor guinea pig and adjusted to the hematocrit of the recipient animal, was injected over a period of 5 minutes. Both types of cells were labelled with different isotopes. Ten minutes before and twenty minutes after the bolus injection of the cocktail, the distribution of cardiac output (radioactive microspheres) was determined. In addition, the distribution of rigidified- relative to that of control red blood cells was calculated for each tissue.

Results.
After the cocktail injection, rigidified red blood cells were trapped selectively in the microcirculation of the liver and spleen, but not in the placentas.

Conclusion.
Decreased red cell deformability appears to have little or no effect on the perfusion of the uteroplacental microvascular bed of this species.

6.
Physiology and placental transfert

Physiologie et transfert placentaire

Calcium dependent ATPases and nucleotide phosphates of the human placental basal plasma membrane

C.H. Smith, L.K. Kelley

Washington University Medical School, St. Louis Children's Hospital, St. Louis, MO 63110 USA

Mineralization of the fetal skeleton requires large quantities of calcium throughout the third trimester of pregnancy. This calcium must be transferred from the maternal to the fetal circulation by the placental syncytiotrophoblast. In all species that have been studied the concentrations of both total and ionized calcium are greater in the fetal than in the maternal circulation. The transport mechanisms of the plasma membranes of the syncytiotrophoblast must also maintain intrasyncytial free calcium at concentrations orders of magnitude lower than those in the plasma or extracellular fluid. It is logical to suppose that the basal (fetal facing) plasma membrane of the syncytium possesses an energy dependent extrusion mechanism or pump.

Using an isolated preparation of the basal plasma membrane we have previously demonstrated an ATP dependent calcium tranporter which is saturated at nanomolar concentrations. (Fisher, et. al. 1987) It requires magnesium in concentrations similar to those of ATP. To study the mechanism of this calcium transport process we desired to establish a method for measuring the associated calcium dependent hydrolysis of ATP. This would permit us to compare its properties with those of other calcium dependent phosphatase activities and devise mechanisms for its purification and detailed investigation. In plasma membranes of red cells and smooth muscle, the magnitude of calcium dependent ATP hydrolysis other than that by the transporter is small permitting direct measurement of calcium transporting ATPase and establishment of its identity with the calcium transporter.

Placental basal membranes were incubated with $[\gamma,^{32}P]$-ATP in the presence and absence of calcium at various concentrations, and the calcium dependent ATP hydrolysis measured by subtraction. The results indicate the presence of at least two saturable components. Since ATP dependent calcium transport is saturable in the nanomolar range we examined concentration dependence in that range. Although only one saturable component in this range is apparent, it soon became clear that many observed properties differed substantially from those of the transporter. Addition of magnesium rather than producing stimulation as with the transporter, caused inhibition of calcium dependent activity. Addition of magnesium at various calcium concentrations stimulated total ATP hydrolysis, but calcium dependence decreased and became zero as magnesium concentration increased.

Conversely, the addition of calcium at various magnesium concentrations stimulated activity and eventually abolished magnesium dependence. These results suggested that rather than a magnesium stimulated, calcium dependent ATPase we were observing an ATPase stimulated by both magnesium and calcium with properties clearly different than those of the calcium transporter.

To further understand the relationship between the high affinity hydrolysis of ATP observed in the absence of added magnesium and ATP dependent calcium transport we compared their nucleotide specificity. Addition of various nucleotides demonstrated a calcium stimulated nucleotide phosphatase activity of broad specificity. All of the triphosphates as well as ADP were effective substrates. Calcium transport, however, was appreciably stimulated exclusively by ATP. ATP and GTP each inhibited hydrolysis of the other and with both substrates ATP was a better competitor than GTP. These observations led us to develop the use of GTP to inhibit the non transport related hydrolysis of ATP allowing the measurement of ATP hydrolysis under conditions which support transport, (Kelley et. al. 1987)

In other cell membranes mercurial inhibitors which react with sulfhydril groups have been used to demonstrate differences between calcium transport and other ATP hydrolyzing activities. We studied the effects of these agents on ATP dependent calcium transport and on calcium dependent ATP hydrolysis in the presence of excess unlabled GTP. Transport was inhibited by many sulfhydril compounds. The effect of PCMBS is typical. In the absence of magnesium PCMBS is essentially without effect on calcium dependent ATP hydrolysis. In the presence of magnesium however, PCMBS inhibits calcium dependent ATP hydrolysis and calcium transport with the same concentration dependence. Thus, measured in the presence of GTP, calcium dependent magnesium stimulated ATP hydrolysis activity has the same behavior towards sulfhydril reagents as does calcium transport.

The magnesium dependence of PCMBS sensitive calcium stimulated ATP hydrolysis further confirms this interpretation. In the absence of PCMBS as we observed earlier, two calcium stimulated components are seen. The first component is not ion specific and its calcium dependent activity is inhibited as magnesium is added, while the second is stimulated by the addition of magnesium giving a biphasic curve. Pretreatment of the membranes with 10 µM PCMBS, however, inhibits the magnesium stimulated component and leaves only the first or nontransport component. Subtraction of the activities measured in the presence and absence of PCMBS gives by definition the PCMBS sensitive magnesium dependent activity. This activity has a magnesium requirement superimposable with that of calcium uptake measured under the same conditions.

To confirm the interpretation suggested by the mercurial inhibitor experiments we studied the effect of rabbit IgG directed against the calcium transporting ATPase purified from human erythrocytes. These experiments were conducted in collaboration with Dr. John Penninston of Rochester, Minnesota who provided the antibodies. As we have recently reported rabbit IgG partially inhibits ATP dependent calcium transport in basal plasma membrane, (Borke et. al. 1989). We have now observed that the same immune IgG partially inhibits ATP phosphatase activity measured in the presence of GTP and magnesium but not that measured its absence. We then extended these studies to use polyclonal and monoclonal antibodies against the erythrocyte transporter provided by Dr. Penninston to identify the homologous subunits in the basal plasma membrane. Using SDS gel electrophoresis followed by transfer to nitrocellulose sheets, we detected antigens by immunostaining. The basal plasma membrane yielded an antibody staining doublet in the region of 140 kDa and the band of lower molecular weight comigrated with the erythrocyte transporter subunit. Furthermore, in collaboration with Dr. Jim Borke of Rochester, Minnesota, we have immunohistochemically identified the transporter ATPase along the basal surface of the trophoblast, (Borke et. al. 1989).

Using gel electrophoresis under acid conditions we demonstrated the formation of a phosphorylated intermediate produced during incubation with $[\gamma,^{32}P]$-ATP. The phosphorylation required calcium and was enhanced by the presence of $LaCl_3$ and saponin. This phosphorylated subunit comigrates with that of the calcium transporter of erythrocyte ghosts.

CONCLUSION

In conclusion, we have identified two or more apparently distinct enzyme components mediating calcium stimulated ATP hydrolysis. One of these is a non-specific divalent cation stimulated nucleotidase. Its activity is higher than that of the calcium transport ATPase. When this high activity is partially inhibited by GTP, it and the transport ATPase can be resolved and their properties distinguished.

The transporter ATPase is calcium specific requiring magnesium and hydrolyzes specifically ATP. It is inhibited by mercurials and by antibodies directed against the erythrocyte calcium transporter. In all these properties it corresponds with those of ATP dependent calcium transport observable in the same membrane and with the properties of the calcium transporter seen in the erythrocyte and in plasma membranes in many other cells. The transporter can be stained immunohistochemically in the basal plasma membrane of both human and rat placenta and it forms phosphorylated intermediates and immunostaining bands of the same molecular weight as those seen in the erythrocyte ghost. We conclude that this activity is the enzymatic expression of the ATP dependent calcium transporter. Its $K_{0.5}$ in the nanomolar range permits it to extrude calcium ion at cytoplasmic concentrations.

The non-specific divalent cation stimulated ATPase has often been confused with the calcium transporter in plasma membranes of the placenta and other. Its role is not established. Description of the properties of such ATPases, distinct from those of the calcium transporter should help in defining their functional role. Furthermore, the ability to measure calcium dependent ATP hydrolysis specifically associated with the calcium transporter should be helpful in determining activity under conditions in which the membrane has been disrupted for example in purification of the transporter or in further defining its properties and regulation in the human placenta.

REFERENCES

Borke, J.L., Caride, A., Verma, A.K., Kelley, L.K., Smith, C.H., Penniston, J.T., and Kumar, R. (1989): Calcium pump epitopes in placental trophoblast basal plasma membranes. Am. J. Physiol 257,341-346.

Fisher, G.J., Kelley, L.K., and Smith, C.H. (1987): ATP-dependent calcium transport across basal plasma membranes of human placental trophoblast. Am. J. Physiol. 252,38-46.

Kelley, L.K., and Smith, C.H. (1987): Use of GTP to distinguish calcium transporting ATPase from other calcium dependent nucleotide phosphatases in human placental basal plasma membrane. Biochem. Biophys. Res. Commun. 148,126-132.

Résumé

L'ATPase calcium dépendante et les phosphatases nucléotidiques de la membrane plasmique basale du syncytiotrophoblaste humain.

C.H. Smith, Professeur de Pédiatrie
Université de Washington
St Louis, U.S.A.

Deux activités phosphatasiques ont été identifiées dans les membranes plasmiques basales (face foetale) du syncytiotrophoblaste humain. L'une dépend de la présence de cation divalent et elle est inhibée par le GTP (phosphatase nucléotidique non spécifique); l'autre, d'activité plus faible, peut être décelée en présence de GTP et correspond à l'activité ATPasique liée au transport de cation divalent. Cette dernière (ATPase calcium dépendante) est annulée spécifiquement par le Calcium, requiert la présence de Magnésium et engendre l'hydrolyse de l'ATP exclusivement. Elle s'exerce à une concentration nanomolaire de Calcium compatible avec les concentrations intracellulaires de cet ion. Les dérivés mercuriques, comme le PCMB, et les anticorps dirigés contre le transporteur de Calcium des membranes d'hématies l'inhibent. En gel d'électrophorèse, une bande est détectée par l'anticorps antitransporteur à la fois dans les membranes basales de placenta et dans les autres hématies à 140 KD. En immunohistochimie, ce transporteur a été localisé dans les membranes basales du trophoblaste.

L'activité ATPasique Calcium-dépendante assure la libération de Calcium vers le compartiment liquidien du foetus dont les besoins pour la minéralisation du squelette sont considérables au troisième trimestre de la grossesse.

Transport placentaire : résumé de l'atelier

J.C. Challier, H.J. Schroder*

Biologie de la reproduction, Université Pierre et Marie Curie, 75005 Paris, France
Universität Frauenklinik, Hamburg, Federal Republic of Germany

Deux catégories de substances ont été examinées dans cet atelier consacré au transport placentaire : celle des minéraux et celle des acides aminés.

La première présentation (MOHAMMED T., SIBLEY C., STULC J., MARESH M. & BOYD R.D.H., Manchester, UK) traite du transport de Potassium dans le placenta de rat. L'acquisition foetale de Potassium entre 20 et 21 jours de gestation est de 0,111 moles.min^{-1}.g^{-1} de placenta. La clairance materno-foetale est cinq fois plus importante. Lors de la perfusion in situ de la circulation foetale du placenta de rat, le flux net en direction du foetus est très faible et à la limite de significativité. En double perfusion, la clairance materno-foetale du Potassium est seize fois plus élevée que celle du complexe Cr-EDTA. Le passage obéit à une cinétique de second ordre et le K_H obtenu est de 7 mM. L'emploi d'Oubaine réduit au tiers le passage du potassium. Une réduction similaire est observée avec le Baryum. L'effet est réversible. Ces résultats montrent que le passage du Potassium dans le placenta de rat est un processus saturable, asymétrique, qui impliquerait la Na-K-ATPase et des canaux potassiques Baryum-dépendants.

Le passage du Magnésium en perfusion in situ (SHAW A.J., MUGHAL M.Z. & SIBLEY C.P., Manchester, UK) chez le rat, indique que ce divalent n'est pas transporté par diffusion. En effet, sa clairance materno-foetale atteint 26 µl.min^{-1}.g^{-1}, soit neuf fois celle du Cr-EDTA, alors que ces deux molécules ont des coefficients de diffusion identiques. De plus, la contribution du pool intra-tissulaire de Magnésium dans le transfert materno-foetal semble négligeable.

STULC J., STULCOVA B. et SVIHOVEC (Prague, Tchécoslovaquie) ont perfusé le placenta de rat simultanément du côté foetal et du côté maternel, pour étudier le passage du Calcium dans le sens mère-foetus et dans le sens foetus-mère. Le passage de Calcium obéit à une cinétique de second ordre de K_H 0,45mM. Dans les conditions physiologiques, le passage du Calcium s'opère donc à vitesse maximale et il est relativement indépendant des variations de son taux sanguin maternel. Le Baryum et le Strontium inhibent le transport du Calcium. Cette inhibition est annulée par l'administration d'excès de Calcium. Le système de transport transplacentaire du Calcium paraît donc peu spécifique.

L'effet de deux inhibiteurs du métabolisme minéral a été étudié (PAGE K.R., ABRAMOVITCH D., DACKE C.G., MAYHEW T., WILLIAMS J.M.A., Aberdeen & Porsmouth, UK) lors de la double perfusion du placenta humain. L'emploi du DNP (10^{-3}M) provoque une libération de Potassium dans les liquides de perfusion et une captation placentaire de Calcium et de Sodium. En présence d'Ouabaine (10^{-5}M), on note l'apparition de Potassium dans les perfusats maternel et foetal, ainsi qu'une libération de Sodium dans le perfusat foetal. Ces résultats semblent indiquer qu'une ATPase membranaire intervient dans le transport placentaire du Calcium, plutôt qu'un échangeur Na/Ca.

La localisation des ATPases membranaires reste très controversée dans le placenta. Chez le porc, l'existence d'un potentiel transmembranaire et d'un courant de court-circuit sensible à l'ouabaine, suggèrent la présence d'une pompe à sodium Na-K dépendante. L'activité ATPasique K-dépendante a été localisée par FIRTH A. (Londres, UK) essentiellement dans l'épithélium chorionique des aréoles qui jouxtent les glandes utérines. Cette activité enzymatique est décelée à la base des cellules, là où la membrane plasmique est très contournée. La signification de cette localisation précise n'est pas connue. Une telle activité pourrait intervenir dans le mouvement des ions et dans celui des liquides sécrétés par les glandes utérines.

Dans la seconde partie, le passage placentaire des acides aminés a été examiné.

Le passage materno-foetal des acides aminés par la vésicule ombilicale a fait l'objet d'une présentation (SCHRODER H.J., SCHOCH C. & LEICHTWEISS H.P., Hambourg, RFA). Le transfert materno-foetal des formes D et L de l'alanine est négligeable. La captation des acides aminés par la vésicule vitelline est cependant significative, excepté pour l'AIB. Le passage des formes L et D de l'alanine et de l'acide aspartique s'effectue par un processus de type compétitif et Sodium-dépendant. Le fait que le D-aspartate soit capté en proportion identique, voire même plus importante que le L-aspartate, a été souligné par les auteurs. Un parallèle peut être fait avec les observations rapportées antérieurement en perfusion dans le placenta humain et confirmées au niveau membranaire dans la présentation de SMITH et al., au cours de cet atelier.

Le passage de la taurine a été examiné au cours de l'incubation de microvésicules placentaires (dérivées de la membrane plasmique maternelle du syncytium) et en perfusion in vitro dans l'espèce humaine (KARL P.I., FISHER S.E., Manhasset, USA). La captation de la taurine par les microvésicules a les propriétés suivantes : (i) stimulation par un gradient de concentration sodique (externe>interne), renforcée par un gradient en Potassium de sens inverse ou dans une plus faible mesure par un gradient en chlorure de même direction, (ii) inhibition "cis" compétitive par la ß-alanine ou l'hypotaurine, et de type "trans" lors d'une charge des microvésicules en taurine ou en hypotaurine. En perfusion, la taurine ne se concentre pas dans la circulation foetale à la différence de l'histidine. Ce dernier acide aminé est par ailleurs sujet à une transtimulation, et non à une transinhibtion, en microvésicules. En conclusion, la captation placentaire de taurine est le fait d'un système symport (stoechiométrie 1:1 avec le sodium), hautement spécifique. L'existence d'une transinhibition pourrait expliquer l'absence de transfert materno-foetal net saturable de la taurine et l'absence de gradient de concentration foeto/maternel de taurine observées en perfusion.

Les résultats obtenus avec les acides aminés anioniques lors d'études de transport portant sur des microvésicules dérivées de membranes plasmiques apicales (côté maternel) ou basales (côté foetal) du syncytiotrophoblaste humain ont été exposés (SMITH C.H., MOE A.J., KELLEY L.K. & HOETZLI S.D., Washington, USA). Les captations de l'aspartate et du glutamate suivent une cinétique de transport à caractère saturable, à haute affinité (d'ordre micromolaire). Leur transport dépend de la présence de sodium et il est inhibé par les dérivés sulfiniques du cystéate et la cystéine. Ces caractéristiques sont celles du système de transport de type X_{AG}, qui est ainsi démontré sur les faces maternelle et foetale du syncytiotrophoblaste. L'absence de transfert materno-foetal de ces acides aminés *in vivo* proviendrait, selon ces auteurs, non pas de l'absence de système de captation, mais de l'absence d'un système approprié d'efflux pour leur libération dans les circulations.

En conclusion, deux idées directrices ont émergé des discussions. A propos du transport des minéraux, la relation entre le transfert transplacentaire, qu'il soit materno-foetal ou foeto-maternel (SHENNAN D.B. & BOYD C.A.R., Placenta, 9, 333-343, 1988) et les systèmes de transport ioniques identifiés récemment sur les membranes plasmiques (SHENNAN D.B. & BOYD C.A.R., BBA, 906, 437-457, 1987) devrait être approfondie. Il paraît en effet urgent de délimiter le rôle de ces derniers, en tant que mécanisme purement cellulaire de transport ou bien en tant qu'intervenant dans le passage transcellulaire. On peut d'ailleurs envisager qu'un passage transplacentaire résulte de l'action coordonnée de plusieurs systèmes de transport membranaires, situés sur des faces différentes du syncytium, ou que certains mécanismes de transport membranaire aient un rôle purement local. Dans le cas des transports d'acides aminés, l'état actuel des recherches a bénéficié des études d'avant-garde effectuées avec des microvésicules. Il se dégage des discussions, que la face maternelle du syncytiotrophoblaste comporte les systèmes A, ASC, L1, L2 et Anioniques, et que la face foetale renferme les mêmes systèmes, à l'exclusion du système ASC. Le système ß, pour la taurine, existe sur la face maternelle. L'existence d'un système de transport des acides aminés anioniques sur les deux faces du syncytiotrophoblaste est en agrément avec les observations faites en perfusion *in vitro*, il explique les concentrations élevées de ces acides aminés observées dans les tissus placentaires *in vitro* et l'hypothèse d'une absence de mécanisme d'efflux de ces acides aminés pour expliquer leur absence de passage dans la circulation foetale mériterait d'être vérifiée. Les relations fonctionnelles pouvant exister entre les systèmes de transport présents sur les faces foetales et maternelles du trophoblaste n'ont pas fait l'objet d'études approfondies, il paraît souhaitable que ce problème soit étudié au niveau cellulaire, en culture par exemple.

Workshop Placental Transport.

Chairmen : Jean-Claude CHALLIER* & Hobe SCHRODER

*University P. & M. Curie, Paris, France.
University-Frauenklinik, Hamburg, FRG.

The placental transport of two series of molecules has been scrutinized during this workshop : minerals and amino acids.

There has been three presentations dealing with mineral transport in the rat placenta. It emerged into notice that Potassium and Magnesium transport cannot be account only by diffusion. This conclusion is based on the values of the ratio of diffusion coefficients of these cations related to Cr-EDTA or on the inhibitory effects obtained. Placental calcium operates at a maximal rates and follows a second order kinetic. Reversible inhibition by other divalent cations implies a certain lack of specificity.

The ATPase inhibitory ouabain and the uncoupler compound DNP produced change in the circulatory release of minerals by the perfused human placenta. These changes indicates a role of an ATPase rather than a Na/Ca exchanger in the placental transport of calcium. A localization of sodium ATPase was attempted in the swine placenta. The activity was observed at the basolateral side of the chorionic epithelial cells of "areolae", opposite the openings of the uterine glands. The reason for such precise location is not presently understood.

The amino acid transport are poorly transported across the perfused guinea pig yolk sac placenta. The uptake of anionic amino acids was competitive and sodium dependent, but apparently lacked stereospecificity. The features of the Taurine transport kinetics have been defined using human placental microvesicles. Similarly, the anionic transport system has been characterized in the basal plasma membrane of the human syncytiotrophoblast.

The general agreement of this workshop was the difficulty to differentiate the transport systems committed in local placental "housekeeping" functions from those actively dedicated to fetal homeostasis.

Gestational studies on the rat placental Na+/H+ exchanger

D.E. Atkinson, J.D. Glazier, C.P. Sibley

Departments of Child Health and Physiology, University of Manchester, St Mary's Hospital, Manchester M13 OJH, UK

A Na^+/H^+ exchanger has been identified in microvillous membrane vesicles prepared from the rat placenta at term(1). Studies on placental Na^+ clearance in this species have shown that as gestation proceeds the rate at which Na^+ is transferred to the fetus increases dramatically(2 & 3). Since the Na^+/H^+ exchanger has been shown to be functionally involved to a small degree in the transfer of Na^+ to the rat fetus(3) we have looked at gestational changes in the activity of this exchanger.

Vesicles were prepared from between 20-50g of rat placenta collected from several animals at each of 15,18,21 & 22 days gestation. Homogenization in ice cold mannitol buffer was followed by Mg precipitation and differential centrifugation as described in(1). Na^+ uptake was then initiated by addition of 0.2ml of vesicles suspended in either control (pH 7.6) or proton gradient (pH 5.5) buffer to 0.8ml of pH 7.6 buffer containing 2.5µCi/ml $^{22}Na^+$. To half of the proton gradient experiments 5×10^{-4}M amiloride was added. Ion exchange columns(4) were used to separate free from intravesicular $^{22}Na^+$. The eluents from the columns were then counted.

	15 DAYS		18 DAYS		21 DAYS		22 DAYS	
	1 Min	30 Min	1 Min	30 Min	1 Min	30 Min	1 Min	30 Min
pH5.5	1.05±.3	1.75±.33	0.82±.1	2.28±.1	0.82±.1	1.82±.2	0.83±.1	1.01±.14
pH5.5+Amil.	0.42±.2*	1.50±.14	0.61±.1	2.14±.5	0.57±.1	1.15±.3	0.37±.1*	0.79±.3
pH7.6	0.42±.1	1.03±.03	0.47±.2*	1.63±.3	0.58±.1	1.62±.4	0.24±.1*	0.72±.07

Table 1. 1 min. uptakes and equilibrium(30mins) values (nmoles Mg protein^{-1}). Means ±s.e. are shown. n=4 in all cases *p<0.05 V pH 5.5 values.

The Na^+/H^+ exchanger in the rat placenta, convincingly shown at 15 and 22 days gestation (less so at 18 and 21 days) does not appear to change activity per mg of vesicle protein over the last third of gestation.

REFERENCES
1) Glazier, J.D., Jones, C.J.P. & Sibley, C.P. (1988) J. Physiol. 403: 55p.
2) Flexnor, L.B. & Gelhorn, A. (1942) Am. J. Obstet. Gynecol. 43: 965-974.
3) Atkinson, D.E., Robinson, N.R. & Sibley, C.P. (submitted paper).
4) Gasko, O.D., Knowles, A.F., Shertzer, A.G., Suolinna, E.M. & Racker, E. (1976) Analyt. Biochem. 72: 57-65.

Epidermal growth factor, placental transfer and secretion in short term dual perfusion of human placental lobules

J.C. Challier, S. Goma, T. Bintein, G. Olive*

*Biologie de la reproduction, Université Pierre et Marie-Curie, et *Pharmacologie biochimique, Hôpital Saint-Vincent-de-Paul, 75014 Paris, France*

A role for the placental EGF receptors of the human placenta has been suspected in the regulation of nutrient transport. In this study, different doses of EGF were perfused in the maternal circulation of human placental lobules. Various parameters were assessed : glucose transfer and uptake, AIB transfer and uptake, lactate release, HCG and HPL release. The control phase (60 min) was followed by the EGF phase (60 min) at concentrations of 10, 66 and 100 ng/ml. Flow rates were carefully controlled in the two phases. In these experiments, we did not observed any significant change for nutrient transport or hormone release. We conclude that EGF is not involved in the short term control of nutrients transport in the human placenta by end pregnancy. Other functions for the placental EGF receptors must be investigated.

Leucine and glutamic acid transport under diabetic conditions in the *in vitro* perfused human placenta

J.U. Hibbard, G. Pridjian, P.F. Whitington

Department of Obstetrics/Gynaecology, University of Chicago, Chicago, Ill USA

We sought to characterize the effects of acute diabetogenic conditions on amino acid transport by in vitro perfused term human placentae. We studied bi-directional and recirculating leucine and glutamic acid transport under normal conditions, hyperglycemia and acidosis. Physiologic concentrations of leucine (50 µM) and glutamic acid (100 µM) were used and [U-14C]-leucine and [U-14C]-glutamic acid as well as [3H]-L-glucose, a reference molecule, were used in tracer amounts. Perfusates were isotonic Krebs-Ringer bicarbonate buffer containing 4% bovine serum albumin oxygenated with 95% O_2:5% CO_2. Normal conditions were: maternal and fetal perfusate glucose 100 mg% and pH 7.40; hyperglycemia was: maternal glucose 400 mg%, fetal 300 mg%; and acidosis was: maternal pH 7.25, fetal 7.20. The t-test was used for statistical comparisons. RESULTS: In non-recirculating perfusions leucine or glutamic acid were on either the maternal or fetal side. The flux (nmol/min/g), clearance and clearance index (CI) in the maternal-to-fetal and fetal-to-maternal directions are presented in the table as means and SEM:

		Maternal to Fetal				Fetal to Maternal		
Leucine	n	Flux	Clearance	C I	n	Flux	Clearance	C I
normal	6	2.58 ± 0.39	0.59 ± 0.09	2.21 ± 0.24	5	3.41 ± 0.47	0.82 ± 0.11	1.58 ± 0.15
hypergly	6	2.39 ± 0.39	0.55 ± 0.09	2.18 ± 0.19	5	2.86 ± 0.30	0.68 ± 0.72	1.75 ± 0.13
acid	6	2.52 ± 0.45	0.58 ± 0.10	2.28 ± 0.21	3	2.32 ± 0.20	0.56 ± 0.05	1.69 ± 0.05
Glu acid								
normal	5	0.64 ± 0.17	0.10 ± 0.03	0.51 ± 0.08	2	3.80 ± 0.43	0.81 ± 0.09	1.29 ± 0.01
hypergly	5	0.87 ± 0.26	0.14 ± 0.04	0.64 ± 0.18	2	3.15 ± 0.52	0.67 ± 0.11	1.10 ± 0.08
acid	4	0.75 ± 0.30	0.12 ± 0.04	0.58 ± 0.10	2	2.87 ± 0.64	0.61 ± 0.14	1.20 ± 0.04

Recirculating experiments with equimolar maternal and fetal leucine were performed under conditions of normality, hyperglycemia and acidosis. No differences were found in the F/M gradients of leucine that were achieved after 3 hours (1.7 (n=5), 1.9 (n=5), and 1.6 (n=2), respectively). CONCLUSIONS: Acute metabolic changes, as seen in diabetic ketoacidosis, when imposed on the human in vitro perfused placenta do not affect transport of leucine or glutamic acid.

Na-dependent amino uptake by human placental microvillous membrane vesicles (MMV) : importance of storage conditions and preservation of cytoskeletal elements

P.I. Karl, S.E. Fisher

Departments of Pediatrics and Research, North Shore University Hospital-Cornell University Medical College, Manhasset, New York, 11030, USA

Human placental MMV were prepared by precipitation of non-microvillous membranes with Mg^{++}, yielding more MMV protein than previously reported. Two aspects of MMV preparation were found to be important to the interpretation of amino acid uptake studies: (1) storage conditions (2) preservation of cytoskeletal elements. In nonfrozen MMV, MeAIB uptake was stimulated by an inward Na^+ gradient and showed "overshoot". The initial Na^+ - dependent uptake rate was concentration dependent with a Vmax of $640 + 80$ pmol. mg^{-1} .30 sec^{-1} and a Km of 0.44 ± 0.07 mM. Na^+ - stimulated Cysteine uptake (119 ± 23 pmol. mg^{-1} .30 sec), previously thought to be very low or absent in the human placenta, was comparable to MeAIB, although there was no "overshoot". Cysteine uptake was partially stimulated by Li^+. In general, storage at either -80°C or -196°C markedly reduced Na^+- dependent uptake of several amino acids, compared to vesicles stored at 4°C. The greatest reduction was seen at -80°C, especially with cysteine. There was no effect of storage temperature on amino acid uptake in the presence of a choline gradient (Na^+ - independent). Within each of the three storage temperature condition, there was no difference in uptake at 12 vs 60 hrs post preparation. Removal of cytoskeletal proteins with KSCN, a chaotropic agent, resulted in greater enrichment of MMV marker proteins, but the preparation lost the capacity for active MeAIB uptake. These data, especially with regard to storage conditions, highlight the importance of precise definition of preparation and storage conditions when interpreting results of amino acid uptake by human placental MMV.

Taurine transport by microvillous membrane vesicles and the perfused cotyledon of the human placenta

P.I. Karl, S.E. Fisher

Departments of Pediatric and Research, North Shore University Hospital-Cornell University Medical College, Manhasset, New York, 11030, USA

Human placental uptake and maternal-to-fetal (M-to-F) net transfer of taurine were evaluated in purified microvillous membrane vesicles (MMV) and the isolated perfused cotyledon. Taurine uptake by MMV was specifically stimulated by an inward Na^+ gradient (Vmax = 24.5 ± 0.6 pmol. mg^{-1} .30 sec^{-1}; Km = 6.2 ± 0.7 uM) with an uptake stoichiometry of 1:1 for taurine: Na^+. In the absence of a Na^+ gradient, an inward Cl gradient also stimulated uptake, although this anion effect was much less than that of Na^+. Na^+- stimulated uptake was enhanced by an outward K^+ gradient. Beta-alanine and hypotaurine competitively inhibited uptake of taurine in MMV. Two-way uptake inhibition studies showed no interaction between taurine and non-beta amino acids. When MMV were preloaded with taurine or hypotaurine, there was trans-stimulation of Na^+ - stimulated taurine uptake. In the perfused placentae, there was no saturable M-to-F net transfer of taurine, despite saturation of tissue uptake from the maternal circulation. During 3 hours of perfusion, no fetal-to-maternal (F-to-M) gradient formed for taurine; yet, a 2:1 gradient simultaneously occurred for histidine (an amino acid which exhibits "trans" stimulation in MMV). This study demonstrates that taurine uptake by the microvillous membrane of the human placenta is highly specific, of high affinity, 1:1 Na^+ - coupled and Cl^- - stimulated. The perfusion experiments suggest that any F-to-M taurine gradient observed in vivo may be due to factors independent of placental transport.

Transport and metabolism of glutamic acid studied in the human placenta by an *in vitro* perfusion method

M. Luckhardt, E. Aegerter, A. Malek, H. Schneider*

*Universitätsfrauenklinik, CH- 3012 Bern, Schanzeneckstrasse 1, Switzerland

In a previous study it was shown that in contrast to other aminoacids there is no transfer for glutamic acid from maternal to fetal side and that glutamic acid is cleared by the placenta from the fetal circulation (1). By means of double-sided in vitro perfusion of an isolated lobulus of the human placenta, some details in metabolism and transport of glutamic acid could be further elucidated. In a first series, the tissue extraction was examined after bolus injection of radioactive glutamic acid as well as differently labelled L-glucose as an extracellular marker. It could be shown that the uptake from the fetal side was distinctly higher than the uptake from the maternal side. This result is in contrast to data of various other aminoacids where a higher uptake from the maternal side could be shown (2). In a second series, the maternal and fetal compartment were perfused in a closed circuit after addition of radioactively labelled glutamic acid and differently labelled alanine on both sides. The concentration changes of both labelled aminoacids on maternal and fetal side showed different patterns. For alanine the concentration on both sides remained constant, while for glutamic acid a considerable decrease in concentration was found on both sides (Fig. 1). The concentration of glutamic acid and glutamine were also de-

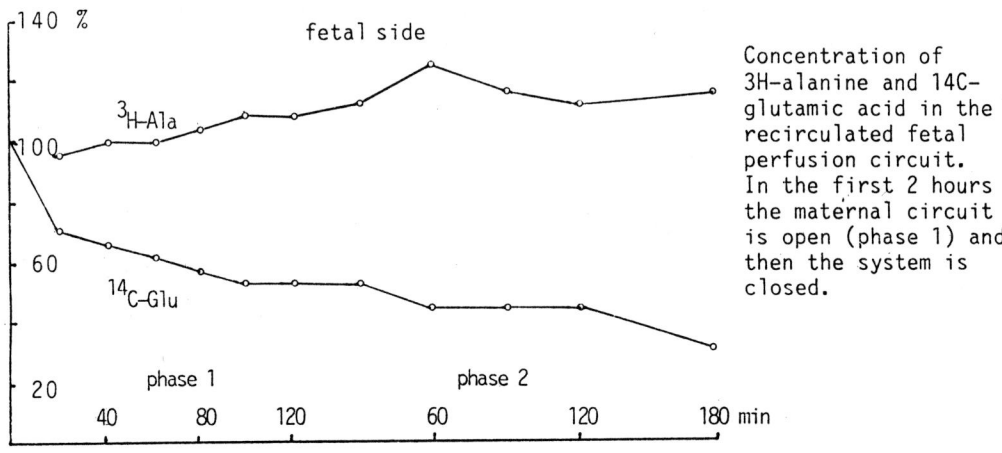

Concentration of ^3H-alanine and ^{14}C-glutamic acid in the recirculated fetal perfusion circuit. In the first 2 hours the maternal circuit is open (phase 1) and then the system is closed.

supported by Schweizerischer Nationalfonds

termined enzymatically and the decrease of glutamic acid as shown with the isotope was confirmed while for glutamine de novo synthesis could be observed. After four hours of perfusion, only 30% of the amount of glutamic acid which was added at the beginning remained and 20% was transformed into glutamine. The remaining 50% was partly recovered from the perfused tissue while the metabolism of the rest so far remains unexplained.

Schneider H, Möhlen KH, Challier JC, Dancis J:
Transfer of glutamic acid across the human placenta perfused in vitro. Br J Obstet Gynecol 1979; 86: 299-306

Schneider H, Proegler M, Sodha R, Dancis J:
Asymmetrical transfer of alpha-aminoisobutyric acid (AIB), leucine and lysine across the in vitro perfused human placenta. Placenta 1987; 8: 141-151

Energy production by the dually perfused term human placenta *in vitro* : ^{31}P nuclear magnetic resonance spectroscopy at 2.0 and 4.7 tesla

A. Malek, R.K. Miller, D.R. Mattison, R. Bryant, M. Panigel, L. Neth

Department of Obstetrics and Gynaecology and Biophysics, University of Rochester, Rochester, NY, and Department of Obstetrics and Gynaecology, University of Arkansas for Medical Sciences, Little Rock, AR; National Center for Toxicology Research, Jefferson, AR, USA

The placenta is responsible for fetal well being, which requires high energy production for active transport and protein synthesis. Previous investigations have suggested that placental ATP remains constant or decreases after initiation of perfusion. Those ATP levels were determined after freeze clamping which, unfortunately, requires termination of perfusion. In contrast, we have assessed placental energy generation by measuring ATP and inorganic phosphate (Pi) during in vitro perfusion with modified M199 tissue culture medium using ^{31}P nuclear magnetic resonance (NMR) spectroscopy at 2.0 Tesla (T) and 4.7 T. At higher magnetic field strengths (4.7 T) adjacent resonance lines are better separated and signal to noise ratios higher. Perfusion conditions included: (i) control, up to 10 hours; (ii) ischemia, produced by halting the perfusion pumps up to 4 hours; and (iii) metabolic inhibition, produced by adding 2,4-Dinitrophenol (DNP) or DNP & Iodacetate (IA) to the perfusate.

SUMMARY OF PLACENTAL ENERGY GENERATION AND ENERGY UTILIZATION

Condition	ATP	Pi	Glucose Consumption	Lactate Production	Hormone Release hCG	hPL	Fetal Volume Loss : ml/h
Control 6 h	+70%	-10%	+3%	-15%	+5%	+8%	<2
Control 10 h	+70%	-20%	-6%	-15%	-11%	+10%	<2
Ischemia	-55%	+30%	*	*	*	*	*
Recovery	0%	-10%	+5%	-10%	-67%	-177%	*
DNP (0.1 mM)	-45%	+60%	-72%	-80%	-73%	-93%	<2
DNP&IA (0.1 mM)	-100%	+55%	-39%	-62%	-51%	-81%	14

* = Not done. Data reported as percent change compared to first 2 hours of perfusion

^{31}P NMR spectra were aquired continuously over approximately 30 min throughout the perfusion. Samples for glucose, lactate, hPL and hCG were obtained at the same time interval. All experiments included at least 2 hours of control perfusion before any alterations were initiated. Transfusion of maternal and fetal perfusate occurred at 2 and 6 hours. During the first 2 hours ATP rapidly increased and Pi fell. Glucose consumption, lactate production and hormone release were constant throughout the control perfusions, indicating adequate cell function, energy generation and utilization. In ischemia, when perfusion was stopped for 4 hours, ATP decreased 55%. During reperfusion ATP rapidly increased to control levels. Energy utilization, as measured by glucose consumption and lactate production, was unchanged after reperfusion. However, release of hPL and hCG were decreased and did not recover during reoxygenation. During perfusion with DNP placental ATP decreased. ATP production was completely blocked when IA was added to DNP. These studies show that ^{31}P NMR spectroscopy is useful for monitoring placental energy production non-invasively and continuously. These studies also suggest that human placental energy generation by the glycolytic pathway is very robust. Supported in part by ES 02774.

Fetal to maternal transport of oxytocin across the dual perfused human placenta

A. Malek, M.J. Meadows, F.C. Miller, E. Blann, D.R. Mattison

Department of Obstetrics and Gynaecology, University of Arkansas for Medical Sciences, Little Rock, AR and Division of Developmental and Reproductive Toxicology, National Center for Toxicological Research, Jefferson, AR, USA

Fetal hormonal signals to the maternal organism may play a role in parturition. For example, the anencephalic human and sheep fetus typically have prolonged gestation. Although the putative fetal signal remains to be defined, oxytocin (OX) has been suggested to fulfill that role. Fetal umbilical artery OX concentrations (15-40 pg/ml) are higher than umbilical vein concentration (4-12 pg/ml) and peripheral maternal concentrations are even lower (1-10 pg/ml). Circulating concentrations of OX are low, or absent, in the anencephalic fetus. In addition, injection of OX into umbilical vessels has been suggested to shorten the third stage of labor and enhance separation of retained placenta. These data suggest that fetal OX may play a role in parturition. If fetal OX is to play a role, one possible route may involve transport across the placenta into the maternal circulation. To explore this question, we have conducted a series of experiments in which OX was injected as a bolus or infused into the fetal circulation and measured in the maternal circulation of the dually perfused term human placenta. In the experiments reported here, ^{3}H OX and ^{14}C inulin (IN) were simultaneously infused into the fetal artery (FA) for 100 minutes. OX and IN concentrations were assessed in MV, FV, MA and FA by liquid scintillation counting. In two experiments little IN or OX was transferred into the maternal circulation. In one experiment substantial amounts of both IN and OX were transferred into the maternal circulation. In the

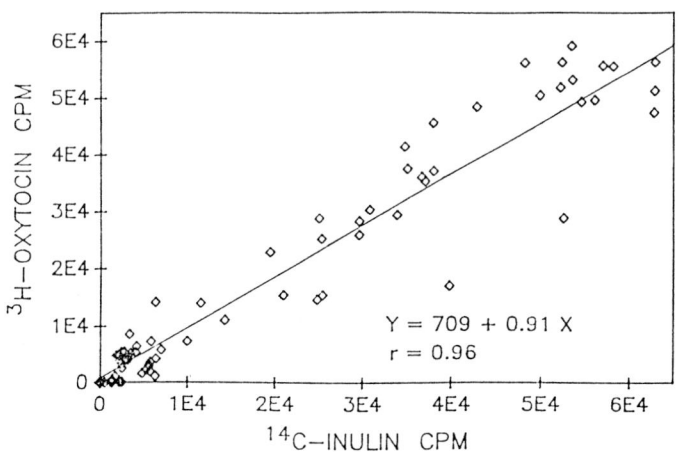

third experiment intermediate amounts of both were transferred into the maternal circulation. If MV (IN) is plotted against MV (OX), the data points form a straight line with intercept not significantly different from 0 and a slope of approximately 1, (r = 0.96). These data suggest that OX transfer from fetal to maternal circulation is not different from IN transfer. If the fetal circulation is intact, little OX would reach the maternal circulation. However, in the presence of fetal to maternal leaks, substantial and undefined quantitative of fetal OX may have access to the maternal circulation. These data suggest that a putative role for fetal OX in the initiation of labor must involve a route other than access to the maternal circulation across the placenta. In addition, given that OX is a hypotensive agent, clinical misadventure may result from umbilical OX injection to shorten the third stage of labor.

Continuous pO_2 monitoring during dual perfusion of the term human placenta

A. Malek, M.J. Meadows, F.C. Miller, D.R. Mattison

Department of Obstetrics and Gynaecology, University of Arkansas for Medical Sciences, Little Rock, AR and Division of Human Risk Assessment National Center for Toxicological Research, Jefferson, AR, USA

In some perfusion systems, maternal-fetal perfusion-perfusion overlap is assessed by O_2 transfer from the maternal to the fetal circuit by intermittent measurement of perfusate pO_2 with a blood gas analyzer. Although this characterizes perfusion-perfusion overlap, it does not allow continuous assessment of O_2 transfer and utilization. Continuous measurement of O_2 transfer with real time display allows rapid identification of perfusions in which maternal to fetal transfer of O_2 falls unexpectedly. In addition, transient effects on placental O_2 transfer and utilization produced by the chemical or perfusion conditions can be studied. We have developed a computerized system to continuously monitor and display pO_2 in real time from the maternal artery and vein (MA, MV) and fetal artery and vein (FA, FV) of the dually perfused term human placenta. Oxygen electrodes (Microelectrodes, Inc. Londonderry, NH) were installed in flow-through chambers in the perfusion system. The signal from the electrodes was digitized (Keithley System 570, Keithley, Cleveland, OH) and acquired by Labtech Notebook (Laboratories Technologies Corp., Wilmington, MA). A Zenith computer with an 8088 CPU, 8087 math coprocessor (Zenith Data Systems, Benton Harbor, MI) and a Mitsubishi EGA display was used to collect, store, analyze and display the pO_2 data in real time during perfusion. Output from the O_2 electrodes was linearly proportional to pO_2 as measured by a Corning 170 pH/blood gas analyzer (Corning Medical and Scientific, Medford, MA) over values ranging from 33 to 502 mm Hg ($r^2 = 0.99$). Term human placenta were obtained following delivery, transported to the laboratory where fetal and maternal perfusion circuits were established within 20 minutes of placental delivery. The pO_2 was captured every other second, displayed and written to the hard disk. Running average pO_2 values (mean $\pm$ SD) were also calculated for MA, MV, FA and FV every minute. Some of the difficulties encountered in this system include: noisy electrodes because of bubbles on the membrane and in the chamber, membrane degradation and electrode warm-up and stabilization time. One of the unique capabilities of the pO_2 electrodes is their ability to respond rapidly to changing conditions. For example, as the fetal circulation begins to breakdown with increasing leaks into the maternal circulation an early change noted, before volume loss in the fetal circuit can be appreciated, is decreasing FV pO_2.

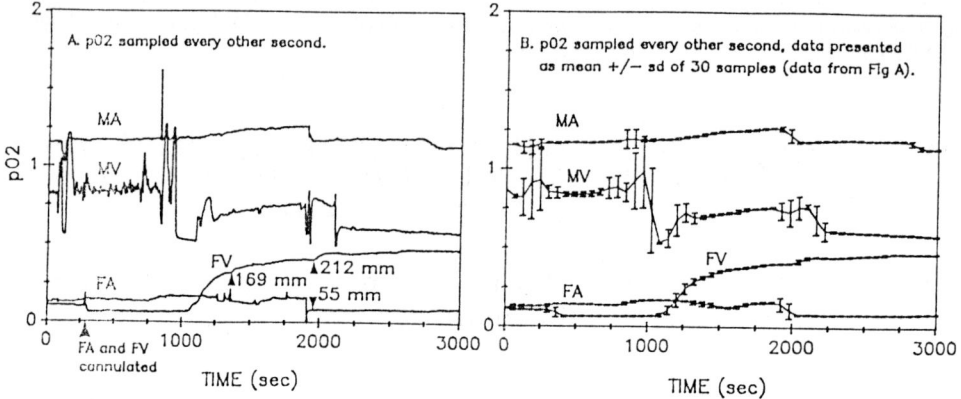

Placental toxicology : retinoids and cadmium

R.K. Miller, W. Faber, Y. Shah, P.J. Wier, R. Perez Di Gregorio, A.A. Levin, P.A. di Sant'Agnese, L. Neth, C. Eisenmann

Department of Obstetrics/Gynaecology, Environmental Health Sciences Center, University of Rochester, Rochester, NY, 14642 USA

The placenta is certainly a conduit for the transfer of toxic agents, but it can be a regulator of transfer via biotransformation/binding. The placenta can also be a site for direct toxic interaction. Two examples of such toxicants will be examined: Retinoids and Cadmium (CD). 13-Cis-Retinoic acid (CRA- Accutane) is used to treat cystic acne and is also teratogenic in man. It is an isomer of all trans-retinoic acid(TRA), a potent animal teratogen which normally is found endogenously as a metabolite of retinol (vitamin A). In rodents, CRA is 10 fold less teratogenic than TRA. In human placental studies in vitro, TRA is converted to CRA, which indicates a potential mechanism for reducing the toxicity of TRA. However, the 1st trimester and term placentae convert CRA to TRA, which suggests a mechanism of bioactivation for the synthetic CRA.

CD, an environmental contaminant and human nephrotoxin, has been shown in rodents to be a placental toxin resulting in fetal lethality. Under in vitro perfusion conditions for periods of 12 hr in the human placenta, CD (0, 10, 20, 100nmoles/ml initial) did not significantly affect the cellular uptake of alpha-aminoisobutyric acid, glucose utilization, lactate production, ATP content or oxygen consumption; however, CD did produce a dose-related decrease in hCG and hPL synthesis/release, reduction in zinc transfer and ultrastructural damage by 4 hours with necrosis by 8 hours at 100 nmoles/ml. The placenta appears to be acting as a filter to concentrate CD and prevent its rapid appearance in the fetal circulation, which results in a potential decreased toxicity to the fetus, but increased toxicity to the placenta. Such studies demonstrate the role of biotransformation in the human placenta to increase or to decrease the toxicity of drugs as well as to produce direct toxicity in the placenta. (Funded in part by NIH ES 02774).

Evidence for carrier-mediated potassium transfer across dually perfused rat placenta

T. Mohammed*, J. Stulc**, C.P. Sibley***, J. Glazier*, R.D.H. Boyd*

*Department of *Child Health and **Physiology, University of Manchester, St Mary's Hospital, Hathersage Rd, Manchester, UK*

Pregnant Sprague–Dawley rats at 21 days gestation were anaesthetized and kept at 37°C in a saline bath. The uterine artery was cannulated and side branches tied so as to perfuse the maternal side of the placenta with Krebs solution containing 20 per cent bovine serum at rate of 1.8 ml min^{-1}. The umbilical vessels were similarly cannulated and perfused with Krebs solution (1 per cent bovine albumin; rate 1.2 ml min^{-1}).

Materno–fetal clearance ($K_{mf} = [v] \cdot Q_f / A$ where v = fetal effluent radioactivity, Q_f = fetal effluent flow, A = maternal perfusate radioactivity) for [^{42}K], [^{51}Cr]-EDTA and in most cases [^{3}H]-Water were determined after allowing 20 min for steady state to be reached. K_{mf} for [^{42}K] and [^{51}Cr]-EDTA were 246 ± 39 and 15 ± 5 µl min^{-1} g^{-1} placenta, respectively (comparable with the in situ singly perfused rat placenta; 309 ± 68 and 5.4 ± 3, respectively).

K_{mf} at maternal perfusate [K$^+$] between 1.6 and 20 mmol l^{-1} showed saturation of [^{42}K] transport (V_{max} = 2.9 ± 0.9 µmol min^{-1} g^{-1} placenta and K_m = 7 ± 3 mmol). K_{mf} [^{3}H]-Water and [^{51}Cr]-EDTA were unaffected. Perfusion with maternal and fetal perfusates containing 1 mmol ouabain or 1 mmol barium chloride reduced K_{mf} [^{42}K] ($P<0.001$) and, less markedly K_{mf} [^{3}H]-Water.

	Before	Ouabain	K_{mf} (µl min^{-1} g^{-1} placenta) After	Before	Barium	After
[^{42}K]	228 ± 28	67 ± 5'	283 ± 25	351 ± 27	109 ±¹	263 ± 47
[^{51}Cr]-EDTA	4 ± 1	5 ± 1	12 ± 5	6 ± 1.3	6 ± 1.6	4 ± 1.3
[^{3}H]-Water	1041 ± 238	617 ± 133'	1047 ± 341	1340 ± 187	880 ± 224	641 ± 312

$n = 5$ ' = $P<0.05$ ¹$P = <0.01$.

Steady state K_{mf} and feto–maternal clearances (K_{mf}) in the same placenta for [^{42}K], [^{51}Cr]-EDTA and [^{3}H]-Water were 233 ± 33, 8 ± 2, 1050 ± 296 and 59 ± 21, 18 ± 8, 574 ± 144, respectively.

Placental potassium transfer in the rat appears to be carrier-mediated, saturable, and to involve Na$^+$, K$^+$-ATPase and barium inhibitable potassium channels.

The uptake of ascorbic acid and its oxidation products in the isolated cotyledon of the human term placenta

H.-P. Leichtweiß, B. Mohar, S. Wohlers, H.J. Schröder

Universitäts-Frauenklinik, Martinistrasse, 52, 2 Hamburg 20, Federal Republic of Germany

We demonstrated earlier that in the guinea pig placenta vitamin C is mainly transported as dehydroascorbic acid (DHA). Using the same method which allows ascorbic acid (AA) and its oxidation products DAA and diketogluconic acid (DKA) to be separated and isolated we now investigated the uptake of these substances in the artificially perfused cotyledon of human term placentae. The uptake was calculated according to the method of Yudilevich after injection of double tracer boli; all uptake values are given in relation to the extracellular marker L-glucose.

Results: Uptake values of DHA are 55.6% ($\pm$ 12.6) on the maternal and 47.6% ($\pm$ 8.3) on the fetal side and exceed those of AA significantly on both sides (11.5% $\pm$6.5 maternal side, 5.9% $\pm$5.2 fetal side). There is no uptake for DKA.

Addition of cytcholasin B (1 mmol/l) decreased the uptake of DAA by 63% on the maternal and 57% on the fetal side.
Reduction of Na^+ (<10 mmol, supplemented by Tris) decreased the DHA uptake to 66% of the controls on the maternal side and to 57% on the fetal side.

The uptake of tracer after application of 14C-AA was reduced by 32% on the maternal side and by 33% on the fetal side when thiourea (1 mmol/l) was added. This indicates that 1/3 of the "AA" represents uptake of DAA which is gained from the AA by rapid oxidation before reaching the cell surface. 2/3 of offered 14C - AA label is taken up as AA. This uptake can be significantly decreased by reduced Na^+ concentration on the maternal side.

Conclusion: as in the guinea pig, in the human placenta vitamin C is preferentially transported as DHA, a minor part as AA. There should exist two transport systems. One is sodium independent for DHA, the other is sodium dependent for DHA and AA.

Inhibitor action on placental mineral metabolism

K.R. Page, D.R. Abramovich*, C.G. Dacke**, T. Mayhew, J.M.A. Williams

*Departments of Physiology, Aberdeen University * Obstetrics/Gynaecology, ** Portsmouth Polytechnic*

We have studied inhibitor action on Na^+, K^+, and Ca^{++} movements between perfusate and tissue in the dually perfused human placental lobule. Krebs ringer (3% Dextran, pH 7.3-7.4) was passed under open circuit, and ion selective electrodes used to measure inflow and outflow concentrations. After an initial perfusion period perfusate containing inhibitor was passed for 15 min. Dinitrophenol (10^{-3} M) caused a significant release of K^+ and increase in Na^+ and Ca^{++} uptake by tissue on both circuits. Ouabain (10^{-5} M) caused a significant release of K^+ on both circuits and had no detectable effect on Ca^{++} or maternal Na^+. A small but significant release of Na^+ occurred on the fetal circuit. Ca^{++} movements showed a dependence on perfusate Ca^{++} concentration similar to that observed in previous closed circuit studies. These results support the view Ca^{++} crosses the placenta from mother to fetus by a membrane bound ATPase system and that a Ca^{++}/Na^+ exchanger is not involved with this process.

Supported by Action Research.

Transfer of betahydroxybutyrate (BOHB) by the *in vitro* perfused human placenta

G. Pridjian, A.H. Moawad, P.F. Whitington

Departments of Obstetrics and Gynaecology and Pediatrics, The University of Chicago, Chicago, IL, USA

BOHB is a ketone body found in high concentration in maternal plasma during diabetic ketoacidosis (DKA), and implicated in associated poor fetal outcome. The *in vitro* perfused human placental cotyledon was used to study its transfer to the fetal compartment at maternal concentrations similar to those in DKA.

RESULTS: Values are expressed as mean ± SEM; clearance index = clearance relative to antipyrine clearance.

[BOHB] (mM)	FLUX (umoles/min)	CLEARANCE (ml/min)	CLEARANCE INDEX
1.0	0.47+.07	0.47+.07	39.14
2.5	1.09+.16	0.43+.06	36.65
5.0	2.57+.43	0.51+.09	40.97
7.5	3.19+.44	0.43+.06	36.44

Open circuit perfusions (n=5) with increasing maternal [BOHB] resulted in a linear increase in the rate of transfer to the fetal side (FLUX); FLUX = 0.452[BOHB] - .018. Neither clearance nor clearance index changed with [BOHB]. In closed circuit perfusions (n=3) with 5 mM BOHB, the fetal/maternal [BOHB] was not different from 1 (1.04+.02, compared to 1.68+.04 for 50 mM leucine under similar conditions). Together these verify transfer by passive diffusion. In open circuit perfusions at 5 mM maternal [BOHB], the cotyledon extracted 5.33+.69 umoles/min (11.4%) from the maternal side, retained 2.76+.81 umoles/min (5.9%) and transferred 2.57+.43 umoles/min (5.5%) to the fetal side. [BOHB] $\geq$10 mM were toxic to the cotyledon.

CONCLUSIONS: 1) Maternal-to-fetal transfer of BOHB by the human placenta is by passive diffusion. 2) [BOHB] double those seen in DKA are toxic to the perfused cotyledon. 3) The placenta may store and/or metabolize some BOHB.

Supported by the American Diabetes Association, N. Illinois Affiliate

Glucose and oxygen consumption, and lactate production of human placentae dually perfused *in vitro* under normoxia, hypoxia, and reoxygenation after a hypoxic period

W. Reiber, H. Nöschel, S. Schröder, B. Müller

Department of Obstetrics and Gynaecology, University of Jena, German Democratic Republic

The aim of the experiments was to investigate metabolic reactions of isolated human placentae under different O_2-conditions, particularly when oxygenation followed a period of hypoxia.

Method: A perfusion system was used with both circuits of a whole placenta being recirculated over two hours with blood-containing perfusate.
Group A: 17 placentae were perfused well oxygenated.
Group B: 10 perfusions under hypoxia.
Group C: 8 placentae were oxygenated after 60 Min. of hypoxic perfusion. 1)

Results and Discussion:
In group B glucose-uptake was slightly increased compared to group A (2,95 ± 1,11 versus 2,59 ± 1,68), while lactate-production was higher (7,69 ± 1,04 versus 3,56 ± 1,53). Oxygen consumption was smaller under hypoxia. Pyruvate showed no significant changes over the two hours in all groups. In group C the results of the first hour resembled group B. After reoxygenation these placentae showed a smaller oxygen consumption compared with group A, but the highest glucose-uptake (4,35 ± 0,72), whereas lactate-production was lowered (5,26 ± 1,86) compared with group B. The smaller oxygen consumption could be explained by a deterioration of mitochondriae and hence a decreased capacity of oxydative metabolism. The higher glucose-uptake could reflect the ability of placental tissue to restore other energy-sources (e.g. proteins) which were affected under the previous hypoxia.

1) all results are given in µmol/g wet weight/h

Uptake and transfer of amino acids in the artificially perfused guinea-pig yolk sac placenta

C. Schoch, H.J. Schröder, H.-P. Leichtweiß

Universitäts-Frauenklinik, Martinistrasse, 52, 2 Hamburg 20, Federal Republic of Germany

It is unknown whether in guinea pigs amino acids may be exchanged across the yolk sac placenta.

The yolk sac vessels of guinea pigs (gestational age >55 days) were perfused in situ (H.Schröder, H.-P.Leichtweiß, EPG meeting Cambridge, 1984) with cell free modified Ringer's solution (95%O_2/5%CO_2, pH=7.4, 38°C) at constant 0.5 ml/min. Catheters were inserted into maternal vessels. We investigated 1) the maternal-fetal transfer of a) 3H2O, b) L-alanine, c) D-alanine and AIBA, and 2) the total uptake (reference: L-glucose) of a) D- and L-alanine, b) AIBA, c) D- and L-aspartate.

ad 1). Radioactive substances were injected into the maternal circulation. Blood and perfusate samples were collected up to 30 min later.
a) Water clearance was 130±7 ul/min (n=30) after 15 min.
b) 28±17% of maternal plasma activity were found after 20 min in fetal perfusate when 3H-L-alanine was injected (n=6). At 30 min less than 3% of maternal 14C-L-alanine levels could be detected at the fetal side (n=4). c) No activity in fetal perfusate was found after maternal injection of D-alanine or AIBA.

ad 2). Labelled amino acids and L-glucose were injected into the fetal arterial tube. Perfusate samples were taken continuously for 10 min.
a) L-alanine uptake was 29±12% (n=6). It was reduced to 9.2±7.4% (n=3, p=.04) with 10 mmol L-alanine and to 21±13% (n=6, NS) in low sodium (5 mmol/l) perfusion fluid. D-alanine uptake was 8.9±5% and reduced to 0.8±2.5% with 10 mmol L-alanine (n=3, p=0.05). Low sodium decreased uptake from 5.3±5% to 0.8±11.4% (n=4, NS). b) AIBA uptake was 2.6±8.1% (n=3).

c) L-apartate uptake was 36.1±12.7%. It was reduced to 3±7% (1 mmol/l L-aspartate) and to 3±5.5% (1 mmol/l D-aspartate, n=4, p<0.03). Low sodium decreased uptake from 37.2±12.7% to 6.7±7.1% (n=5, p<.02). Uptake of D-aspartate was 42.8±15%. It was reduced to 8.6±12% (1 mmol/l L-aspartate, n=4, p<0.03) and to 6.8±5% (1 mmol/l D-aspartate). Low sodium decreased uptake from 49.8±14.2% to 10.1±6.2% (n=5, p<0.01).

Conclusion: the transfer of D-/L-alanine or AIBA from mother to fetus is negligible. Uptake of D-/L-alanine and -aspartate is mediated by processes (metabolism, membrane carrier) of low stereospecifity. D-/L-aspartate uptake depends on sodium.

D-glucose uptake and transfer in the artificially perfused guinea-pig yolk sac placenta

H.J. Schröder, W. Elwers, H.-P. Leichtweiß

Universitäts-Frauenklinik, Martinistrasse, 52, 2 Hamburg 20, Federal Republic of German

It is not known whether in guinea-pig nutrients like glucose or amino acids may be exchanged through the yolk sac placenta between mother and fetus. The yolk sac vessels of guinea pigs (gestational age > 55 days) were perfused (H. Schröder, H.-P. Leichtweiß, EPG meeting, Cambridge, 1984) with cell free TC199 (+ 20 g/l Dextran T40, + 2 ng/l bovine albumin, 95 per cent O_2/5 per cent CO_2, pH = 7.4, 38°C at constant 0.3 ml/min. Arterial and venous catheters were inserted into maternal vessels. We studied (1) the transfer of (a) triated water and (b) D- and L-glucose from mother to fetus, and (2) the uptake (U_{tot}, U_{max}) of D-glucose and 3-O-methyl-D-glucose from the fetal perfusate.

(Ia) $[^3H]_2O$ was injected into the dam at the end of an experiment. Maternal blood and perfusate venous samples were collected at 2 min intervals. The mean water clearance at 15 min was 82 ± 15 (s.d.) µl/min (n = 6).
(b) $[^3H]$-D- and $[^{14}C]$-L-glucose were injected into the mother. The mean D- and L-glucose clearance after 15 min was less than 1.2 µl/min (n = 10).

(2) $[^3H]$-D-glucose or $[^3H]$-0-methyl-D-glucose with $[^{14}C]$-L-glucose were injected into the arterial tube. Perfusate samples were taken continuously for 10 min. (a) Total D-glucose uptake was 11.2 ± 4.9 percent whereas total O-methyl-D-glucsoe was significantly less (8.8 ± 41 per cent, n = 8).

(b) With cytochalasin B (1 x 10^{-4} mmol/l) total D-glucose uptake decreased significantly from 10.8 to 4.9 per cent (paired t-test, n = 5).

(c) Total D-glucose uptake (reference: L-glucose) decreased with increasing chemical D-glucose concentration. U_{max} values were used to calculate unidirectional D-glucose fluxes. K_m (5.6 mmol/1) and V_{max} (0.5 mmol/min) were derived from the Hofstee plot (n = 4).

Conclusion : maternal glucose hardly moves from the maternal circulation to fetal tissues. D-glucose uptake from the fetal circulation is carrier mediated and due to membrane bound and metabolic processes.

Magnesium transfer across the *in situ* perfused rat placenta

A.J. Shaw, M.Z. Mughal, C.P. Sibley

Departments of Child Health and Physiological Sciences, University of Manchester, St Mary's Hospital, Manchester, UK

Pregnant Sprague-Dawley rats were anaesthetised (110 mg/Kg sodium thiobutabarbitone) on day 21 of gestation (term = 23d.), uterus opened and the fetal circulation of one placenta perfused with a Mg free Krebs Ringer containing 3.5% dextran (pH 7.4, 37°C) at approximately 0.5 ml/min.(Q). ^{51}Cr-EDTA was injected as a diffusional marker into a cannulated maternal jugular and samples of arterial blood (A) subsequently taken from a carotid cannula at 2, 12, 24 and 36 min. for determination of plasma ^{51}Cr-EDTA (gamma spectrometry) and Mg (atomic absorption spectrometry). Fetal perfusate effluent (v) (4 min. collections) was similarly analysed. Mean clearance (K = v.Q/A. placental wt.) was calculated for the first three effluent periods (during which K was steady). Diffusion coefficients (D) were determined in a 0.9% NaCl, 0.6% CaCl$_2$, 1% agar medium by the method of Berhe et al (1987). Mean $\pm$ SEM K, D, and K/D values for Mg and Cr-EDTA were:

Solute	K (n = 21) (μl min^{-1} g^{-1})	D (n = 6) (cm2 x 10^6 sec^{-1})	K/D (cm g^{-1})
Mg	26.7 $\pm$ 2.0	6.9 $\pm$ 0.3	64.8
Cr-EDTA	3.2 $\pm$ 0.2	6.8 $\pm$ 0.3	7.8

Five experiments were also performed using a bilateral perfusion system in which the maternal uterine artery was perfused with a Krebs Ringer containing Mg (0.5 mmol/l). K_{Mg} and $K_{Cr-EDTA}$ estimated with this preparation were 51.5 $\pm$ 6.8 and 8.7 $\pm$ 3.5 μl min^{-1} g^{-1} respectively. Removing Mg from the maternal perfusate reduced Mg appearance in the fetal perfusate by 87 $\pm$ 4% within 8 min. This suggests that K_{Mg} obtained in the singly perfused preparation largely reflects Mg transfer from the maternal circulation and not simply elution of a placental Mg pool. The finding that $K/D_{Mg} > K/D_{Cr-EDTA}$, suggests that maternal-fetal Mg transfer by the perfused rat placenta is not only by diffusion.

Reference: Berhe, A., Bardsley, W.G., Harkes, A. and Sibley, C.P. (1987) Placenta 8, 365-380.

Transcellular transport of Ca²⁺ across the perfused human placental lobule

J. Štulc, B. Štulcovà, J. Švihovec, M. Břešťák

Department of Pharmacology and Department of Obstetrics and Gynaecology, Faculty of Pediatrics, and Department of Physiology, Faculty of Medicine, Charles University, Prague, Czechoslovakia

Transcellular and paracellular components of the maternal-fetal (m-f) transport of Ca, Pi and K across the intact human term placenta were calculated from the published data. Assuming that there is no significant transcellular transport in the fetal-maternal (f-m) direction, the transcellular component is only a fraction of the total m-f flux, most of the flux being paracellular.

M-f transport of ^{45}Ca and ^3H-L-glucose (an extracellular marker) was measured across the dually perfused isolated human placental lobule. The transcellular component of the Ca transport was estimated from the transfer of the two tracers and from the ratio of their diffusion coefficients in water. The transcellular component was about a half of the total m-f transport. The transcellular transport of ^{45}Ca decreased reversibly during perfusion with NaCN (1 mM) or when Ca^{2+} concentration in the perfusate was elevated to 2.85 mM from the control level of 1.1 mM. It increased when Ca^{2+} concentration was reduced to 0.3 mM.

Problems in studying transport across the human term placenta, arising from a slow growth rate of the human fetus near term, are discussed.

7.
Placental structure and pathology

Structure placentaire et pathologie

Ultrastructural correlates of placental substance transfer : a state-of-art-review

R. Leiser, T. Egloff

Institute for Animal Anatomy, University of Bern, Länggass-Strasse 120, CH-3001 Bern, Switzerland

INTRODUCTION

The placenta as a mediator organ between mother and conceptus is engaged in innumerable functions. These can be related to materno-fetal substance exchange, synthesis and secretion of substances, catabolic and regenerative processes, materno-fetal contact development, etc.

However - after approximately 35 years of electron microscopic research - relatively few of these functions can be correlated with ultrastructural morphology, i.e., directly observed or related to cytochemical, immunocytochemical or other methods.

In this review some findings of ultrastructural phenomena in the chorion frondosum of chorioallantoic placenta shall be presented and explained. Moreover, a preview of future research concerning this subject will be given. We are very grateful to the placentologists from all over the world, who kindly sent us their negatives, slides and copies, which now allow us to illustrate this review on a broader base.

FACILITIES IN MATERNO-FETAL SUBSTANCE EXCHANGE OR TRANSFER

1. Thinning, fusion and partial loss of layers of the interhemal membrane

The **diffusional transplacental exchange of substances** between mother and conceptus partly depends on the thickness and on the stratification of the **placental interhemal membrane** or placental barrier (Metcalfe et al., 1967; Björkman, 1968; Steven, 1975; King, 1982a; Ramsey, 1982; Faber and Thornburg, 1983).

During implantation, this interhemal membrane includes in all species the six layers based on Grosser's (1927) classification: the maternal endometrium is composed of the capillary endothelium, the connective tissue and the uterine epithelium, whereas the fetal chorion includes the trophoblast, the connective tissue and the capillary endothelium (see Scheme 1). Soon after implantation, the interhemal membrane changes in a very species-specific way by thinning, fusion and partial loss of layers in order to promote substance exchange to fulfill the increasing needs of the developing and growing conceptus (Steven and Morriss, 1975; Mossman, 1987).

- **Thinning** of the interhemal membrane down to 1-2 μm can be observed in many species (cf. Dantzer et al., 1988; Scheme 1). In the pig, for instance, the interhemal membrane is thinned by capillaries, which indent the trophoblast and the uterine epithelium as so-called "intraepithelial capillaries" (Amoroso, 1952; Fig. 1a). However, without a real reduction of placental layers, the pig placenta remains one of the epitheliochorial type (Björkman, 1968, 1970; Fig. 2a). Particularly the **trophoblast** can also be thin in some places, containing a few organelles only. As a "nephropneumoide zone" with easy gaseous exchange (for review see Ludwig, 1981) this part of the trophoblast is called epithelial plate in the woman (Bremer, 1916; Fig. 1d). In addition, the maternal **capillary endothelium** can show fenestrations by local thinning in the pig (Friess et al., 1980; Fig. 1a) as well as in the fetal capillary endothelium of some hemochorial rodent placentas (King and Hastings, 1977; King, 1982a; Fig. 1b).

- **Fusion** of the **trophoblast** and the **uterine epithelium** (Scheme 1) can be observed in the epitheliochorial placentation of ruminants as a mainly non-degenerative process. The so-called trophoblast binucleate cells migrate either to the materno-fetal contact zone or into the uterine epithelium (Wooding, 1984; Fig $1c_1$; Scheme 2). Substance transfer by trophoblast granules is obviously facilitated by direct penetration into the cleared up uterine epithelial cells. As a partly degenerative process, fusion of the two epithelia might be a form of transitory stage in the development of the early hemochorial placenta in the rabbit (Larsen, 1961; Enders and Schlafke, 1971; for review see Schlafke and Enders, 1975) and in the woman (Larsen, 1980; Fig. $1c_2$).

The basal laminas of endothelium and trophoblast, especially serving as a filter for macromolecules (King, 1988), are locally closely related. Fusion of the two basal laminas (Scheme 1) is known, e.g., in the pig (Friess et al., 1980; Fig. 2a). However, it is not a fusion between different layers, but one of subelements in the layer of fetal connective tissue.

- **Partial loss** of layers of the materno-fetal interhemal membrane develops in species which show a trophoblast characterized by invasive, degenerative and lytic capabilities acting on the endometrial tissue (Grosser, 1927; Amoroso, 1952; Björkman, 1968; Schlafke and Enders, 1975).

Concerning the endometrial tissue, only the uterine epithelium becomes dissolved, as shown in the endotheliochorial placental type (Björkman, 1973; Scheme 1), e.g., in the carnivores (Fig. $2b_1$). The break down of all **three maternal layers** (uterine epithelium, maternal connective tissue, maternal endothelium), allowing the trophoblast a direct contact to the maternal blood, characterizes the hemochorial placental type in man (Fig. 1d) and rodents (Fig. $2b_2$). The **trophoblast** itself may be reduced from three to one layers as shown by Enders (1965) concerning the hemochorial placental type: from hemotrichorial in many rodents (Fig. 1b), to hemodichorial in the rabbit, to hemomonochorial in the Guinea pig and in man (Scheme 1; Figs. $2b_2$, 1d). In the woman, however, two trophoblastic layers appear in some places: the inner layer, the cytotrophoblast, decreases in extent during pregnancy and regenerates the outer layer bordering the maternal blood, the syncytiotrophoblast (cf. Boyd and Hamilton, 1970; Wynn, 1973; Kaufmann, 1981).

Scheme 1. Scheme of the chorioallantoic interhemal membrane of placenta demonstrating the development from preimplantation to implantation and placentation (left to right). The different possibilities of placental typification from epitheliochorial to endotheliochorial and to hemo(mono-/di-/tri-)chorial are exemplified by progressive lysis of maternal tissue (bottom left to right).

The numbers refer to the structural phenomena and to the Figures in this study as well:
Thinning (**1a, 1b**)/ fusion (**1c**)/ partial loss (**1d**) of layers of interhemal membrane.
Epitheliochorial (**2a**)/ endotheliochorial (**2b₁**)/ hemochorial (**2b₂**) enlargement of cell membranes.
Uterine epithelial (**3a**)/ trophoblastic (**3b**) endocytosis
Uterine epithelial (**4a**)/ trophoblastic (**4b**) transcytosis

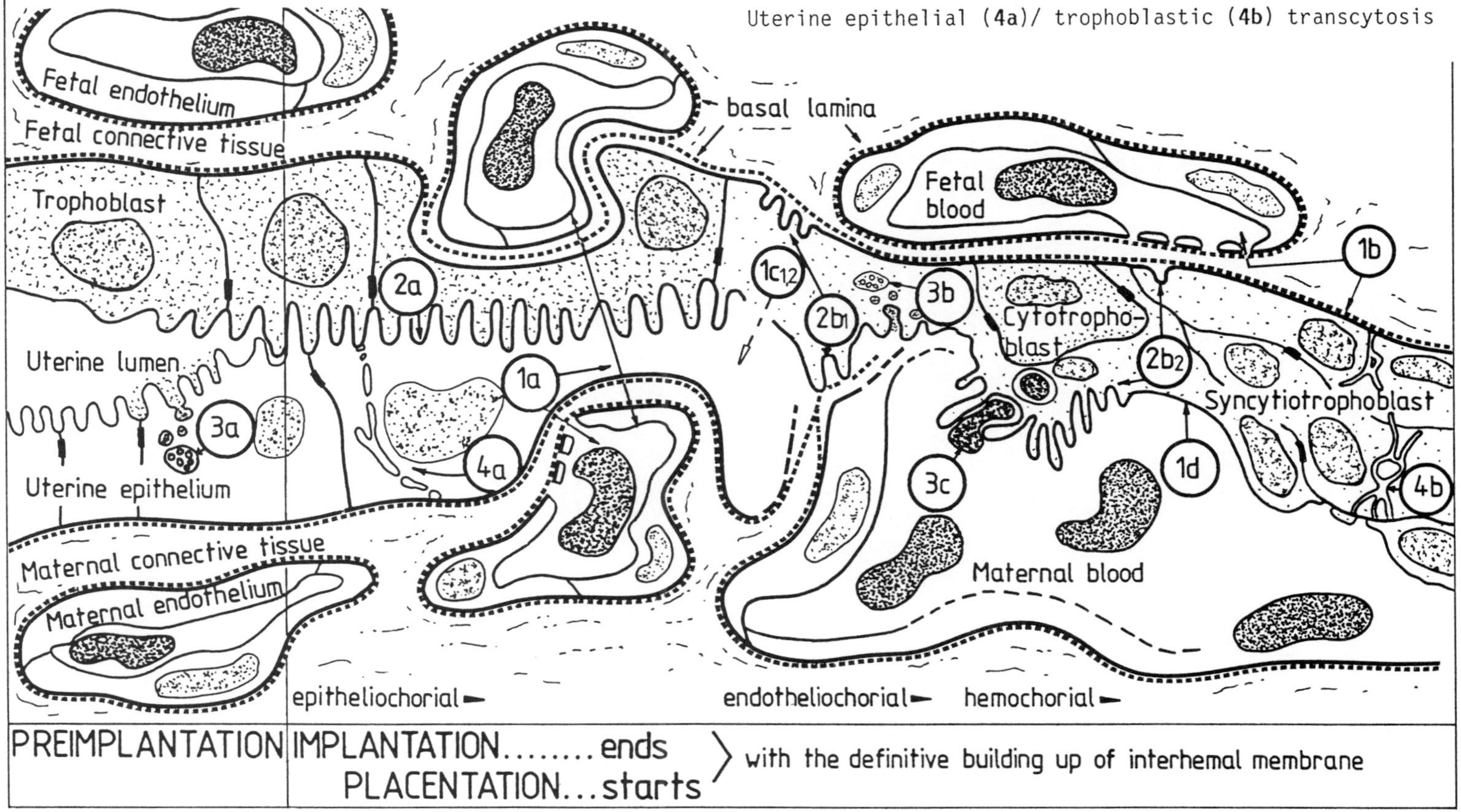

PREIMPLANTATION | IMPLANTATION......ends ⟩ with the definitive building up of interhemal membrane
| PLACENTATION...starts

epitheliochorial ▶ | endotheliochorial ▶ | hemochorial ▶

221

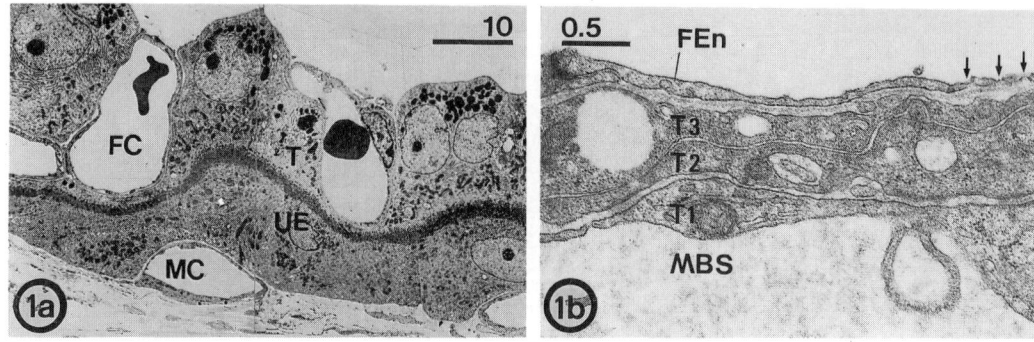

Fig. 1a. Epitheliochorial pig placenta with "intraepithelial capillaries" on maternal (MC) and fetal side (FC), 110 days post coitum (d.p.c.). Uterine epithelium (UE), trophoblast (T). Courtesy of AE Friess, Bern (Switzerland).

Fig. 1b. Placental interhemal membrane of hemotrichorial **Lemmus lemmus**. The maternal blood space (MBS) borders on a three-layered syncytiotrophoblast (T1, T2, T3), of which T1 shows pores. The **arrows** point to fenestrations of the fetal capillary endothelium (FEn). Courtesy of BF King, Davis (USA).

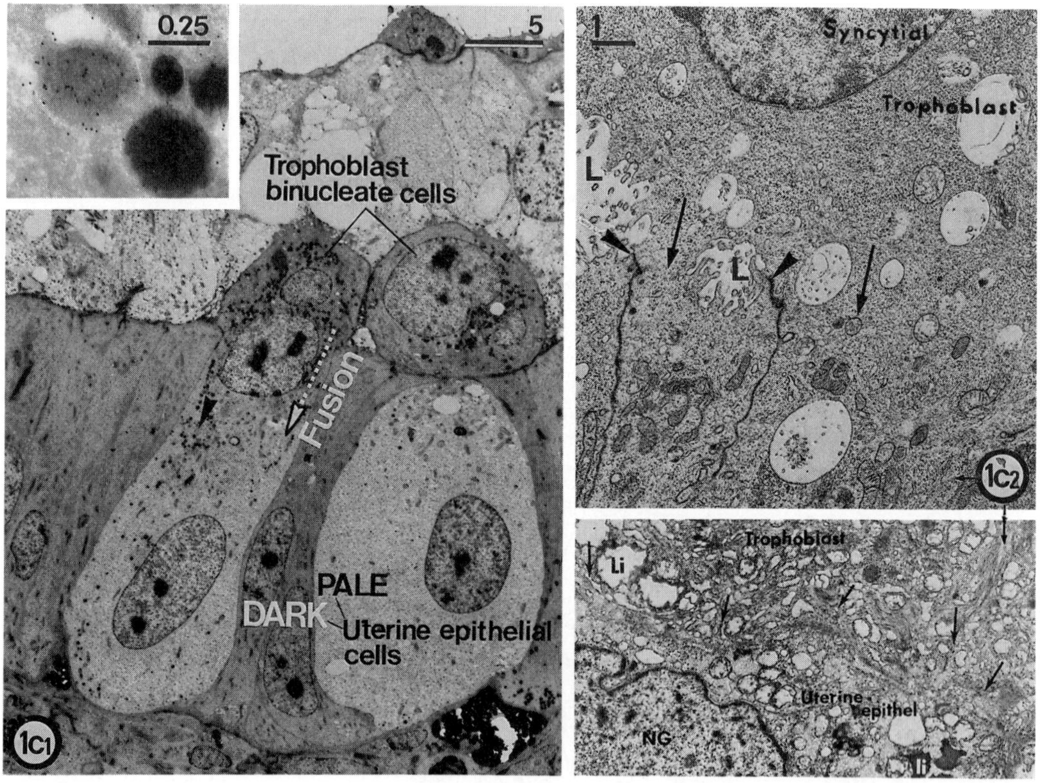

Fig 1c$_1$. Migration of trophoblast binucleate cells and fusion with pale uterine epithelial cells in the sheep placenta, 16 d.p.c. Dark trophoblast granules penetrate the uterine epithelial cells (**arrowhead**). Inset: Trophoblast secretory granules with gold particles showing the localization of placental lactogen. Anti-oPL/gold-labelling. Courtesy of FBP Wooding, Cambridge (England).

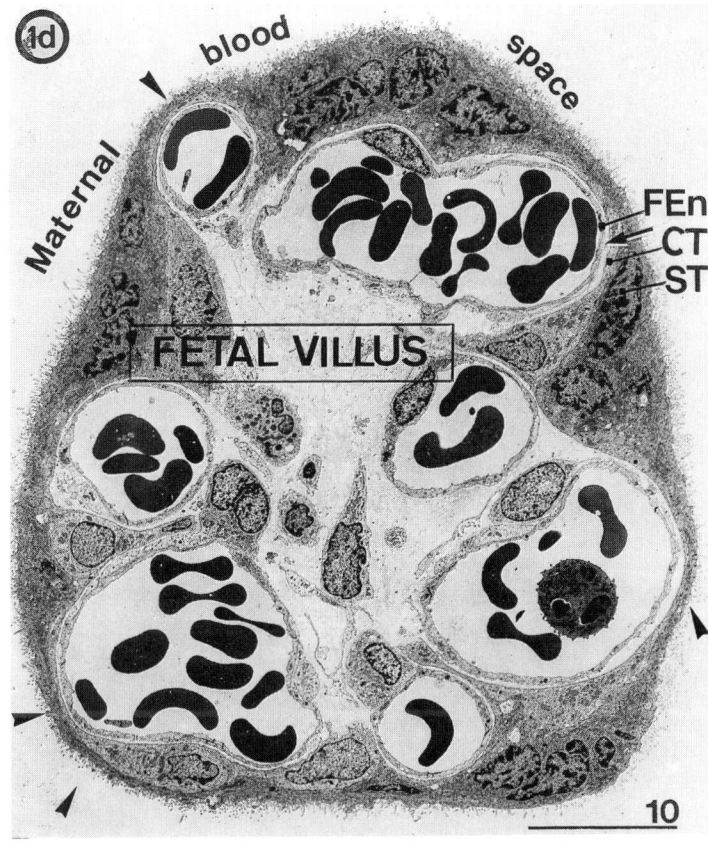

Fig. 1d. The hemochorial placental interhemal membrane of woman, as shown on the fetal terminal villus, consists of three layers: dark syncytiotrophoblast (**ST**)/clear cytotrophoblast (**CT**), connective tissue (**arrow**), fetal endothelium (**FEn**). The **arrowheads** indicate parts of particularly thin syncytiotrophoblast, called epithelial plates. Courtesy of P Kaufmann, Aachen (FRG).

<- see page before

Fig. 1c$_2$. Top: Fusion of syncytiotrophoblast and 2 uterine epithelial cells (**arrows**) during implantation in the rabbit, 7 d.p.c. Isolated uterine lumen (**L**), "blindly ending" tight junctions (**arrowheads**). Courtesy of AC Enders and S Schlafke, Davis (USA). **Bottom:** Advanced trophoblast-uterine epithelial fusion with little remnants of cell membranes (**arrows**) during implantation in the woman. Lipid droplets of trophoblast (**Li**) and uterine epithelium (**li**). Nucleus of gland cell (**NG**). Courtesy of JF Larsen, Copenhagen (Denmark).

2. Cell membrane enlargement of the contacting uterine epithelium and trophoblast

Cell membranes are phase boundaries. Therefore, enlargement of cell membranes means an increase of **qualified substance transfer** and, hence, a decrease of barrier function, especially in the uterine epithelium and trophoblast of the interhemal membrane (Björkman, 1968, 1973; King, 1982a; Hasselager, 1986).

The uterine epithelium is provided with apical microvilli before implantation already. During implantation, these microvilli may induce the **trophoblast** to develop apical microvilli as well (Leiser, 1975; Scheme 1). The interdigitating microvillous apical cell membranes of both epithelia enlarge the absorptive uterotrophoblastic surface, e.g., of the late epitheliochorial pig placenta, by a factor of 8 to 11 (Hasselager, 1986; Fig. 2a). In the epitheliochorial placenta this type of absorptive surface lasts during the whole pregnancy. In animals with a break down of endometrial tissue during late implantation (see above) only the trophoblast shows apical enlargement of the cell membrane by means of microvilli,

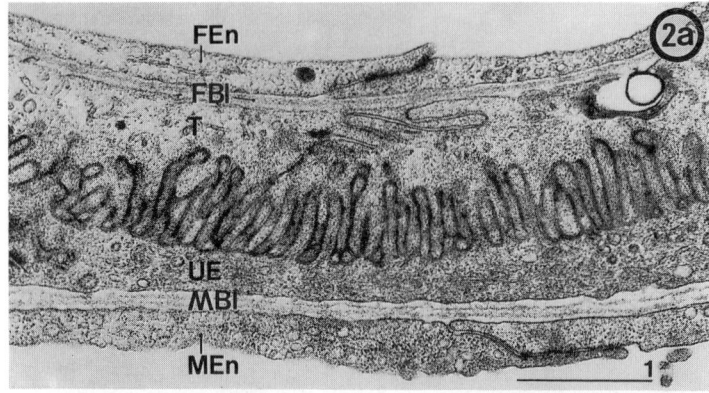

Fig. 2a. Epitheliochorial pig placenta: interhemal membrane, 6 layered and measuring 2 µm, 110 d.p.c. Maternal endothelium (**MEn**), maternal (double) basal lamina (**MBl**), uterine epithelium (**UE**), trophoblast (**T**), fetal (fused) basal lamina (**FBl**), fetal endothelium (**FEn**). Courtesy of AE Friess, Bern (Switzerland).

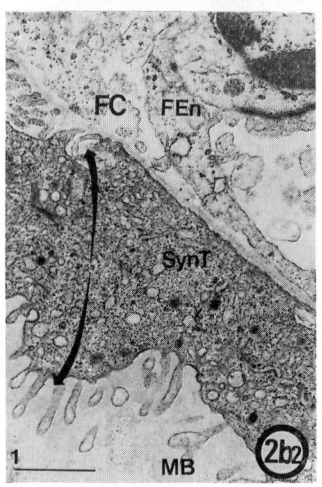
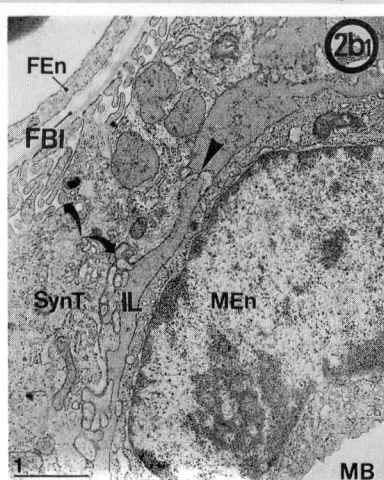

Fig. 2b$_1$. Endotheliochorial cat placenta: interhemal membrane, 4 layered, near term. Maternal endothelium (**MEn**), syncytiotrophoblast (**SynT**) with apical and basal processes, fetal (double) basal lamina (**FBl**), fetal endothelium (**FEn**). The maternal connective tissue or interstitial layer (**IL**) is locally interrupted by direct "endothelio-chorial contact" (**arrowhead**). Maternal blood (**MB**).

Fig. 2b$_2$. Hemomonochorial Guinea pig placenta: interhemal membrane, near term. Syncytiotrophoblast (**SynT**) showing apical and basal microvilli (**double arrow**) faces maternal lacunar blood space (**MB**). Fetal connective tissue (**FC**), fetal endothelium (**FEn**).

folds or blunt processes. They are rather moderate in the endotheliochorial type, e.g., the cat (Leiser and Enders, 1980a; Leiser, 1982; Fig. 2b$_1$), or variable in the hemochorial type, e.g., the Guinea pig (Kaufmann and Davidoff, 1977; Fig. 2b$_2$). Variable microvillous enlargement also appears on the basal cell membrane of the endotheliochorial and hemochorial trophoblast (Scheme 1; Figs. 2b$_1$, 2b$_2$), probably influencing here the ionic transport (King, 1982a).

3. Endocytosis as shown by (micro)pinocytosis and (hetero)phagocytosis

Endocytosis of fluid material (soluble molecules), **pinocytosis**, is ultrastructurally evident on the apical and basal unit membranes of all epithelia/endothelia of the placental barrier (Scheme 1). Substances become mostly connected to receptors (receptor ligand complexes) and then internalized via coated pits of the cell membrane (Goldstein et al., 1979). The coated pits transform into coated vesicles (Ockleford and Whyte, 1977), multivesicular bodies and endosomes carrying their content into the lysosomal apparatus, the Golgi zone and even to the basolateral cell membrane (King, 1982a, 1988).

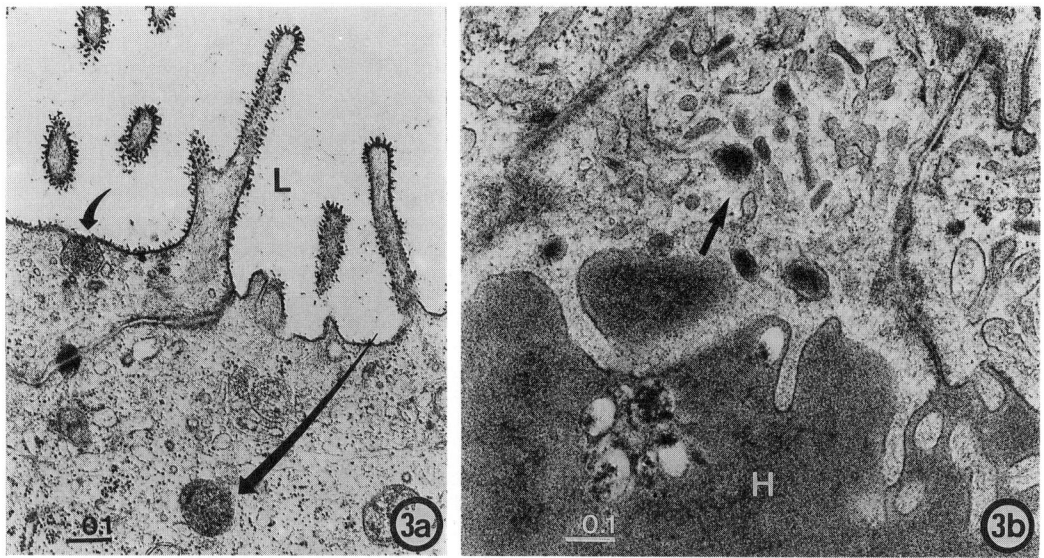

Fig. 3a. Cat uterine epithelium at preimplantation, 10 d.p.c., with peroxidase-traced micropinocytosis (**arrows**) from the uterine lumen (**L**).

Fig. 3b. Cat cytotrophoblast, near term, with vesicle-bound pinocytosis (**arrow**) of dark hemolysate (**H**) from the naturally existent placental hematoma.

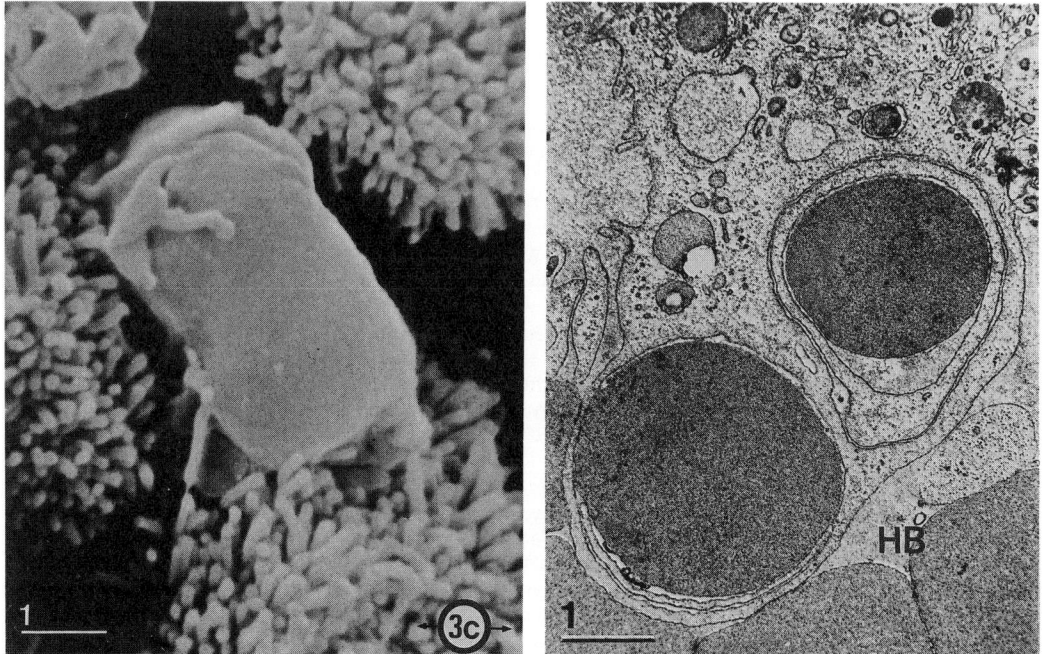

Fig. 3c. Cat cytotrophoblast, 58 d.p.c., showing erythrophagocytosis. Corresponding scanning electron microscopy (**left**) and transmission electron microscopy (**right**). Hematomal blood space (**HB**).

Particularly on the apical part of preimplantative rat **uterine epithelium** (Scheme 1) uterine secretory products become reabsorbed (Psychoyos and Mandon, 1971) by "pinopods" and vesicles or "dome-shaped bulges" (Enders and Nelson, 1973; Parr and Parr, 1974). This has been proved by tracing with ferritin (Parr and Parr, 1974), peroxydase or other tracers (Enders and Nelson, 1973), as shown, e.g., in the cow (Guillomot et al., 1986) and in the cat (Leiser, 1982; Fig. 3a). This reabsorption is vaguely understood. It may help in passing over information from the conceptus to the uterine epithelium or may enhance the viscosity of the embryotrophe in order to get an adhesive film for embryo attachment (Enders and Nelson, 1973). Pinocytosis is also very common in the apical cell part of the **trophoblast** (Scheme 1), i.e., among others in man as made visible by tracing with peroxydase-conjugated Immunoglobulin G (King, 1982b), in the cat with ferritin (Leiser and Enders, 1980a) or by the naturally existent hemolysate from the placental hematoma (Leiser and Enders, 1980b; Fig. 3b).

Endocytosis of solid material (aggregated molecules), **phagocytosis** (Scheme 1), can be observed in the preimplantative **uterine epithelium**, resembling an "exaggerated pinocytotic process" as mentioned above (see also Enders and Nelson, 1973). The apical **trophoblast**, however, is most conspicuously phagocytotic when invading the endometrial tissue. This is rare in the epitheliochorial placenta, e.g., in the equidae during migration of the endometrial cup cells (see Scheme 2; Allen et al., 1973), but is very common in the endotheliochorial and hemochorial placental type (Enders and Schlafke, 1971; Schlafke and Enders, 1975; Scheme 1). The trophoblast is capable of erythrophagocytosis as well. This allows the embryo to get the needed iron from the placental hematomas, e.g., in the epitheliochorial sheep and goat (Burton, 1982; Myagkaya et al., 1984) or in the endotheliochorial cat as evident in Figure 3c (Leiser and Enders, 1980b; Malassiné, 1982).

4. Substance exchange by intracellular pathway (transcytosis)

The intracellular placental pathway by **transfer tubules** (and dense bodies) is designed for direct transport across the cells without a degradation of the content (proteins) or it may take part in the transport of substances to the Golgi complex for synthesis of compounds suitable for fetal utilization (King and Enders, 1971; Parr, 1980; Dantzer, 1982; Scheme 1). Transfer tubules are branched, have a diameter of approximately 0.1 µm and can be observed, e.g., in the **endometrial epithelium** of the pig placental areola (Fig. 4a). Here, they are part of the lysosomal system, detectable by the reaction of acid phosphatase. They contain ferritin as well, intracellularly stored in an intermediate form for the synthesis of uteroferrin (Dantzer and Nielsen, 1984; placental iron transfer reviewed by van Dijk, 1988).

The **transtrophoblastic channels** of the **syncytiotrophoblast** (Scheme 1) have been detected by experimental transfer data in the rabbit (Stulc et al., 1969) and in man (Kaufmann, 1984), whereas their ultrastructural occurence could be shown in the Guinea pig by using a lanthanum hydroxide tracer (Fig. 4b). These tubular and ramified, membrane-lined channels have a diameter of 15 to 25 nm and extend from the apical or basal epithelial surface into the epithelial matrix forming a rather dense meshwork. After Kaufmann et al. (1987) they may function as water filled routes across the syncytiotrophoblast which allow unrestricted materno-fetal diffusion of water soluble, lipid insoluble molecules with an effective molecular diameter of about 1.5 nm.

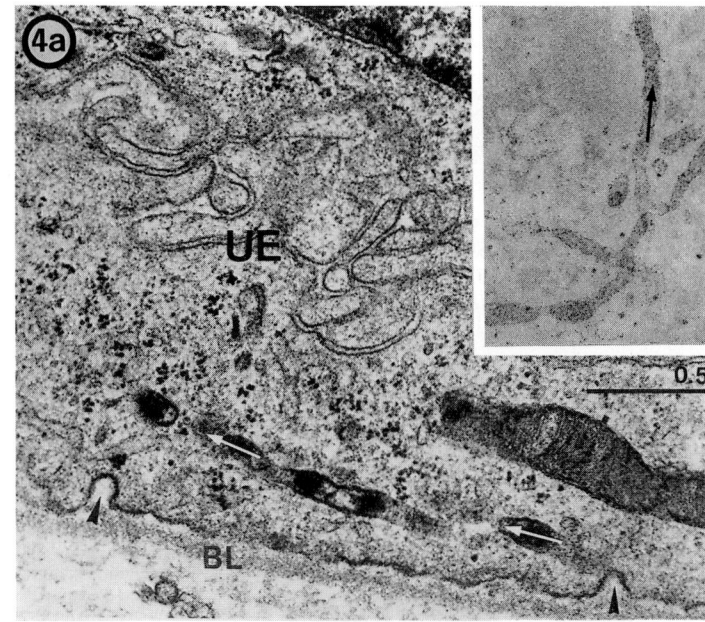

Fig. 4a. Pig uterine epithelium (UE) of placental areola, 111 d.p.c., with transfer tubules (**arrows**) originating from the basal unit membrane (**arrowheads**).
Basal lamina (BL).
Inset: System of ramified transfer tubules stained with periodic acid-thyocarbohydrazide-silver-proteinate.
Courtesy of V Dantzer, Copenhagen (Denmark).

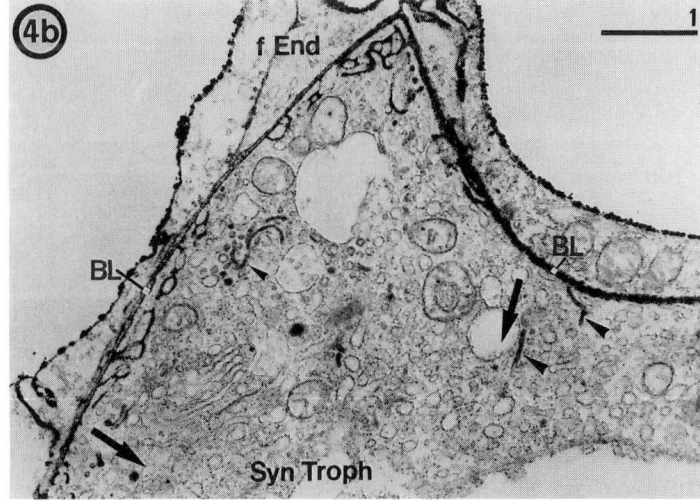

Fig. 4b. Guinea pig placenta: interhemal membrane, near term. Lanthanum hydroxide tracer penetrated (**arrows**) from intercellular clefts of fetal capillary endothelium (fEnd) and basal lamina (BL) into trans-trophoblastic channels (**arrowheads**) of syncytiotrophoblast (Syn Troph). Courtesy of P Kaufmann, Aachen (FRG).

5. Substance transfer by migration of epithelial cells

Trophoblast giant cells or **binucleate cells** develop during implantation of the embryo in bovidae, as shown in the cow (Leiser, 1975; Wooding, 1982a). These cells migrate towards the contact zone of trophoblast and uterine epithelium, fuse with now cleared up uterine epithelial cells (Fig. $1c_1$) and finally reach the basal lamina of the uterine epithelium (Wooding, 1982a; Wooding and Wathes, 1980; Scheme 2). In this way, secretory granules, enclosing placental lactogen (Wooding, 1981), are transported from the level of chorion to the plasmalemma bounding the endometrial stroma. There they may exocytose, delivering the granule contents close to the blood system.

Scheme 2. Substance transfer through interhemal membrane by migration of trophoblast cells.

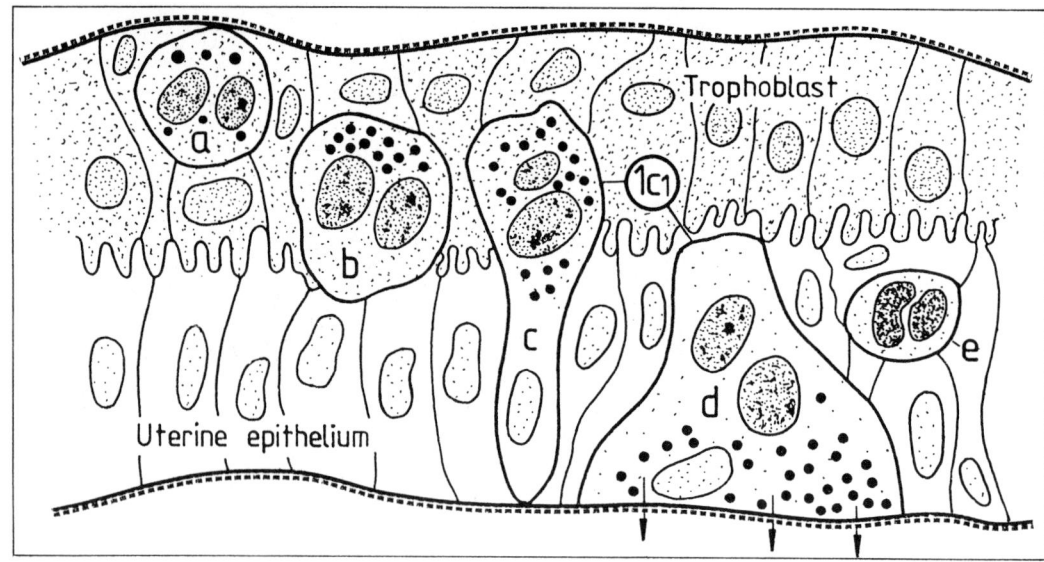

BOVIDAE: Migration of throphoblast binucleate cells by fusion with uterine epithelial cells (**a** to **d**), degeneration of the "hybrid" cells (**e**), release of placental lactogen from secretory granules (**arrows**). Modified from Wooding and Wathes (1980) **J. Reprod. Fert**, 59, 425-430.

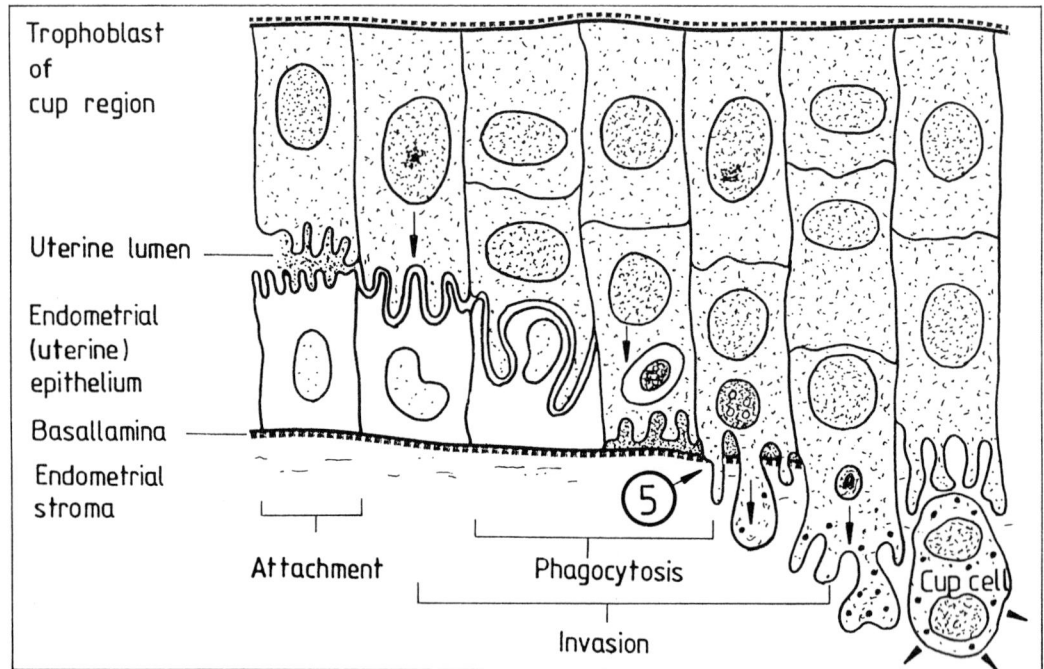

EQUIDAE: Formation and migration of trophoblast cup cells (top to bottom) delivering eCG into the endometrial stroma (**arrowheads**). The numbers in the scheme refer to the corresponding Figures. Modified from Allen et al.(1973) **Anat. Rec.** 177, 485-502.

Fig. 5. Horse placenta, 37 d.p.c. Trophoblast cup cell processes penetrate the remnants of the basal membrane of endometrial epithelium (**arrowheads**) and endometrial stroma cells (**bottom**). Courtesy of WR Allen, Cambridge (England).

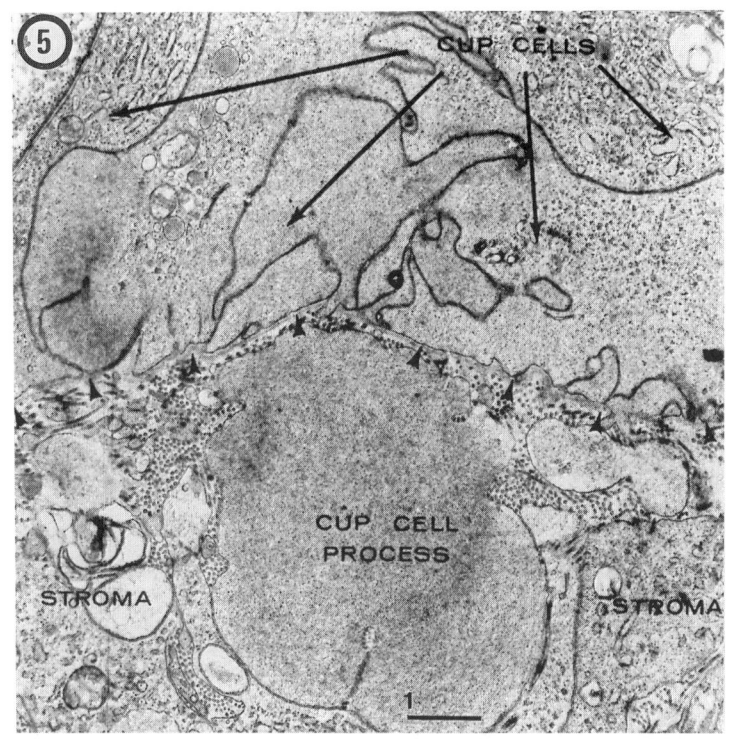

In the equidae, e.g., the horse between day 36 and 80 of pregnancy (Allen et al., 1973), **trophoblast** cells of the girdle or cup region (the annulate chorion girdle at the junction of the allantochorion and regressing yolk sac) become attached to the uterine epithelial cells. They proliferate and with the help of numerous processes invade the uterine epithelium by phagocytosing uterine epithelial cells (Scheme 2). Unlike the bovidae (see above), these migrating trophoblast cells pass the basal membrane and penetrate into the endometrial stroma (Fig. 5). Here, they proliferate again, become enlarged and roundish and are called girdle or cup cells. These cup cells, identifiable by tubules and secretory granules (see also Fig. 6b), may synthesize and release equine chorionic gonadotropin (eCG) (Allen, 1982).

ULTRASTRUCTURAL SIGNIFICANCE OF SUBSTANCE SYNTHESIS (METABOLISM) AND SECRETION (EXOCYTOSIS)

6. Pregnancy proteines and proteohormones

Protein synthesis and secretion are common phenomena in the epithelia of placental interhemal membrane (Nelson et al., 1978; Bell, 1988; Martal et al., 1988). However, there are rather few ultrastructural correlates. Only some of these proteinaceous substances are represented here:

- **Uteroglobin** (Hegele-Hartung and Beier, 1986; Fig. 6a), is the first found **pregnancy protein**. It is apically secreted by the **endometrial epithelium** of the pregnant preimplantative or pseudopregnant rabbit with a maximum from day 4 to 6 post coitum (for reviews see Beier, 1978; Beato et al., 1983).

Ultrastructurally, this protein can be located with gold particles after labelling with PAP (peroxidase-antiperoxydase stained with anti-uteroglobulin from the sheep) (Hegele-Hartung and Beier, 1986). Uteroglobulin is progesteron-dependent and there are strong indications to the fact that it works as an inhibitor of the phospholipase A2, an enzyme catalyzing the synthesis of a precursor of prostaglandin $F2_\alpha$ (Miele et al., 1987).

- Synthesis of **proteohormones** can take place in the **cytotrophoblast** as well as in the **syncytiotrophoblast. Equine chorionic gonadotropin** (eCG) synthesis in the girdle or cup cells, from cellular trophoblast origin (see also Scheme 2), is suggested to take place in a tubular system adjacent to the cell membrane (Hamilton et al., 1973; Fig. 6b). In man, however, **human chorionic gonadotropin** (hCG) has been distinctly located in the syncytiotrophoblast (Kawagoe et al., 1978; Morrish et al., 1987; Suemizu et al., 1988). In their recent study, Morrish et al. (1987) used a monoclonal ß-hCG antibody/protein A-gold technique (Fig. 6c). Here, contrary to earlier investigations (Dreskin et al., 1970; Kawagoe et al., 1978), gold particles appear along the classical secretory granule-exocytosis-pathway in placentas at 8 - 10 weeks of gestation, but not at term (Morrish et al., 1987).

- **Placental lactogen** (or chorionic somatomammotrophin) is a peptide hormone with trophic and hormone-like effects in both mother and fetus (Porter, 1980; Freemark et al., 1987). Recently, it has been detected with gold particles using immunocytochemical methods in, e.g., the human syncytial **trophoblast** (Morrish et al., 1988) and in the sheep trophoblast binucleate cells (Wooding, 1981; Morgan et al., 1987; inset of Fig. $1c_1$). During migration of the binucleate cells from the level of trophoblast epithelium to the basal membrane of uterine epithelium (see migration: Scheme 2) the granules become mature containing an increasing amount of gold particles.

7. Steroidhormones

The ultrastructural equivalent of steroidhormones can be detected by characteristic manifestations, such as form, size, amount of organelles and by cytochemical reactions in both epithelia of interhemal membrane, endometrial epithelium and trophoblast (Wynn, 1973; Burgess and Tam, 1978). As shown in Table 1, estrogen dominance generally increases the amount of smooth ER and enlarges the Golgi zone (Dantzer and Svenstrup, 1986), whereas progesterone results in distinct smooth ER and Golgi zone, combined with the occurence of more and larger mitochondria with tendency to tubular cristae, and the appearance of lipid droplets and glycogen deposits (Secchi and Lecaque, 1984).

Fig. 6b. Horse placenta, 50 d.p.c. Trophoblast cup cell (**bottom**), facing glandular epithelium (**top**), shows a system of tubules presumably containing eCG. Courtesy of WR Allen, Cambridge (England).

Fig. 6c. Human syncytiotrophoblast, first trimester. Positive protein A-gold staining of granules, using monoclonal ß-hCG antibody (**arrows**), point to the existence of hCG in dark granules probably dischargeing into the maternal blood space (**MB**). Lipid droplet (**L**), mitochondria with tubular cristae (*). Courtesy of DW Morrish, Edmonton (Canada).

Fig. 6a. Rabbit uterine epithelium, pseudo-pregnant. Uteroglobin, visualized by gold particles immunolabelled to protein A, is packed into vesicles/secretory granules at the transface of the Golgi zone (**bottom**); secretory granules become discharged into the uterine lumen (**top**). Courtesy of C Hegele-Hartung, Aachen (FRG).

<- see page before

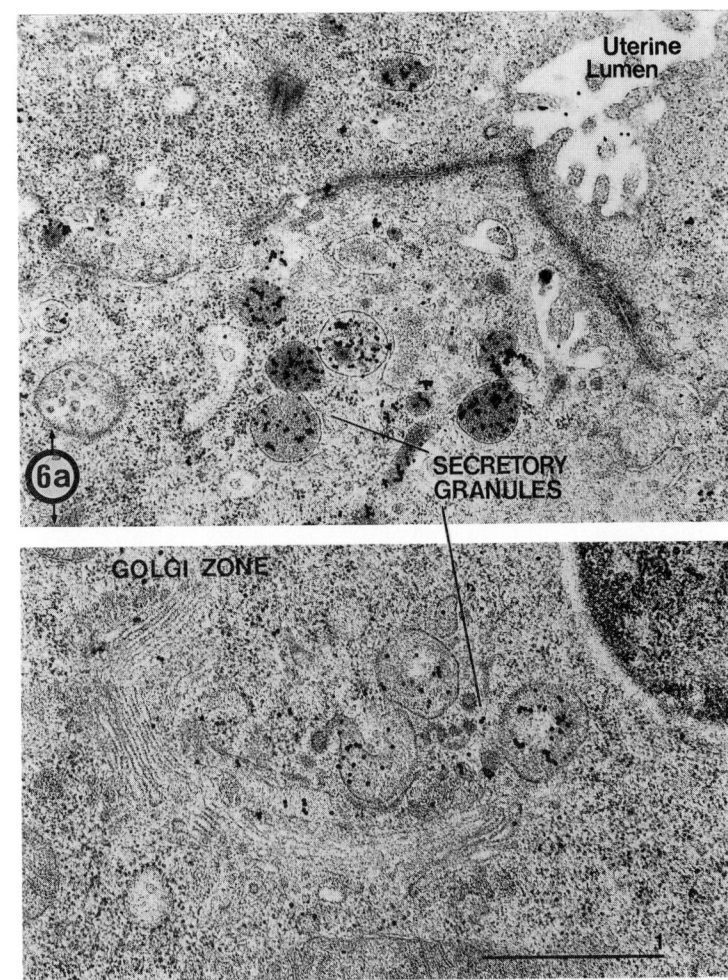

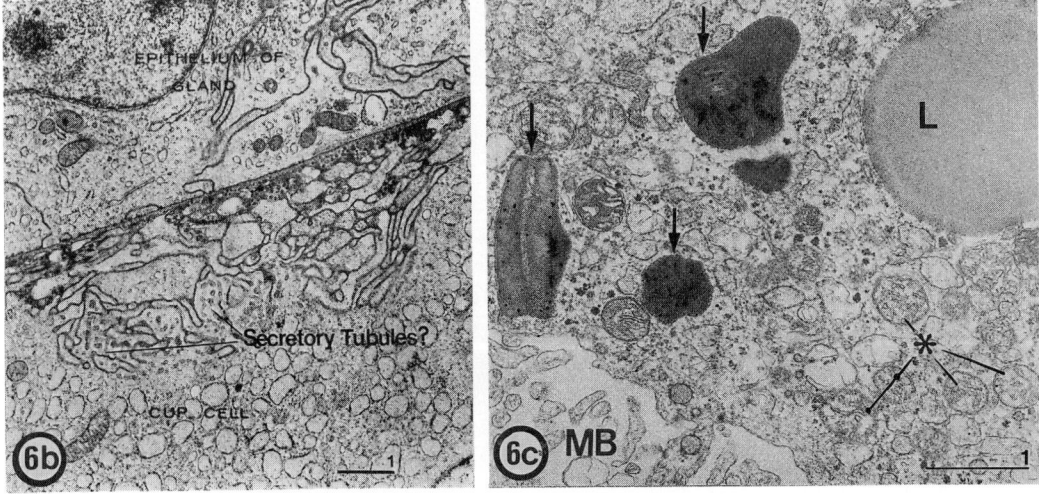

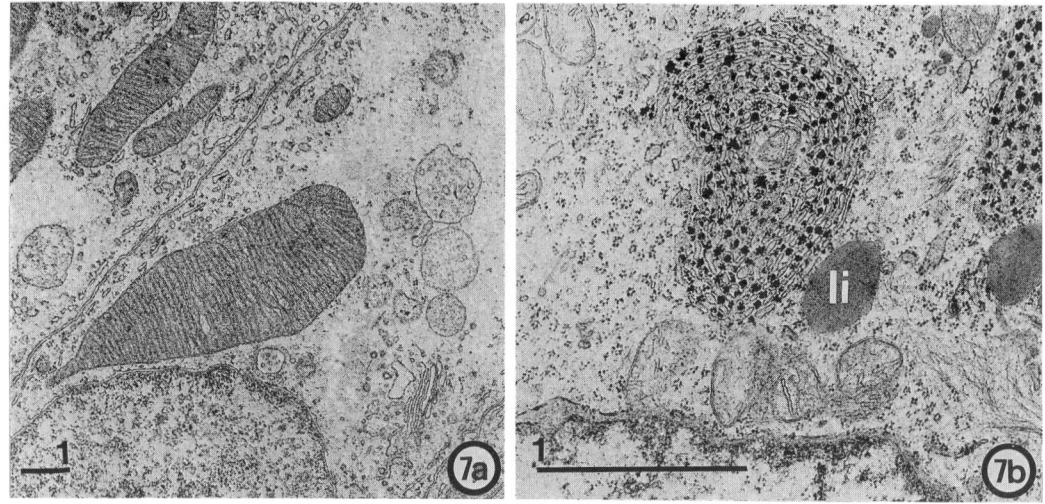

Fig. 7a. Cow uterine epithelium at implantation, 23.5 d.p.c. Giant mitochondrion near the Golgi complex indicates progesterone dominance.

Fig. 7b. Cat uterine epithelium with two glycogen bodies and lipid droplets (li) point to a progesterone condition. Postimplantation 14 d.p.c.

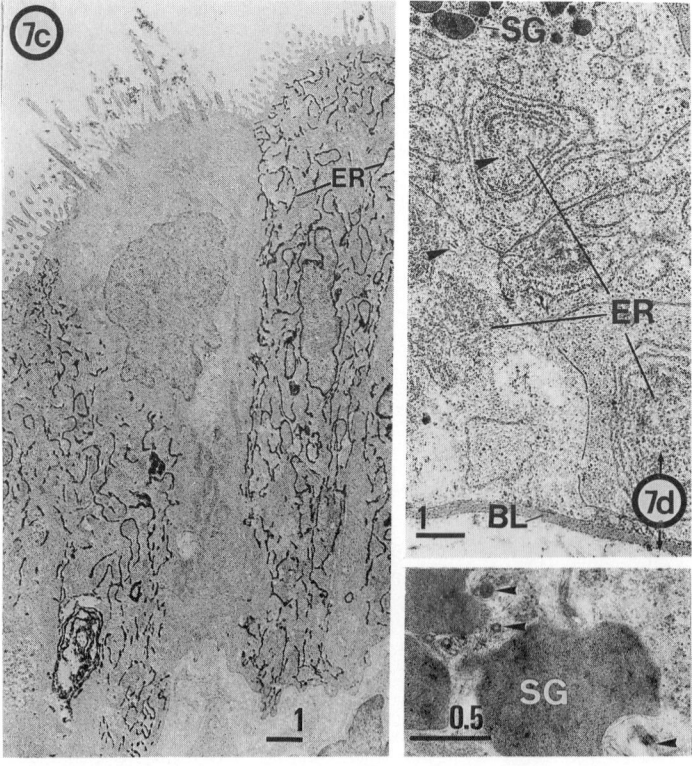

Fig. 7c. Human uterine epithelium, late proliferative phase. Peroxidase activity, related to estrogen synthesis, is evident on the ER and on nuclear envelopes of non-ciliated but not of ciliated cells. Courtesy of Y Ishikawa, Kanazawa (Japan).

Fig. 7d. Pig trophoblast of chorioallantoic fossae, 56 d.p.c. Digitonin-3ß-hydroxysterol complexes, magnified at bottom (**arrowheads**), signify occurence of steroid hormones. Basal lamina (BL), whorls of endoplasmatic reticulum (ER), secretory granules (SG). Courtesy of E Czarnowska, Warsaw (Poland).

Table 1. Increase of size and amount of organelles by dominance of estrogen or progesterone.

estrogen	progesterone
smooth ER	smooth ER
Golgi complex	Golgi complex
	mitochondria (with tubular cristae)
	lipid droplets
	glycogen bodies

In the following, some ultrastructural characteristics of steroid hormone synthesis are shown:

In the endometrial epithelium

Giant mitochondria near the Golgi zone during implantation (day 23½ post coitum) refer to progesterone dominance in the cow (Fig. 7a).

A so-called **glycogen body** in the cat's postimplantative phase (Malassiné, 1974) - a system of smooth ER which spirals between singly aligned glycogen particles - is typically related to lipid droplets and expresses the progesterone condition (compare Secchi et al., 1987; Fig. 7b).

The cytochemical activity of **uterine peroxidase** can be observed in the ER and nuclear envelope, but not in the secretory granules and Golgi zone of the non-ciliated or microvillous cells during the human late proliferative phase (Ishikawa et al., 1988; Fig. 7c). This activity, present in the nuclear envelope, may catalyze the metabolism involving DNA and estrogen (Metzler and Zeeh, 1986).

In the trophoblast

Digitonin 3β-hydroxysteroid complexes indicate the presence of 3β-hydroxy-sterols beeing located in contact with granules and near the smooth ER whorls of porcine **trophoblast cells** covering the fossae of the allantochorion. 3β-hydroxysteroid-dehydrogenase is involved in dehydroepiandrosterone metabolism which takes part in the course of **steroid** (estrogen/progesterone) **biosynthesis** (for review see Czarnowska, 1988; Fig. 7d).

In the **syncytial trophoblast**, e.g., of man (Wynn, 1973; Kaufmann, 1981; Morrish et al., 1987; Fig. 6c), there is a typical organellic pattern significant for steroidogenesis, such as smooth ER, mitochondria with tubular cristae, lipid droplets and secretory granules. Beyond this, however, this syncytium might catalyze many other substances whose ultrastructural correlates are not yet known. - This reflects the tremendous pluripotency of the placenta itself.

CONCLUDING REMARKS Placental structure is known from relatively few mammal species only. In view of the wide variability in structure already observed, it would not be surprising to find still other modifications of the interhemal membrane in other species in the future. The primary challenge facing research efforts in this area will be to relate these structural modifications to functional or physiological activities of the placenta. More experimental and in-vitro studies combined with new techniques, i.e., immunocytochemistry, autoradiographic tracing, X-ray energy dispersive microanalysis, etc., will be successful.

Acknowledgements. We wish to thank Mr Daniel Lisi for his excellent schematic drawings and Mr Simon Koenig for his experienced photographical assistance.

REFERENCES

Allen WR (1982) Immunological aspects of the endometrial cup reactions and the effect of xenogeneic pregnancy in horses and donkeys. **J. Reprod. Fert.,** Suppl 31, 57-94.

Allen WR, Hamilton DW, Moor RM (1973) The origin of equine endometrial cups. II. Invasion of the endometrium by trophoblast. **Anat. Rec. 177,** 485-502.

Amoroso EC (1952) Placentation. In: **Marshall's Physiology of Reproduction,** 3rd ed., vol 2, pp 127-311. Parkes AS (ed.), Longmans, Green and Co, London.

Beato M, Arnemann J, Menne C, Müller H, Suske G, Wenz M (1983) Regulation of the expression of the uteroglobin gene by ovarian hormones. In: **Regulation of gene expression by hormones,** McKerns KW (ed.), pp 151-175. Plenum Press, New York.

Beier HM (1978) Physiology of uteroglobin. In: **Novel aspects of reproductive physiology,** Spilman CH, Wilks JW (eds), vol 8, pp 219-248. SP Medical and Scientific Books, New York, London.

Bell SC (1988) Secretory endometrial/decidual proteins and their function in early pregnancy. **J. Reprod. Fert. Suppl. 36,** 109-125.

Björkman N (1968) Contributions of electron microscopy in elucidating placental structure and functions. **Int. Rev. gen. exp. Zool. 3,** 309-371.

Björkman N (1970) **An atlas of placental fine structure.** Baillière Tindall and Cassell, London.

Björkman N (1973) Fine structure of the fetal-maternal area of exchange in the epitheliochorial and endotheliochorial types of placentation. **Acta anat. 86,** Suppl. 61, 1-22.

Boyd JD, Hamilton WJ (1970) **The human placenta.** W Heffer and Sons, Cambridge (England).

Bremer JL (1916) The interrelations of the mesonephros, kidney and placenta in different classes of mammals. **Am. J. Anat. 19,** 179-209.

Burgess SM, Tam WH (1978) Ultrastructural changes in the Guinea pig placenta, with special reference to organelles associated with steroidogenesis. **J. Anat. 126,** 319-327.

Burton GJ (1982) Placental uptake of maternal erythrocytes: A comparative study. **Placenta 3,** 407-434.

Czarnowska E (1988) The ultrastructure of the trophoblast fossal regions in the pig placenta during pregnancy. **Anat. Histol. Embryol. 17,** 207-225.

Dantzer V (1982) Transfer tubules in the porcine placenta. **Biblthca anat. 22,** 144-149.

Dantzer V, Nielsen MH (1984) Intracellular pathways of native iron in the maternal part of the porcine placenta. **Europ. J. Cell Biol. 34,** 103-109.

Dantzer V, Svenstrup B (1986) Relationship between ultrastructure and oestrogen levels in the porcine placenta. **Anim. Reprod. Sci. 11,** 139-150.

Dantzer V, Leiser R, Kaufmann P, Luckhardt M (1988) Comparative morphological aspects of placental vascularization. **Troph. Res. 3,** 235-260.

Dreskin RB, Spicer SS, Greene WB (1970) Ultrastructural localization of chorionic gonadotropin in human term placenta. **J. Histochem. Cytochem. 18,** 862-874.

Enders AC (1965) A comparative study of the fine structure of the trophoblast in several hemochorial placentas. **Am.J. Anat. 116,** 29-68.

Enders AC, Schlafke S (1971) Penetration of the uterine epithelium during implantation in the rabbit. **Am. J. Anat. 132,** 219-240.

Enders AC, Nelson DM (1973) Pinocytotic activity of the uterus of the rat. **Am.J. Anat. 138,** 277-300.

Faber JJ, Thornburg KL (1983) **Placental physiology.** Structure and function of fetomaternal exchange. Raven, New York.

Freemark M, Comer M, Korner G, Handwerger S (1987) A unique placental lactogen receptor: implications in fetal growth. **Endocrinology 120,** 1865-1872.

Friess AE, Sinowatz F, Skolek-Winnisch R, Träutner W (1980) The placenta of the pig. I. Finestructural changes of the placental barrier during pregnancy. **Anat. Embryol. 158,** 179-191.

Goldstein JL Anderson RGW, Brown MS (1979) Coated pits, coated vesicles and receptor-mediated endocytosis. **Nature 279,** 679-685.

Grosser O (1927) **Frühentwicklung, Eihautbildung und Placentation des Menschen und der Säugetiere.** JF Bergmann, München.

Guillomot M, Betteridge KJ, Harvey D, Goff AK (1986) Endocytic activity in the endometrium during conceptus attachment in the cow. **J. Reprod. Fert. 78,** 27-36.

Hamilton DW, Allen WR, Moor RM (1973) The origin of equine endometrial cups. III. Light and electron microscopic study of fully developed equine endometrial cups. **Anat. Rec. 177,** 503-518.

Hasselager, E (1986) Surface exchange area of porcine placenta: morphometry of anisotropic interdigitating microvilli. **J. Microscopy 141,** 91-100.

Hegele-Hartung C, Beier HM (1985) Immunocytochemical localization of uteroglobin in the rabbit endometrium. **Anat. Embryol. 172,** 295-301.

Hegele-Hartung C, Beier HM (1986) Distribution of uteroglobin in the rabbit endometrium after treatment with an anti-progesterone (ZK 98.734): an immunocytochemical study. **Human Reprod. 8,** 497-505.

Ishikawa Y, Hirai K-I, Uchida H (1988) Ultrastructural localization of human uterine peroxidase. **J. Electron Microsc. 37,** 287-293.

Kaufmann P (1981) Reife Plazenta. In: **Die Plazenta des Menschen,** Becker V, Schiebler TH, Kubli F (eds), chapt. 3, pp 51-100. Georg Thieme Verlag, Stuttgart, New York.

Kaufmann P (1984) Influence of ischemia and artificial perfusion on placental ultrastructure and morphometry. **Contr. Gynec. Obstet. 13,** 18-26.

Kaufmann P, Davidoff M (1977) The Guinea-pig placenta. **Adv. Anat. Embryol. Cell Biol. 53,** 1-91.

Kaufmann P, Schroeder H, Leichtweiss HP, Winterhager E (1987) Are there membrane-lined channels through the trophoblast? A study with lanthanum hydroxide. **Troph. Res. 2,** 557-571.

Kawagoe K, Sugase M, Machinami R (1978) Ultrastructural evidence of human chorionic gonadotropin on trophoblastic surface. **Acta Histochem. Cytochem. 11,** 187-194.

King BF (1982a) Comparative anatomy of the placental barrier. **Biblthca anat. 22,** 13-28.

King BF (1982b) Absorption of peroxidase-conjugated immunoglobulin G by human placenta: an in vitro study. **Placenta 3,** 395-406.

King BF (1988) Permeability pathways in the placenta - Some morphological perspectives. 11th Rochester Troph. Conf. with Europ. Plac. Group, Oct. 1988, Rochester. Abstract booklet, p 138.

King BF, Enders AC (1971) Protein absorption by the Guinea pig chorio-allantoic placenta. **Am. J. Anat. 130,** 409-430.

King BF, Hastings RA (1977) The comparative fine structure of the interhemal membrane of chorioallantoic placentas from six genera of myomorph rodents **Am. J. Anat. 149,** 165-180.

Larsen JF (1961) Electron microscopy of the implantation site in the rabbit. **Am. J. Anat. 109,** 319-334.

Larsen JF (1980) Human implantation and clinical aspects. **Prog. reprod. Biol. 7,** 284-296.

Leiser R (1975) Kontaktaufnahme zwischen Trophoblast und Uterusepithel während der frühen Implantation beim Rind. **Anat. Histol. Embyol. 4,** 63-86.

Leiser R (1982) Development of the trophoblast in the early carnivore placenta of the cat. **Biblthca anat. 22,** 93-107.

Leiser R, Enders AC (1980a) Light- and electron-microscopic study of the near-term paraplacenta of the domestic cat. I. Polar zone and paraplacental junctional areas. **Acta anat. 106**, 293-311.

Leiser R, Enders AC (1980b) Light- and electron-microscopic study of the near-term paraplacenta of the domestic cat. II. Paraplacental hematoma. **Acta anat. 106**, 312-326.

Ludwig KS (1981) Vergleichende Anatomie der Placenta. In: **Die Placenta des Menschen**, Becker V, Schiebler TH, Kubli F (eds), chapt. 1, pp 1-12. Georg Thieme Verlag, Stuttgart New York.

Malassiné A (1974) Localisation ultrastructurale et cytochimique de glycogène dans le placenta de chatte: présence de "corps glycogénique". **C.R. Acad. Sci. D 278**, 629-632.

Malassiné A (1982) Scanning and transmission electron-microscopic observations of the hemophagous organ of the cat placenta. **Biblthca anat. 22**, 108-116.

Martal J, Chêne N, Charlier M, Charpigny G, Camous S, Guillomot M, Reinand P, Bertin J, Humblot P (1988) Protéines trophoblastiques. **Reprod. Nutr. Dévelop. 28**, 1655-1672.

Metcalfe J, Bartels H, Moll W (1967) Gas exchange in the pregnant uterus. **Physiol. Rev. 47**, 782-838.

Metzler M, Zeeh J (1986) Different metabolism by mammalian and horseradish peroxydases in-vitro of steroidal estrogens and their catechol metabolites. **J. steroid. Biochem. 24**, 653-655.

Morgan G, Wooding FBP, Brandon MR (1987) Immunogold localization of placental lactogen and the SBU-3 antigen by cryoultramicrotomy at implantation in the sheep. **J. Cell Sci. 88**, 503-512.

Morrish DW, Marusyk H, Siy O (1987) Demonstration of specific secretory granules for human chorionic gonadotropin in placenta. **J. Histochem. Cytochem. 35**, 93-101.

Morrish DW, Marusyk H, Bhardwaj D (1988) Ultrastructural localization of human placental lactogen in distinctive granules in human term placenta: Comparison with granules containing human chorionic gonadotropin. **J. Histochem. Cytochem. 36**, 193-197.

Mossmann HW (1987) **Vertebrate fetal membranes**. Mac Millan Press Ltd.

Myagkaya GL, Schornagel K, Van Veen H, Everts V (1984) Electron microscopic study of the localisation of ferric iron in chorionic epithelium of the sheep placenta. **Placenta 5**, 551-558.

Nelson DM, Enders AC, King BF (1978) Cytological events involved in protein synthesis in cellular and syncytial trophoblast of human placenta. An electron microscope autoradiographic study of [^{3}H] Leucine incorporation. **J. Cell Biol. 76**, 400-417.

Ockleford CD, Whyte A (1977) Differentiated regions of human placental cell surface associated with exchange of material between maternal and fetal blood. Coated vesicles. **J. Cell Sci. 25**, 293-312.

Parr MB (1980) Endocytosis in the uterine epithelium during early pregnancy. **Prog. reprod. Biol. 7**, 81-91.

Parr MB, Parr EL (1974) Uterine luminal epithelium: Protrusions mediate endocytosis, not aprcrine secretion, in the rat. **Biol. Reprod. 11**, 220-233.

Porter DG (1980) Fetomaternal relationship: The actions and the control of certain placental hormones. **Placenta 1**, 259-274.

Psychoyos H, Mandon P (1971) Etude de la surface de l'épithelium uterin au 5e jour de la gestation. **C.R. Acad. Sci. Paris D 272**, 2723-2725.

Ramsey E (1982) **The placenta**. Human and animal. Praeger Publishers, New York.

Schlafke S, Enders AC (1975) Cellular basis of interaction between trophoblast and uterus of implantation. **Biol. Reprod. 12**, 41-65.

Secchi J, Lecaque D (1984) Effects of progestins and antiprogestins on mitochondria in uterine glandular cells in the rat. A quantitative investigation. **Cell Tiss. Res. 238**, 247-252.

Secchi J, Lecaque D, Tournemine C, Philibert D (1987) Early glygogenesis in the uterine glandular cells of the rabbit induced by progestins: a quantitative

investigation. **Cell Tiss. Res. 248**, 359-364.
Steven DH (1975) Anatomy of the placental barrier. In: **Comparative Placentation**, Steven DH (ed.), chapt 2, pp 25-57. Academic Press, London, New York, San Francisco.
Steven DH, Morriss G (1975) Development of the foetal membranes. In: **Comparative Placentation**, Steven DH (ed.), chapt 3, pp 58-86. Academic Press, London, New York, San Francisco.
Stulc J, Friedrich R, Jiricka Z (1969) Estimation of the equivalent pore dimensions in the rabbit placenta. **Life Sci. 8**, 167-180.
Suemizu H, Osamura Y, Watanabe K (1988) Ultrastructural localization of HCG and subunits in the human placenta and choriocarcinoma cell lines - morphological approach to the secretion pathway. **Acta Histochem. Cytochem. 21**, 265-271.
Van Dijk JP (1988) Regulatory aspects of placental iron transfer - a comparative study. **Placenta 9**, 215-226.
Wooding FBP (1981) Localisation of ovine placental lactogen in sheep placentomes by electron microscope immunocytochemistry. **J. Reprod. Fert. 62**, 15-19.
Wooding FBP (1982a) Structure and function of placental binucleate ('giant') cells. **Biblthca anat. 22**, 134-139.
Wooding FBP (1982b) The role of the binucleate cell in the ruminant placental structure. **J. Reprod. Fert. (Suppl.) 31**, 31-39.
Wooding FBP (1984) Role of binucleate cells in fetomaternal cell fusion at implantation in the sheep. **Am J. Anat. 170**, 233-250.
Wooding FBP, Wathes DC (1980) Binucleate cell migration in the bovine placentome. **J. Reprod. Fert. 59**, 425-430.
Wynn RM (1973) Placental ultrastructure. **Obstet. Gynecol. Ann.**, Meredith corporation 1-24.

Addendum:
Miele L, Cordella-Miele E, Mukherjee AB (1987) Uteroglobin: Structure, molecular biology, and new perspectives on its function as a phospholipase A_2 inhibitor. **Endocr. Rev. 8**, 474-490.

Résumé

ASPECTS ULTRASTRUCTURAUX DU TRANSPORT PLACENTAIRE

Le travail se propose de rechercher les caractéristiques ultrastructurales, significatives des fonctions placentaires, chez divers mammifères. Les échanges ou transferts foeto-maternels sont facilités par différentes spécialisations anatomiques. L'un de ces facteurs est la diminution de l'épaisseur de la "membrane placentaire interhémale". Cet amincissement concerne le trophoblaste mais également l'endothélium des capillaires maternels et foetaux suivant les espèces. Dans le cas du placenta épithéliochorial de ruminant, la fusion entre des cellules trophoblastiques et des cellules de l'épithélium utérin donne naissance à des cellules binuclées, lieux de synthèse des hormones lactogènes. Ce processus permet le transport de ces hormones vers la circulation sanguine maternelle. Dans certains placentas le trophoblaste est envahissant, l'on observe alors la disparition de certaines assises de la "membrane placentaire interhémale" : épithélium utérin, tissu conjonctif, endothélium maternel. Le trophoblaste peut également se réduire à une assise trophoblastique dans le cas du placenta hémomonochorial. Dans les placentas épithéliochoriaux l'interdigitation des microvillosités apicales du trophoblaste et de l'épithélium utérin augmente la surface d'échange foeto-maternelle. L'endocytose placentaire s'effectue par micropinocytose, endocytose par récepteur et erythrophagocytose. Le placenta présente également des images de synthèses et de sécrétions. Les hormones protéiques peuvent être synthétisées dans le cytotrophoblaste (équine chorionic Gonadotropin) ainsi que dans le syncytiotrophoblaste (hCG, hCS). Des structures caractéristiques de la stéroïdogénèse sont également observables.

Localization of cytoskeletal proteins in chorionic villi

C.D. Ockleford, F.M. Bradbury, I. Indans

Department of Anatomy, University of Leicester Medical School, P.O. Box 138, University Rd., Leicester LE1 9HN, UK

The biological concept of signalling is primarily a physiological one. A specific or special sign is given by one element which effects a change in another element. When reduced to molecular terms the study of signalling becomes <u>inter alia</u> the study of a series of interactions frequently involving a cell or group of cells sensing an environmental change, usually affecting the secretion of peptide containing molecules which bind (become ligands) to receptors at target sites in the same neighbouring or distant cells. These are classified as autocrine, paracrine or endocrine events respectively. The receiving cells then change their behaviour in complex ways possibly altering their growth, differentiation or physiology.

All such events as the generation or receiving of signals require transient or lasting changes in the molecular and organellar compositions of the cells and tissues which support the generation and receipt of signals.

Earlier work has told us a considerable amount about the binding of ligand to membrane receptors in the human placental syncytiotrophoblast and the membrane associated molecular architecture which arises to enclose ligand and receptor (Fine and Ockleford, 1984). We know that this procedure alone can concentrate some molecules extensively (Ockleford and Dearden, 1984).

There is reason to believe that important signalling molecules are handled in similar ways by cells and therefore the cell biological study of the machinery involved in receipt, internalisation, processing, recycling and transfer of a variety of molecules by cells assumes importance.

Many of these processes are motile and some depend upon guidance or targetting systems. Generally underpinning these processes in a variety of cell types and locations is a complex cytoskeletal organisation which is relatively well understood in cells in culture but rather poorly described in most whole tissues.

We and others have undertaken the task of mapping elements in human placental chorionic villi (Ockleford, Wakely & Badley, 1981; Ockleford & Wakely, 1982; Ockleford, 1987; Ockleford & Addai, 1988), molar villi (Ockleford et al. 1983; Ockleford et al. 1989), cultured trophoblast (Ockleford & Wakely, 1982; Butterworth & Loke, 1985; Addai, 1987 ; Butterworth et al. 1985; Loke et al 1986) and choriocarcinoma in culture (Ockleford, Dearden & Badley, 1984) during a period in which preparation methods and microscopic techniques have been improving. Our initial data showed best resolution in the cultured cell systems but with the advent of confocal epifluorescence microscopy we are now able to achieve better resolution in suitable sections of whole villi. These studies have shed light on the contribution which cytoskeletal elements make to the elaborate histological organisation of the chorionic villus tree (Kaufmann, Stark & Stegner 1977; Panigel 1986; Kaufmann, Luckhardt, Schweikhart & Cantle, 1987; Burton 1987).

The villous core

The structure of the core region of chorionic villi essentially contains mesenchymally derived cells and extracellular matrix. As expected the cells of the core tissue react with antivimentin antibodies (Fig. 1a).

In the endothelial cells of the cores of villi this anti-vimentin specific fluorescence is especially strong. The observation of 10nm filaments confirm the presence of intermediate filaments composed of vimentin (Ockleford, Wakely, Badley & Virtanen, 1981 ; Fig.1b).

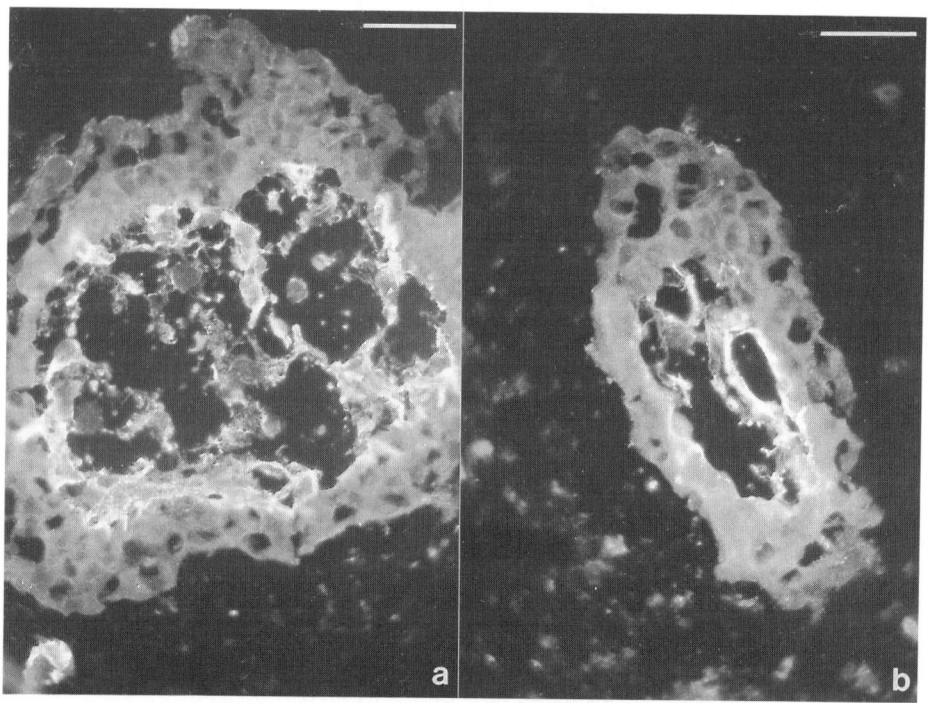

Fig. 1 a) An indirect immunofluorescence micrograph showing the distribution of anti-vimentin reactive elements in a first trimester chorionic villus. The trophoblast shows no fluorescence above background levels whereas there is strong staining of cells in the villous core. Scale bar = 50μm.

b) A similar preparation to that shown in figure 1a. The villous core shows a section through a small vessel. The endothelium of the vessel is strongly reactive. Scale bar = 50μm.

Of more interest is the reaction pattern which we have obtained with antibodies having specificity for the intermediate filament protein desmin which is typical of muscle cells. This is positive for an extensive population of cells in the villous core and is not simply restricted to the smooth muscle of the media of larger vessels (Fig. 2).

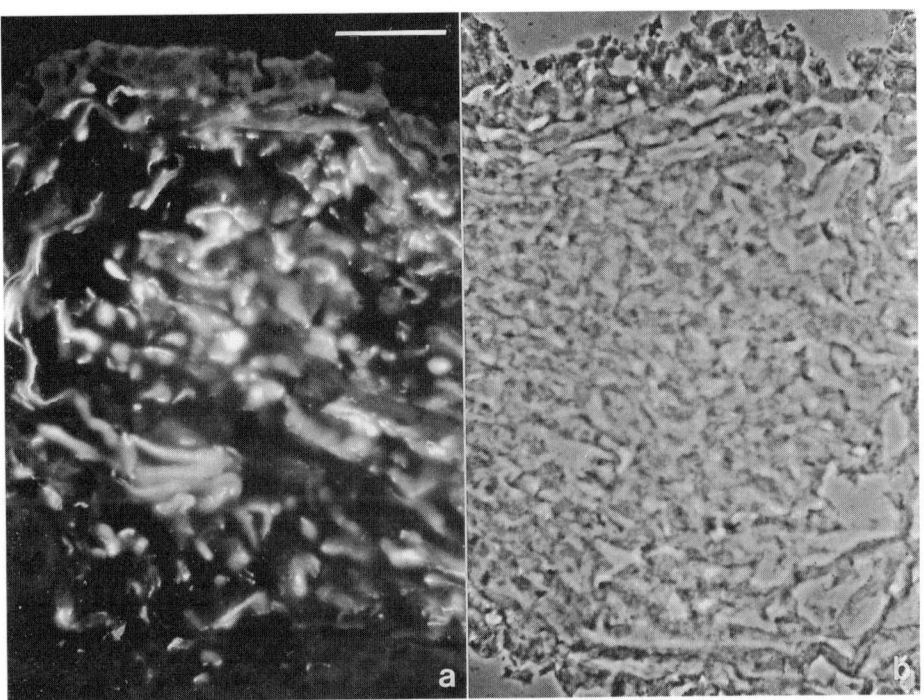

Fig. 2. The indirect immunofluorescence micrograph (a) shows the presence of anti-desmin reactive cells in the villous core but not in the trophoblastic epithelium. Frame (b) shows the same field in phase contrast. Scale bar = 50μm.

The possibility that an extensive population of myofibroblasts populates the villous core has been raised recently on the basis of the observation that they contain dipeptidyl peptidase IV (Feller, Schneider, Schmidt & Parwaresch, 1985). The finding of a muscle specific intermediate filament population in many cells thought previously to be simply mesenchymal or fibroblastic seems to support the concept of extensive potential for contractility in the villous core and outside the vessels raising the possibility that the placenta acts as an auxillary fetal heart.

The trophoblastic epithelium
Concerning syncytiotrophoblast, evidence has been gathered describing an apical layer of cytoplasm containing concentrations of material immuno-reactive with anti-actin and anti-tubulin antibody. In electron micrographs these are seen as regions where microfilaments and microtubules are found together.

An analysis of syncytiotrophoblast has been undertaken using Jamin Lebedeff interference contrast microscopy a technique which has the power to reveal differences in the dry mass and refractive index of different regions of cells and tissues (Ockleford, 1990). These can be observed qualitatively (Fig. 3) but may also be quantitated (Table 1).

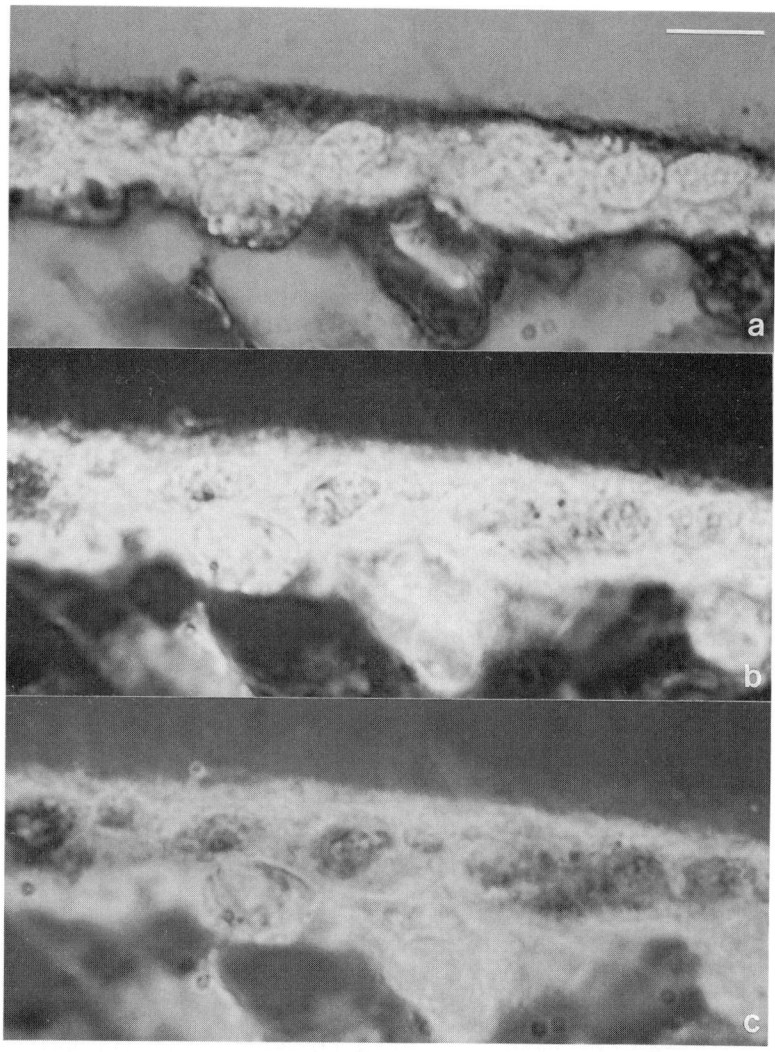

Fig.3 Three panels show the same portion of trophoblastic epithelium prepared as a frozen section unstained and uncontrasted. The preparation has been observed using polychromatic illumination under a Jamin-Lebedeff interference microscope. Alterations in the setting of the microscope allow the differences in refractive index between different parts of the specimen to be observed selectively as different colours. In frame a) the colour contrast between the central syncytiotrophoblast cytoplasm and the apical and basal regions can clearly be seen. Scale bar = 10μm.

	AVERAGE SPECIMEN THICKNESS(μm)	AVERAGE REFRACTIVE INDEX	DRY MASS(GMS/μm^3)
A	4.02	1.3332	8.29×10^{-16}
B	4.29	1.3605	1.52×10^{-13}
C	4.94	1.3920	3.27×10^{-13}
D	5.29	1.3929	3.33×10^{-13}
E	5.05	1.4094	4.24×10^{-13}
F	4.27	1.3624	1.63×10^{-13}
G	4.44	1.3626	1.64×10^{-13}
H	32.31	9.6132	1.56×10^{-12}
I	4.62	1.3733	2.24×10^{-13}
J	0.48	0.0259	1.44×10^{-13}

Table 1 This table shows the quantitative measurements of dry mass, refractive index, and average specimen thickness at different sites in a frozen section of a chorionic villus.
Measurements were made using monochromatic light with the aid of a Zeiss Jamin Lebedeff interference microscope.
{A = Core fluid; B = Brush border; C = Syncytial apical cytoplasm; D = Syncytial basal cytoplasm; E = Syncytial nuclei; F = Fibroblast cytoplasm; G = Cytotrophoblast cytoplasm; H = Σ_x; I = x; J = SD.}

To summarise the relevant conclusions in this context, dry mass is shown to be concentrated in the apical cytoplasm of the syncytium as one might predict from the findings of the structural protein polymer forming microtubules and microfilaments. Mass is, however, also concentrated in the basal cytoplasm of the syncytium and as a band arching over where cytotrophoblast cells contact syncytiotrophoblast. Since desmosomes are seen on electron microscopy in this basal region and these are associated with 10nm intermediate filaments (Fig. 4) it is logical to assume that at least some of the dry mass is contributed by the keratin containing intermediate filaments.

As expected indirect immunofluorescence microscopic observations of 3μm frozen sections of first trimester placenta reacted with anti-keratin antibodies to show strong basal fluorescence. However, in examining such preparations with the Biorad Lasersharp MRC 500 confocal epifluorescence microscope (White, Amos & Fordham 1987) a distinct apical band of immuno-reactive cytoplasm was observed which indicated that the upper dry mass rich layer also contains intermediate filaments of keratin (Fig.5). Now in any <u>cellular</u> epithelium this would be expected because the upper lateral intercellular membranes bear junctional complexes which usually contain desmosomes and rooted in these are networks of 10nm keratin filaments contributing to a series of mechanically integrated terminal webs.

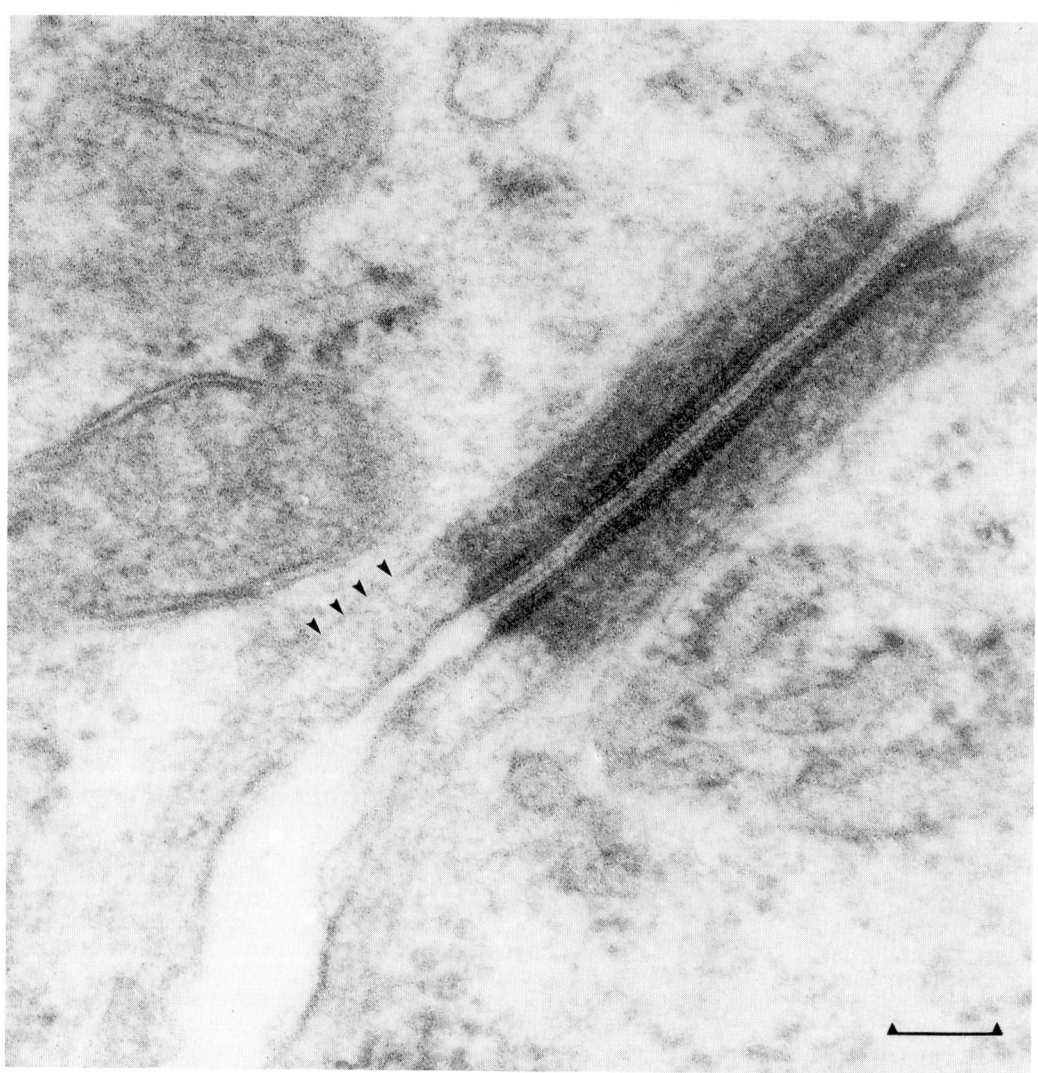

Fig.4 This transmission electron micrograph shows the presence of a classical desmosome in the region of interaction between a cytotrophoblast cell and the syncytiotrophoblast. The opposing membranes from the syncytium (above) and the cytotrophoblast (below) come closer and are held rigidly parallel within the desmosome. The desmosomes contain several well characterised components including desmoglein, desmoplakin and desmocollin. Keratin containing intermediate (10nm) filaments (arrowheads) are rooted in the desmosome. Scale bar = 100nm.

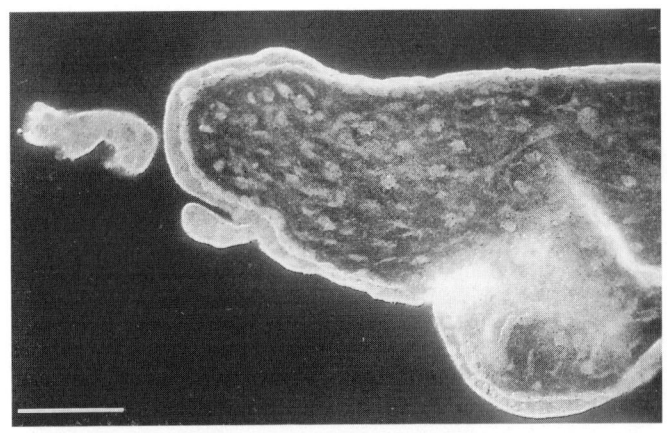

Fig.5 This confocal epifluorescence microscopic image shows the overall distribution of keratin containing syncytiotrophoblast cytoplasm which is concentrated in basal and apical bands. Scale bar = 100µm.

However in syncytium there are no lateral intercellular membranes as we have confirmed previously by microinjection of fluorescent probes into its cytoplasm (Gaunt, Addai and Ockleford, 1986). This raises the interesting question of how the apical 10 nm filaments in the syncytiotrophoblast are co-ordinated.
Now ultrastructural observations of "persistent desmosomes" thought to be the relics of the fusion of cytotrophoblasts with the overlying syncytium have frequently been made (Cavicchia, 1971; De Virgilis et al 1982; Fig, 10a of Dearden and Ockleford 1983). This population is clearly distinct from the population frequently observed in ultrastructural studies of the basal parts of the epithelium (e.g. Fig.4) A new perspective on their distribution and frequency in apical syncytium has recently been contributed by Beham, Denk and Desoye (1988). They used anti-desmoplakin indirect immunofluorescence to show a punctate staining pattern showing the populations of these organelles in the apical cytoplasm and in the junction region between cytotrophoblast and syncytiotrophoblast.

This evidence supports the suggestion that there are sub-apical vesicles or tucks of apical membrane bearing these desmosomes which integrate the apical keratin intermediate filaments (Bradbury & Ockleford, 1990).

Our 0.1µm optical sections obtained using the confocal microscope also show strands of anti-keratin fluorescence rich cytoplasm connecting basal and apical cytoplasm at intervals (Fig. 6).

The Biorad confocal epifluorescence microscope applied to whole villi prepared using anti-keratin in indirect immunofluorescence protocols shows by Z-axis reconstruction of serial optical sections how the upper surface of the villus is partially flattened against the overlying coverslip. By carefully examining optical sections in the upper cortical region a network pattern of anti-keratin immunofluorescence is observed en face (Fig. 8). It is similar to one previously described for anti-tubulin staining. There is good reason for believing that networks of microtubular and intermediate filaments may be positionally inter-related in other situations because drugs which cause depolymerisation of microtubules may

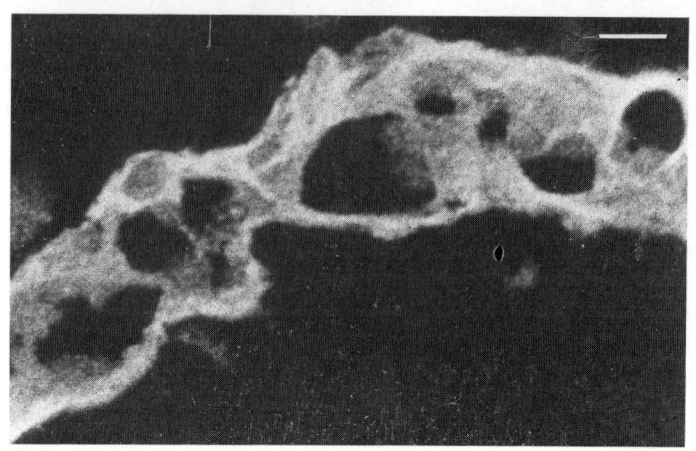

Fig.6 This confocal anti-keratin immunofluorescence micrograph shows strands of keratin containing cytoplasm passing between the basal and apical layers of the syncytium. It also shows a fine strand of keratinaceous material terminating end on and perpendicular to another larger strand. Scale bar = 10μm.

One obvious possible advantage of this configuration is the mechanical integration of basal and apical cytoskeletal domains as shown in schematic form (Fig. 7).

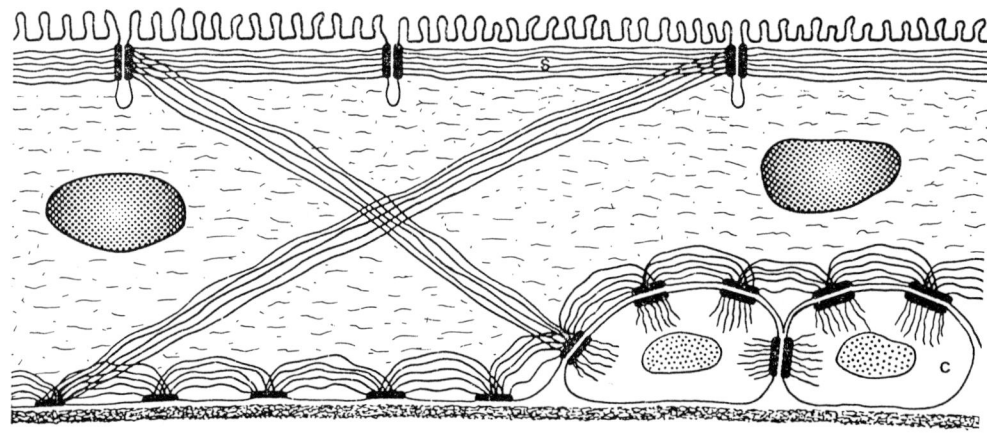

Fig.7 This diagram summarises the distribution of keratin bundles which we have observed using indirect immunofluorescence microscopy. The essential features of interest are the basal and apical concentration of filaments, and the interconnecting strands which pass between the two layers and permit mechanical integration. The absence of lateral intercellular membranes in this atypical (syncytial) epithelium appears to be compensated by apical membrane tucks or vesicles bearing desmosomes.

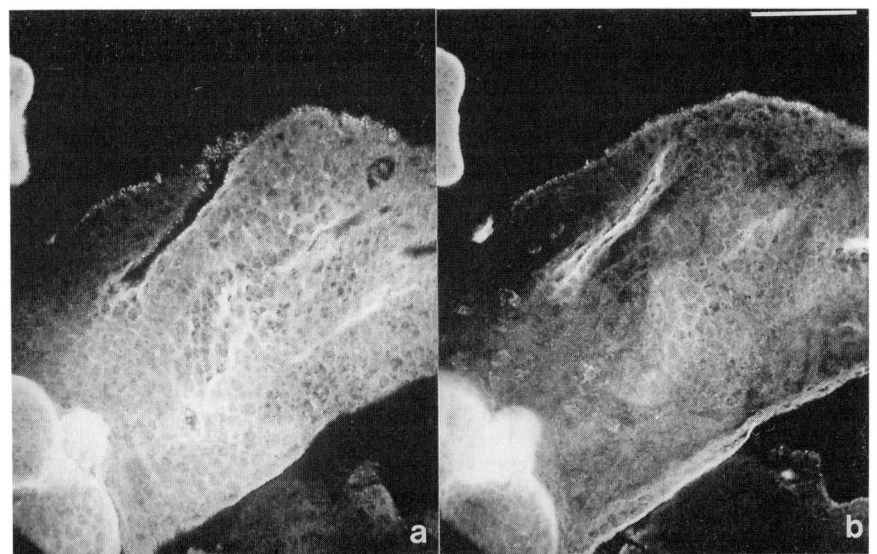

Fig.8 A whole chorionic villus was prepared for indirect immunofluorescence localisation of keratin and compressed under a coverslip. The superficial flattened surface was then optically sectioned at 0.1µm to produce these serial en face views of extremely thin layers of cortical cytoplasm which are 5µm apart. The keratin bundles appear to be arranged in an open lattice configuration. Scale bar = 100µm.

result in **relocation** of intermediate filaments in other cells (Summerhayes, Wong and Chen, 1983).

It is tempting to interpret this information in terms of strands of keratin filaments attaching to membrane invaginations bearing desmosomes at key vertex points arranged over the syncytioskeletal layer. Higher resolution analysis of strands crossing the syncytial epithelial cytoplasm show a termination of one keratin containing bundle on another larger bundle running perpendicular to it.
Therefore, it is possible 1) That keratin fibres may interact with each other laterally to form bundles 2) They may interact in end to side fashion 3) in some as yet undefined way they may interact with microtubules 4) They may attach through desmosomes to membrane. To investigate these relationships in detail and to establish any numerical coincidence between desmosome bearing membrane and vertex points will require careful analysis of dual labelling experiments.

References

Addai, F. K. (1987) Living human placental mitochondria. Ph.D. Thesis. University of Leicester.
Beham, A., Denk, H. & Desoye, G. (1988) The distribution of intermediate filament proteins, actin and desmoplakin in human placental tissue as revealed by polyclonal and monoclonal antibodies. Placenta, 9, 479-492.
Bradbury, F.M., & Ockleford, C.D. (1990)
A confocal and conventional study of the intermediate filaments in chorionic villi. Submitted for publication.

Burton, G. J. (1987) The fine structure of the human placental villus as revealed by scanning electron microscopy. Scanning Microscopy, 1(4), 1811-1828.
Butterworth, B.H. & Loke, Y.W. (1985) Immunocytochemical identification of cytotrophoblast from other mononuclear cell populations isolated from first trimester human chorionic villi. J.Cell Sci., 76, 189-197
Butterworth, B.H. Khong, T.Y., Loke, Y.W. & Robertson, W.B. (1985) Human cytotrophoblast population studied by monoclonal antibodies using single and double biotin - avidin - peroxidase immunocytochemistry. J.Histochem. Cytochem. 33, 977-983.
Cavicchia, J. C. (1971) Junctional complexes in the trophoblast of the human full term placenta. Journal of Anatomy 108, 339-346.
Dearden L., Ockleford, C.D. & Gupta, M. (1983) Structure of human trophoblast correlation with function. In Biology of Trophoblast Y.W.Loke and A.Whyte eds., Elsevier Science Publishers 69-110.
De Virgilis, G., Sideri, M., Fumagalli, G., & Remotti, G. (1982) The junctional pattern of the human villous trophoblast. A freeze-fracture study. Gynecologic and Obstetric Investigation, 14, 263-272.
Feller, A.C., Schneider, H., Schmidt, D. & Pawaresch, M.R. (1985) Myofibroblast as a major cellular constituent of villous stroma in human placenta. Placenta, 6, 405-415.
Fine, R.E. & Ockleford, C.D. (1984) Supramolecular cytology of coated vesicles. Int.Rev.Cytol. 91, 1-42.
Gaunt, M., Addai, F. K. & Ockleford, C. D. (1986). Microinjection of human placenta. I. Techniques. Placenta, 1, 315-324.
Kaufmann, P. Stark, J. & Stegner, H. E. (1977). The villous stroma of the human placenta 1. The ultrastructure of the fixed connective tissue cells. Cell Tiss. Res., 177, 105-121.
Kaufmann, P., Luckhardt, M., Schweikhart, G. & Cantle, S. J. (1987) Cross-sectional features and three-dimensional structure of human placental villi. Placenta, 8, 235-247.
Loke, Y.W. Butterworth, B.H. Margetts, J.J. & Burland, K. (1980) Identification of cytotrophoblast colonies in cultures of human placental cells using monoclonal antibodies. Placenta, 7, 221-231.
Ockleford C.D. (1987) Fine structure of chorionic tissue in health and disease. In : Chorion Villus Sampling. eds. Liu, D.T.Y., Symonds, E.M. & Golbus, M.S. Chapman and Hall, London, 17-28.
Ockleford, C. D. (1989) A quantitative interference light microscope study of human placental chorionic villi. Journal of Microscopy. To be published.
Ockleford, C. D. & Addai, F. K. (1988). Maintenance and development of the structure of chorionic villi in healthy placenta and hydatidiform mole. In: Placental and Endometrial Proteins : Basic and Clinical Aspects. eds. Y. Tomada, S. Mizutani, O. Narita and A. Klopper. V.S.P., Utrecht, The Netherlands. 133-136.
Ockleford C.D. & Dearden, L. (1984) The acquisition of immunity by the fetus. In Immunological Aspects of Reproduction in Mammals. ed. D.B.Crighton, Butterworths, London. 251-263.
Ockleford, C. D. & Wakely, J. (1982). The skeleton of the placenta. Vol. 2 in: Progress in Anatomy, vol. 2, eds. Harrison, R. J. and Navaratnam V., Cambridge University Press, 20-47.
Ockleford, C.D. Dearden, L. & Badley, R.A. (1984) Syncytioskeletons in chorocarcinoma in culture. J. Cell Sci, 66, 1-20.
Ockleford, C. D., Wakely, J. & Badley, R. A. (1981). Morphogenesis of human placental chorionic villi: cytoskeletal, syncytioskeletal and extracellular matrix proteins. Proc. Roy. Soc., 202, 305-316.

Ockleford, C.D., Reti, L.L., Calvert, J.P. & Badley, R.A. (1983) Histopathological diagnosis of hydatidiform mole. Journal of Pathology 140, 51-67.

Ockleford, C.D., Wakely, J., Badley, R.A. & Virtanen, I. (1981) Intermediate filament proteins in human placenta. Cell Biology International Reports, 5, 762.

Ockleford, C. D., Barker, C., Griffiths, J., McTurk, G., Fisher, R. & Lawler, S. (1989) Hydatidiform mole: an ultrastructural analysis of syncytiotrophoblast surface organisation. Placenta, 10, 195-212.

Panigel, M. (1986) Anatomy and morphology. In Clinics in Obstetrics and Gynaecology 13, No. 3, The Human Placenta ed. T.Chard, W.B.Saunders, London pp 421-446.

Summerhayes, I. C., Wong, D. & Chen, L.B. (1983) Effect of microtubules and intermediate filaments on mitochondrial distribution. J. Cell Sci. 61, 87-105.

Wang, T. & Schneider, J. (1987) Cellular junctions on the free surface of human placental syncytium. Archives of Gynecology, 240, 211-216.

White J. G., Amos, W. B., & Fordham, M. (1987). An evaluation of confocal versus conventional imaging of biological structures by fluorescence light microscopy. J. Cell Biol., 105, 41-48.

Acknowledgements

Figures 1-8 are reproduced courtesy of Cambridge University Press and The Journal of Anatomy. Figure 3 and Table 1 are reproduced courtesy of the editor of the Journal of Microscopy. They will be appearing in Ockleford C.D. (1989) as indicated in the references.

CDO and FMB thank DG12 of the EEC and the MRC of Great Britain for grant support.

Résumé

Localisation des protéines du cytosquelette dans la villosité chorionique.

L'organisation structurale du placenta est d'une grande complexité. L'anisométrie du tronc villositaire chorionique est paticulièrement remarquable. L'association, en des lieux précis, de protéines de structure dans le cytoplasme et dans la matrice extracellulaire semble déterminer la forme de la villosité. A l'origine, la morphogénèse de la villosité semble dépendre principalement des protéines du cytosquelette mais par la suite l'effet stabilisateur des protéines fribreuses extracellulaires devient important.
Les techniques de microscopie par immuno-épifluorescence indirecte et de microscopie confocale (microscope confocal laser Biorad Lasersharp MRC 5000) ont permis une meilleure observation de la distribution des filaments de kératine dans des coupes à congelation de trophoblaste.

Il semble que les filaments de kératine participent à la couche de cytosquelette de la face apicale mais également de la face basale du syncytiotrophoblaste. L'observation en microscopie confocale de coupes immunofluorescentes met en évidence des fibres intracytoplasmatiques de kératine, reliant par intervalles les couches de kératine des faces apicales et basales. De ce fait elles doivent permettre l'intégration mécanique des couches basales et apicales du cytosquelette de syncytiotraphoblaste. Cette distribution des filaments de kératine est en rapport avec la mise en évidence par microscopie électronique de jonctions "tight" dans les zônes apicales et basales du cytoplasme. L'utilisation de l'immunofluorescence directe permet également d'accroître notre connaissance sur les filaments intermédiaires à vimentine et à desmine. La desmine est associée à la paroi des vaisseaux sanguins, elle n'est pas présente dans les villosités terminales. La vimentine se situe dans le coeur du tronc villositaire, principalement dans les cellules endothéliales de la paroi des vaisseaux. Sa présence est également signalée dans d'autres cellules mésodermiques du coeur villositaire, des cellules du système réticulé et des fibroblastes.

Proliferation and differentiation of Langhans cells

H. Arnholdt, U. Löhrs

Institut für Pathologie, Medizinische Universität zu Lübeck, Ratzeburger Allee 160, D-2400 Lübeck, Federal Republic of Germany

The trophoblastic proliferation rate in normal placentae, examined immunohistochemically by monoclonal antibody Ki67 or by incorporation of bromodeoxyuridine (BrdU) utilizing monoclonal anti-BrdU, decreases from early gestation to term more markedly when using BrdU, which labels S-phase cells, indicating that the cell cycle time is longer in term placentae. Ki67 and BrdU indices are slightly increased in pre-eclampsia and in intrauterine growth retardation. The portion of differentiated Langhans cells, measured immunohistochemically by their content of cytokeratin 8/18, decreases from 90 % during the first trimester to ca 40 % at term. In contrast to the differentiated cells the number of cytokeratin negative, undifferentiated cells remains constant. Differentiated Langhans cells are increased in diabetes mellitus and pre-eclampsia. In searching for mechanisms regulating these processes we localized EGF receptors not only on the syncytial surface, but also at the cell membrane of some Langhans cells, indicating a possible influence of EGF on these proliferating cells. After short in vitro stimulation by EGF its receptors can also be located at the basal plasma membrane of the syncytium. We interpret this as a transcellular transfer of receptor bound EGF and as evidence for a possible role of EGF in the regulation of trophoblastic proliferation and differentiation.

Histochemical and ultrastructural research of the mesenchymal tissue in the placenta of pig

Z. Bielanska-Osuchowska, A. Kunska

Department of Histology and Embryology, Faculty of Veterinary Medicine, Agricultural University, Warsaw, Poland

The placentae of pig from 18 to 112 day post coitum have been investigated with histochemical and ultrastructural methods. The chorio-allantois is formed as result of the progressive fussion of mesodermal epithelium of chorion and allantois from 18 day pc. The difference between chorionic and allantoic mesenchyme is evident in spite of the fusion into one layer. The chorionic mesenchyme contains the significant amounts of acid glycosoaminoglycans, the allantoic mesenchyme is rich in neutral ones up to 80 day pc. Mesenchyme adjacent to the regular and irregular areaolae undergoes specific differentiation. Apart from typical mesenchymal cells the presence of rounded cells with vacuoles nad heterogenic granules was observed. About the 90 day pc the changes in mesenchymal tissue take place, in extracellular matrix as well as in fibril content. The glycosoaminoglycans can play a mechanical role giving to chorio-allantois rigidity and resistance to pressure and they can participate in the transfer of materials through the placenta and induce the morphogenesis of the folds.

Correlative scanning and transmission electron microscopy of individual placental villi

G.J. Burton, K.W. Thurley

Department of Anatomy, University of Cambridge, Cambridge, UK

Scanning electron microscopy allows extensive areas of the villous surface to be viewed at high resolution with relative ease and speed. Many surface features have been described in the past, but often their interpretation or internal structure has been in doubt. In order to clarify some of these uncertainties correlative microscopy has been performed. After critical-point drying individual intermediate or terminal villi from placentae at varying stages of gestation were stuck onto drawn-out glass pipettes using epoxy adhesive. After scanning microscopy the villi were embedded in resin and sectioned for light and transmission electron microscopy. By using the pipette tip as a marker it was possible to orientate the villus so that areas of interest could be sectioned in the optimal plane. In this way it was possible to identify the different components of villous sprouts at intervals along their length, and to correlate these findings with varying surface appearances. Conversely in term placentae the technique was used to confirm the surface appearances of features such as vasculo-syncytial membranes in perfusion-fixed tissue.

This technique should assist in the interpretation of scanning electron micrographs of both normal and abnormal villi, and may therefore extend the role of scanning microscopy as an investigative technique in placental research.

Calcium in pig placenta : histochemistry and microanalysis

V. Dantzer, M.H. Nielsen*, S. Boisen**, W. Prosbst***

Department of Anatomy, Royal Veterinary and Agricultural University, Copenhagen, Denmark
* Institute of Pathology, University of Copenhagen, Denmark
** Research in Pigs and Horses, National Institute of Animal Science, Foulum, Denmark
*** Carl Zeiss, Oberkochen, Federal Republic of Germany

Physiological calcium in the diffuse epitheliochorial pig placenta was localized by a modified histochemical method using 1% K^+-pyroantimonate in all preparatory steps (1). The presence of calcium in native tissue was verified on cryofixed, ultrathin cryosectioned and cryotransferred material for X-ray energy dispersive microanalysis (Jeol). The calcium content in electron-dense antimonate precipitates was verified on 40 nm thinn sections by electron spectroscopic imaging and electron energy loss spectoscopy (Zeiss). Calcium in the amniochorion, in the allantochorion and in the paraplacenta including the blind ends was determined from day 26-112 of gestation.
Large aggregations of spherite-like bodies in the mesenchyme contained calcium as verified by microanalysis on cryofixed material by the analysis. Calcium was localized in dense areas in mesenchymal cells. The preliminary microanalysis of the tissue prepared with antimonate verified the presence of both calcium and antimonate in the electron-dense precitates located at the lateral plasma membranes and intracellularly in the trophoblast. In the following the location of antimonate precipitates are described from the maternal to the fetal side of the porcine interhemal placenta barrier. Precipitates were located in the maternal epithelium closely related to the folded basal plasma membrane, in large light bodies located basally and occasionally in membranous whorls of large lysosomes. Precipitates were consistently seen at the contact area formed by interdigitating microvilli of maternal and fetal epithelium. In the trophoblast precipitates were seen close to the folded lateral plasma membrane, in the small and large apical vesicles and in the cisternae of the Golgi complex.
The calcium content varied between the different compartments, amniochorion, allantochorion and paraplacenta including the blind ends of the fetal membranes. In allantochorion there was a 7 to 10 fold increase of calcium after day 33 and up to day 60 reaching a level of 4.1% of protein. The higest amount of calcium in the two other compartments was 1,4% of protein at day 50. For the rest of the gestation time the levels were below 1% of protein.
The reliability of calcium localization by a histochemical method using antimonate is demonstrated in well fixed tissue. The results elucidate the intra- and extracellular pathways of calcium in an organ, the porcine placenta, where the transfer of calcium is high due to the fetal demand. The occurrence of spherite-like inclusions in the mesenchym, predominantly from day 30 to day 60, is consistent with a high calcium content in the allantochorion at these stages. The inclusions seem to serve as a temporary physiological storage of loosely bound calcium.

1. Dantzer, V. & Nielsen, M.H.: Localization of calcium in the pig placenta. J. Ultrastruct. & Mol. Struct. Res., 100, 296, 1988.

Supported by the Danish Agricultural and Veterinary Research Council. Grant no. 13-3945.

Distribution of some enzymes in implantation site of pregnant rats*

R. Demir, I. Üstünel

Department of Histology and Embryology, Faculty of Medicine, University of Akdeniz, Antalya, Turkey

The present study aims to describe the distribution and activity of ATP'ase, 5'-nucleotidase, alkaline and acide phosphatases in implantation site of the pregnant rats. Implantation sites from rats at days 12,13,14,15,16 and 17 of pregnancy were sectioned in coronal plane at 10 micros in a microtome-cryostat at $-20°C$. The histochemical methods were aplied for alkaline and acide phosphatases, ATP'ase. 5-nucleotidase activities. The methods used for the demonstration of above enzymes activity revealed the cellular enzymatic reactions observed in different level of the tissue species. The results are summarized on a (+) to (+++) scale according to the gestational age.

Granulated metrial gland cells (GMGCs) are found in the mesometrial decidua, mesometrial triangle, between basal plate of decidua and myometrial layers, in the maternal blood species of the labyrinth from day 12 of pregnancy.

Table I : Distribution and relative activity of ATP'ase, 5'nucleotidase and phosphatases in inplantation site of pregnant rats.

	Alkaline Phosphatase	Acide Phosphatase	ATP'ase	5'-Nucleotidase
Decidua	+++	++	+++	+
GMGCs	++	+	+	+
Mesometrial triangle	+++	++	++	+
Placental labyrinth	+++	+++	+++	+++

(+) = Slightly positive, (++) = Positive, (+++) = Strongly positive.

Examination of cryostat sections of rat placental labyrinth showed that implantation site and fetal membranes reacted in different enzymatic activity (Table I). Decidual and placetal labyrinth vessels were stained for alkaline phosphatase.

ATP'ase activity products appeared at intra- and extra-cellulary in the most area of implantation site and fetal membranes but not in the metrial smooth muscle wall and GMGCs. Acide phosphatase activity varied fairly within the different areas of implantation site. 5'-nucleotidase was slightly present in decidual vessels endothelium but fetal placental labyrinth vessels positive reacted this enyzme localy.

---

* This study was partly supported by Research Fund of Akdeniz University. Grant no.87-01-0103-18.

Light and electron microscopical observations on cellular interaction during initial stages of implantation and trophoblastic invasion in rats*

R. Demir, I. Üstünel, N. Demir

Department of Histology and Embryology, Faculty of Medicine, University of Akdeniz, Antalya, Turkey

Recently attention of investigators is focused on the initation stages of implantation. It is known that the implantation morphologically consists of the establisment of a contact between the blastocyst and the uterinal tissues. During this embryo-maternal adhesion the physiological interaction between them is changing and becoming more differentiated (Denker, In: Novel Aspects of Reproductive Physiology, pp.181-182, 1978). The aim of this study has been the presentation of cellular interaction during the initial stages of implantation process. Implantation sites obtained from rats were studied on day 5,6,7,8,10 and 12 pregnancy to determine the trophoblastic and decidual cells. The cells with cytoplasmic glycoprotein granules appear in decidua during initial stages of pregnancy, are larger irregularly shaped, binucleated and are called as granulated metrial gland cells (GMGCs). Recently confirmed as a bone marrow derived cell, GMGCs precursores pass to the uterus (Stewert, J.Anat. 139, 1984) where they show a differentiation toward the mature from which are identifiable by their numerous cytoplasmic granules. These cells appear to migrate to the placental labyrinth, interact with traphoblast a few days later after implantation. Morphological evidence has suggested that they differentiate from B or T lymphocytes (Clark et al, Am.J.Reprod.Immunol. 5, 1984).

Granulated metrial gland cells (GMGCs) are found in the mesometrial decidua, mesometrial triangle, between basal plate of decidua and myometrial layers, in the maternal blood species of the rat labyrinth from day 12 of pregnancy. GMGCs appeared to interact with adjacent trophoblast, stromal cells and were closely related to mononuclear blood cells. Both the GMGCs and fetomaternal mono- or polynuclear lymphocytes strongly stained with only alcian blue. In conclusion, these cells might play a functional role in the immunological protection of the fetal allograft and a relationship between GMGCs and the development of the fetal placenta and they might play a role in controlling the direction affection of the implantative trophoblast, like a stoping barrier, during the development of the fetal placenta. These cells, perhaps in conjuction with their associated stromal and trophoblastic cells, could participate in removal and remodelling of the extracellular stromal material, interact with other cellular components of the decidua and to stope the trophoblastic invasion.

---

* This study was partly supported by Eczacıbaşı' Scientific Research and Fund of Reward. Grant No. 6211.02/88.

Localization of prostaglandin synthase in human and ovine placental membranes

R.A. Jacobs, J. Oostherhuis, P. Libby, V. Han, J.R.G. Challis

Departments of Obstetrics and Gynaecology and Physiology, University of Western Ontario, Lawson Research Institute, St Joseph's Health Centre, London, Ontario, Canada

Production of prostaglandins by the utero-placental unit is fundamental to pregnancy and parturition. Prostaglandin synthase (PGHS) is a key enzyme in the conversion of arachidonic acid to prostaglandins. We used immunohistochemical techniques to locate the presence of PGHS in placenta and membranes from human tissues collected from either spontaneous labour at term (n=4) or caesarean section at term (n=4). In addition, placenta and membranes were collected from sheep at d45 (n=3), d65, (n=5), d100 (n=7) and labour (n=7; term=d145). Tissues were fixed in Bouin's solution for 4h, washed with several changes of 70% ethanol, embedded in paraffin and sectioned at 3u. The avidinbiotin method (Vectastain) was used for staining, Diaminobenzedine was used as the chromagen and cells were counterstained with haematoxylin. A polyclonal antibody (donated by W. Smith, Michigan State University) was raised in rabbits against PGHS extracted from ram seminal vesicles. Sections from ovine seminal vesicles were used as positive controls and pre-absorption of the primary antibody with ovine PGHS resulted in a loss of staining in tissues. Staining appeared in the subepithelial layers of amnion in both species, but was more intense in sheep. Chorion from human and sheep exhibited staining in trophoblastic cells and in the endothelium of blood vessels. The decidua from human patients stained intensely in both subepithelium and around blood vessels. A similar pattern was observed in the ovine endometrium, but staining was more intense when associated with blood vessels. Staining was apparent in the subepithelial cells of allantois from sheep. No staining was observed in the placenta from either species. There was no significant difference in the pattern of staining between tissues obtained from spontaneous delivery or caesarean section. In the sheep, the intensity of staining appeared to increase with advancing gestational age. In conclusion, we have demonstrated the presence of PGHS in membranes, but not placenta. PGHS is not found in epithelial cells of sheep or human membranes, but is strongly associated with subepithelial cells and blood vessels.

Morphological aspects of transferrin endocytosis by human trophoblast cells

B.F. King, G.N. Fry, G.C. Douglas

Department of Human Anatomy, School of Medicine, University of California, Davis, California 95616, USA

The process of maternal-fetal iron transport involves the endocytic uptake of maternal transferrin (Tf) by trophoblast and subsequent release of iron from the Tf. In order to examine the morphological pathways utilized by trophoblast during Tf procesing, we carried out electron microscopic examination of isolated human trophoblast cells exposed for varying lengths of time to peroxidase - conjugated diferric Tf. Cells were incubated with HRP-Tf for 4 hours at 4°C to maximize binding. Cells were washed to remove unbound label and then reincubated at 37°C for varying lengths of time. Cells were then fixed, incubated in DAB and prepared for EM examination. After incubation for 4 hours at 4°C HRP-Tf was localized on the apical surface microvilli and in coated pits. Upon warming to 37°C HRP-Tf was rapidly internalized in coated vesicles and shortly appeared in endosomes and multivesicular bodies in the cytoplasm. With longer times of incubation HRP-Tf continued to be localized in both large and small vesicles in the cytoplasm. Previous experiments utilizing ^{125}I-labelled Tf (Douglas and King 1988) have shown that Tf was released from cells after short incubation times, but the morphological pathway by which Tf recycling occurred was difficult to determine. In cells exposed to the ionophore monensin for short periods of time no obvious perturbations of the cells or HRP-Tf localization was noted. After longer exposures to monensin, reaction product accumulated intracellularly. These studies demonstrate the potential usefulness of this experimental in vitro system for studying Tf and iron metabolism. Supported by NIH grant HD 11658.

Localization of endogenous immunoglobulin-G in the human placenta

L. Leach, B.M. Eaton, J.A. Firth*, S.F. Contractor

Department of Obstetrics and Gynaecology, Charing Cross and Westminster Hospital Medical School, London, UK
*Department of Anatomy and Cell Biology, St Mary's Hospital Medical School, London, UK

The human fetus acquires passive immunity by transplacental transfer of immunoglobulins of the IgG class. To further ultrastructural studies on possible modes of transfer we have attempted to localise endogenous IgG using ultrathin frozen sections and an indirect immunoelectron methods which utilises colloidal gold.

Immunoreactivity to the rabbit anti-human IgG antibody was seen in the syncytiotrophoblast, interstitial space and fetal endothelium of the placenta. In the syncytiotrophoblast, immunoreactivity was seen at tips, sides and bases of microvilli, in coated and uncoated pits, within coated vesicles and multivesicular bodies in the apical side and within vesicles in the basal side. No immunoreactivity was seen in trans Golgi or CURL regions. Immunoreactivity was observed in the interstitial spaces between the syncytiotrophoblast and the fetal endothelium. Both abluminal and luminal caveolae of the endothelium were found to show immunoreactivity. No immunoreactivity was found in intercellular clefts of the endothelium.

The observed pattern of immunoreactivity is constant with receptor-mediated endocytosis of IgG in the syncytiotrophoblast and vesicular routing across the fetal endothelium.

Interference microscopic mass mapping of human first trimester chorionic villi

C.D. Ockleford

Department of Anatomy, University of Leicester Medical School, P.O. Box 138, University Rd, Leicester LE1 9HN, UK

Using a Jamin-Lebedeff interference microscope an analysis of frozen sections of human first trimester chorionic villi reveals regional differences with subcellular resolution. The evidence indicates compositional differences between villus cell types and shows that the syncytiotrophoblast is differentiated into at least 3 layers, one of which corresponds positionally to the previously described syncytioskeletal layer.

Quantitative measurements have been made of specimen thickness, refractive index and dry mass of regions in the tissue. Local differences in syncytiotrophoblast have been noted with respect to the content of a population of organelles with distinctive optical properties. These may correspond to stored forms of steroid hormone or their precursors.

A «top stage» scanning ultrastructural study of healthy and premalignant trophoblast

C.D. Ockleford, C. Barker, J. Griffiths, G. McTurk, R. Fisher*, S. Lawler*

University of Leicester Medical School, P.O. Box 138, University Rd, Leicester LE1 9HN and
*The Institute of Cancer Research, Royal Marsden Hospital. Fulham Rd, London SW3 6JJ, UK

A scanning ultrastructural examination of a series of 31 hydatidiform mole and 12 healthy placental specimens of similar gestational age has revealed a variety of surface architectures more common in molar tissue. Characteristic paddle-shaped sprouts, ridging of the syncytial maternal oriented surface and microgibbosities are described. These structures are explicable in terms of organellar hyperplasia of cortical cytoskeletal elements found in healthy tissue. These observations constitute specific morphological evidence of involvement of cytoskeletal elements in a condition where aberrant growth control leads to hyperplasia; a further indication that cytoskeletal elements may mediate transformation.

Increased resolution has allowed the description of detailed features such as "caveolar collars" on healthy and molar trophoblast surfaces. These structures are relevant to understanding of receptor mediated endocytosis.

The effects of hypoxia and reoxygenation on barrier thickness of placental villi maintained in organ culture

P. Ong, G.J. Burton

Department of Anatomy, University of Cambridge, Cambridge, UK

For many years it has been assumed on the basis of the work of Tominaga and Page (1966, Am. J. Obs. Gyn. 94, 679-691) that placental villi are able to adapt to hypoxia in vitro by thinning of the villous membrane over a period of a few hours. Furthermore the authors claimed these effects were fully reversible over an equally short time-course upon reoxygenation.

In order to investigate this adaptation further, villi from 10 normal term placentae were maintained in organ culture under 6% O_2 and 21% O_2 respectively for up to 20 hours, with tissue being removed at 5 hourly intervals. A further group of villi were cultured under hypoxic conditions for 10 hours, and then transferred to normoxic conditions for a further 10 hours. Sections 1 um thick were analysed stereologically, and estimates of the volume fractions of the villous components and of the arithmetic and harmonic mean barrier thicknesses obtained.

Barrier thickness was found to be strongly negatively correlated with the volume fraction of the fetal capillaries ($r = -0.67$), although the latter did not vary significantly with either time or oxygen tension. In order to remove this bias caused by collapse of the fetal circulation in organ culture, capillary volume fraction was introduced as a covariate in the subsequent analyses. Two-way analysis of variance revealed that the harmonic mean barrier thickness was significantly lower under hypoxic rather than normoxic conditions ($F = 5.93$, $P\ 0.05$). However an identical pattern was observed with the arithmetic mean, suggesting this reduction was caused by involution of the trophoblast. This was confirmed by the finding that after 10 hours of reoxygenation again both the arithmetic and harmonic mean barrier thicknesses showed a tendency to increase towards the values for the normoxic controls.

These results indicate that any observed thinning of the barrier in hypoxic organ culture is most likely caused by degeneration of the trophoblast rather than a structural adaptation.

8.
Placental pathology : clinical aspects
Pathologie placentaire : aspects cliniques

Induction of germ cell alkaline phosphatase gene expression in human malignant trophoblasts

J.H. Chou, S. Watanabe, T. Watanabe

Human Genetics Branch, National Institute of Child Health and Human Development, National Institutes of Health, Bethesda, Maryland 20892, USA

The existence of multiple alkaline phosphatase (AP) genes in humans has been demonstrated by genetic and biochemical analyses (Harris, 1980; Stigbrand et al., 1982). Recently, four AP genes, placental (Knoll et al., 1988), placental AP-like or germ-cell (Knoll, et al., 1987; Millan and Manes, 1988), intestinal (Henthorn et al., 1988), and tissue-nonspecific liver/bone/kidney (Weiss, et al., 1988), were isolated and characterized. The structures of placental and intestinal AP genes were established by comparison of the sequences of the corresponding genes with the cDNA sequences (Knoll et al., 1988; Henthorn et al., 1988). Both genes are composed of 11 exons and 10 introns, with the introns located at the same positions in both genes. Moreover, placental and intestinal APs have been purified, and the purified enzymes have been extensively characterized (Sussman et al., 1968).

The placental AP-like germ-cell isozyme was discovered by Nakayama and coworkers in 1970 and has subsequently been found in trace amounts in the testis and thymus and in elevated amounts in germ-cell tumors (Rasmuson et al., 1984). However, due to the high allelic polymorphism observed in human placental AP, the existence of the fourth human AP gene was not confirmed until the germ-cell AP gene was isolated and characterized (Knoll et al., 1987; Millan and Manes, 1988; Watanabe et al., 1989). Millan and Manes (1988) cloned and sequenced the germ-cell AP gene, synthesized a peptide that was unique to the AP encoded by the gene, and finally generated antibodies that reacted specifically with the seminoma placental AP-like isozyme (but not the placental AP). However, These authors did not identify or characterize the germ-cell AP mRNA which is essential for the unequivocal verification of functional genes.

We analyzed AP synthesis and regulation in human malignant trophoblasts (choriocarcinoma) which produce a placental AP-like enzyme immunologically similar to the placental AP, but differing from the placental isozyme by a greater sensitivity to heat, EDTA, and the uncompetitive inhibitor L-leucine (Sakiyama et al., 1978), properties shared by the germ-cell AP (Rasmuson et al., 1984).

Moreover, choriocarcinoma cells contain a 2.6-kb AP mRNA whose expression was greatly induced by sodium butyrate (Watanabe et al., 1989). Placental AP mRNA is a 2.8-kb polynucleotide. S1 nuclease mapping using probes derived from placental AP cDNA and ribonuclease protection analysis employing probes derived from the germ-cell AP gene demonstrated that choriocarcinoma express a functional germ-cell AP gene. This was confirmed by the isolation and characterization of a germ-cell AP cDNA clone from a choriocarcinoma cDNA library. The germ-cell AP mRNA is 2487 b in length which contains one major transcription initiation site. Our results showed that malignant transformation of the placenta suppresses expression of the placental AP, but activates expression of the germ-cell AP.

The germ-cell AP gene is also composed of 11 exons and 10 introns, and the predicted primary amino acid sequences of placental and germ-cell APs are 98% homologous (Millan and Manes, 1988; Watanabe et al., 1989). The amino acid sequences of mature placental and germ-cell APs as deduced from cDNA sequences (Kam et al., 1985; Millan, 1986; Watanabe et al., 1989) and the amino- and carboxyl-termini analysis of mature proteins (Ezra et al., 1983; Micanovic et al., 1988; Ogata et al., 1988) indicate that both APs are composed of 484 amino acid residues (Fig. 1). Placental AP differs from the choriocarcinoma germ-cell AP in 9 amino acid residues and differs from the germ-cell AP of Millan and Manes (1988) in 10 amino acid residues. If the two germ-cell APs represent allelic variants, the amino acid substitutions in positions 15, 67, 68, 84, 241, 254, and 429 may account for the increased sensitivity of germ-cell ALP to heat, L-leucine and EDTA (Fig. 1). Site-directed mutagenesis analysis should allow us to identify the amino acid residue (s) responsible for the observed differences in physicochemical properties between the placental and germ-cell APs. These studies will yield valuable information concerning the tertiary structure of the AP and the interactions of key amino acid residues with the active site of AP. The characteristics of mammalian APs are their sensitivities to amino acids as inhibitors. Our structural studies are important to our understanding of the physiological functions of these APs.

```
MQGPWVLLLL GLRLQLSLGI IPVEEENPDF WNRQAAEALG AAKKLQPAQT AAKNLIIFLG      41
DGMGVSTVTA ARILKGQKKD KLGPEITHLAM DRFPYVALSK TYSVDKHVPD SGATATAYLC    101
GVKGNFQTIG LSAAARFNQC NTTRGNEVIS VMNRAKKAGK SVGVVTTTRV QHASPAGTYA    161
HTVNRNWYSD ADVPASARQE GCQDIATQLI SNMDIDVILG GGRKYMFPMG TPDPEYPDDY    221
SQGGTRLDGK NLVQEWLAKH QGARYVWNRT EILDASLDPS VTHLMGLFEP GDMKYEIHRD    281
STLDPSIMEM TEAALLLLSR NPRGFFLFVE GGRIDHGHHE SRAYRALTET IMFDDAIERA    341
GQLTSEEDTL SLVTADHSHL FSFGGYPLRG SSIFGLAPGK ARDRKAYTVL LYGNGPGYVL    401
KDGARPDVTE SESGSPEYRQ QSAVPLDGET HAGEDVAVFA RGPQAHLVHG VQEQTFIAHV    461
MAFAACLEPY TACDLAPRAG TTDAAHPGPS VVPALLPLIA GTLLLLGTAT AP            484
```

Fig. 1. The amino acid sequence of germ-cell AP. The complete amino acid sequence of human choriocarcinoma germ-cell AP was deduced from cDNA and gene sequences (Watanabe et al., 1989). The amino- and carboxyl-terminal signal peptides are underlined, the active site Ser at position 92 is indicated by an asterisk, and nulceotides that differ from the placental AP are boxed.

REFERENCES

Ezra, E., Blacher, R., and Udenfried, S. (1983): Purification and partial sequencing of human placental alkaline phosphatase. Biochem. Biophys. Res. Commun. 116, 1076-1083.
Harris, H. (1980) The Principal of Human Biochemical Genetics. Elsevier, Holland.
Henthorn, P.S., Raducha, M., Kadesch, T., Weiss, M.J., Harris, H. (1988): Sequence and characterization of the human intestinal alkaline phosphatase gene. J. Biol. Chem. 263, 12011-12019.
Kam, W., Clauser, E., Kim, Y.S., Kan, Y.W., and Rutter, W.J. (1985): Cloning, sequencing, and chromosomal localization of human term placental alkaline phosphatase cDNA. Proc. Natl. Acad. Sci. U.S.A. 82, 8715-8719.
Knoll, B.J., Rothblum, K.N., and Longley, M. (1987): Two gene duplication events in the evolution of the human heat-stable alkaline phosphatase. Gene 60, 267-276.
Knoll, B.J., Rothblum, K.N., and Longley, M. (1988): Nucleotide sequence of the human placental alkaline phosphatase gene: evolution of the 5' flanking region by deletion/substitution. J. Biol. Chem. 263, 12020-12027.
Micanovic, R., Bailey, C.A., Brink, L., Gerber, L., Pan, Y.-C.E., Hulmes, J.D., and Udenfriend, S. (1988): Aspartic acid-484 of nascent placental alkaline phosphatase condenses with a phosphatidylinositol glycan to become the carboxyl terminus of the mature enzyme. Proc. Natl. Acad. Sci. U.S.A. 85, 1398-1402.
Millan, J.L. (1986): Molecular cloning and sequencing analysis os human placental alkaline phosphatase. J. Biol. Chem. 261, 3112-3115.
Millan, J.L., and Manes, T. (1988): Seminoma-derived Nagao isozyme is encoded by a germ-cell alkaline phosphatase gene. Proc. Natl. Acad. Sci. USA 85, 3024-3028.
Ogata, S., Hayashi, Y., Takami, N., and Ikehara, Y. (1988): Chemical characterization of the membrane-anchoring domain of human placental alkaline phosphatase. J. Biol. Chem. 263, 10489-10494.
Rasmuson, T., Jeppsson, A., Stigbrand, T. (1984): Placental and placental-like alkaline phosphatases in sera from healthy adults and cancer patients. In Human Alkaline Phosphatases (Stigbrand, T., and Fishman, W.H., eds) pp. 309-315.
Sakiyama, T., Robinson, J.C., and Chou, J.Y. (1978): Multiple forms of alkaline phosphatase in untreated and 5-bromo-2'deoxyuridine-treated choriocarcinoma cells. Arch. Biochem. Biophys. 191, 782-791.
Stigbrand, T., Millan, J.L., and Fishman, W.H. (1982): The genetic basis of alkaline phosphatase isozyme expression. Isoenzymes 6, 93-117.
Sussman, H.H., Small, P.A., Jr., and Cotlove, E. (1968): Human alkaline phosphatase. Immunonochemical identification of organ-specific isoenzymes. J. Biol. Chem. 243, 160-166.
Watanabe, S., Watanabe, T., Li, W.B., Soong, B.-W., and Chou, J.Y. (1989): Expression of the germ cell alkaline phosphatase gene in human choriocarcinoma cells. J. Biol. Chem. 264, 12611-12619.
Weiss, M.J., Ray, K., Henthorn, P.S., Lamb, B., Kadesch, T., and Harris, H. (1988): Structure of the human liver/bone/kidney alkaline phosphatase. J. Biol. Chem. 263, 12002-12010.

Résumé

Expression du gène de la phosphatase alcaline de cellules germinales dans les cellules de choriocarcinome humain.

Comparée à son isozyme placentaire, la phosphatase alcaline (ALP) du choriocarcinome humain se caractérise par une plus grande sensibilité à l'EDTA et à l'inhibition par la L-leucine. De plus les formes entièrement maturées et non glycosylées de l'ALP de choriocarcinome migrent plus rapidement que l'enzyme placentaire correspondante en électrophorèse sur gel de polyacrylamide SDS.

A la différence du placenta humain normal qui exprime un mRNA d'ALP de 2.8 kb, les cellules de choriocarcinome expriment un mRNA d'ALP de 2.6 kb. L'administration de butyrate de sodium aux cellules de choriocarcinome augmente fortement les taux à l'équilibre du mRNA d'ALP du choriocarcinome entrainant une augmentation de l'activité et de la biosynthèse du l'enzyme.

Les analyses à l'aide de la nucléase S1 et de sondes dérivées de cNDA d'ALP placentaire ainsi que des essais de protection à la ribonucléase utilisant des sondes dérivées de gène d'ALP de cellules germinales démontrent que les cellules de choriocarcinome expriment le gène d'ALP de cellules germinales. Le gène d'ALP de cellules germinales code l'isozyme ALP-"like" placentaire qui est principalement exprimé dans le thymus, le testicule et les tumeurs de cellules germinales. Les structures des exons internes (de II à X) des gènes d'ALP de cellules germinales ont été déterminées précedement en se basant sur leur similitude avec les gènes de l'ALP placentaire. Cependant, les frontières des exons I et XI (3'exon) des gènes de l'ALP de cellules germinales n'ont pas été définies du fait de la divergence de séquences entre les deux gènes dans les régions 5' et 3'. Par des expériences de protection à la ribonucléase et d'extension à l'aide d'une amorce il a été montré que l'exon I de ce gène est long de 119 bp et que le mRNA d'ALP de cellules germinales contient un site majeur d'initiation de transcription. L'isolement et la caractérisation de clones de cDNA d'ALP de cellules germinales issus d'une librairie de cDNA de choriocarcinomes traités par le butyrate montrent que le mRNA d'ALP de cellules germinales est d'une longueur de 2487 b et que l'exon XI de ce gène est long de 1135 bp.

Demonstration of HIV-antigens in birth placentae and therapeutic abortions

E. Jimenez, M. Unger, F. Eitelbach, Z. Huang, G. Wagner, M. Vogel, I. Grosch-Wörner*, A. Schäfer**

*Institute of Pathology, Department of Pediatric Pathology and Placentology, Spandauer Damm 130, and * Pediatric Clinic (KAVH), Heubnerweg, and ** Obstetric Clinic, Pulsstrasse, Universitätsklinikum Rudolph Virchow (Charlottenburg), Berlin 19, Federal Republic of Germany*

OBJECT OF STUDY

The diagnosis of intrauterine HIV infection usually takes at least 6 months after birth, mainly because of persistence of maternal antibodies crossing the placenta and because of the probably low viral load in the infected newborn (Scott 1989). Since the placenta constitutes the major link between mother and fetus in intrauterine infection we tried to demonstrate HIV associated antigens 1. in birth placentae as diagnostic contribution in the immediate postpartal assessment of the neonate's risk of infection, 2. in placentae and selected fetal organs of therapeutic abortions in HIV exposed pregnancies to determine the earliest time and frequency rate of HIV infection. Most pathology laboratories receive formalin-fixed tissues. Therefore our goal was to develop a routinely applicable immunohistochemical method for the demonstration of HIV antigens in formalin-fixed, paraffine-embedded placentae and fetal tissues. For other formalin-fixed tissues (i.e. the brain) efficient procedures for the demonstration of HIV associated P24 are already available (Artigas et al. 1989).

MATERIAL AND METHODS

Material. Table 1 summarizes our examination material. 48 of the neonates whose placentae were examined by us are part of the retrospective and prospective Berlin HIV Perinatal Study (Grosch-Wörner et al. 1987).

HIV-exposed	Formalin-fixed	Cryostat Material
Birth Placentae (36th - 40th wk.*)	56	32
Abortions (7th - 27th wk.)		
Placentae	29	16
Fetus	19	12
* wk. = week of gestation		

Table 1: Examination material of 85 HIV exposed pregnancies

Methods. As immunoenzymatic detection system for p24 served the APAAP method (Cordell et al. 1984), the other 3 HIV associated antibodies were stained with the Avidin Biotin Complex (HSU et al. 1981) after blockage of endogenous avidin-binding activity according to Wood and Warnke (1981).

Antibody Versus	Source	Detection System
p24	DuPont Niedrig	APAAP
p17	DuPont	ABC
p18	Seromed Niedrig	ABC
gp41	DuPont	ABC

Table 2: Summary of applied immunohistochemical methods

We used 2 positive and 5 negative control systems (table 3).

Control Tissues	
Positive	Negative
HIV - infected H9 cell line (Pauli, RKI)	„Saline" controls (n = 920)
	H9 cell line (Pauli, RKI)
HIV - lymphadenopathy	Birth placentae (HIV-negative mothers with heroin abuse) (n = 49)
	Normal placentae (n = 20)
	Fetal thymus and spleen (HIV-negative abortions) (n = 10)

Table 3: Summary of Control Tissues

RESULTS

Table 4 comprises the results obtained.

Positive Reaction (Formalin-fixed Material)	anti-p24	anti-p17	anti-p18	anti-gp41
Birth Placentae	13/56	2/56	1/10	–/10
Abortions				
Placenta	12/29	1/29	–/10	–/10
Thymus	2/8	–/3	–/8	–/2
Spleen	2/10	–/3	–/3	–/2
Brain	–/10	–	–	–
Liver	(2)/14	–/2	–/3	–/2
Lymph nodes	–/13	–/1	–/2	–
Lung	–/11	–/1	–/5	–

Table 4: Reactivity of various HIV-associated monoclonal antibodies in placentae and fetal organs of HIV-exposed pregnancies

Anti-p24 yielded the highest number of positive results. In birth placentae as well as placentae from abortions the antigens were localized in Hofbauer cells. Earliest detection of p24 antigen occurred in a placenta of the 13 th week of gestation (Fig. 1a). Also in 2 cases there was a positive staining reaction for p24 in thymus and spleen as well as questionably positive staining histiocytes in the liver.

The demonstration of HIV antigens by the 3 other monoclonals (mabs) succeeded only in 3 placentae although all mabs reacted with our 2 positive controls.

DISCUSSION

HIV-antigen containing Hofbauer cells could be detected in the villous stroma as early as the 13th week of gestation. The Hofbauer cell, therefore, plays a central role in the intrauterine infection chain. Its antigen profile (positive for CD 1 and CD 4, negative for CD 8) defines it as antigen-presenting cell (Jimenez et al. 1989). In analogy to intrauterine rubella infection (Töndury et al. 1966) it can be assumed that the virus localized in infected maternal cells within the intervillous space reaches the villous stroma after trophoblastic damage. Another site of viral transfer could be the maternofetal interface at the basal plate. Yet the demonstration of HIV proteins in Hofbauer cells does not necessarily mean that the fetus is infected too. Considering the short life-span of the placenta the infection may remain dormant in Hofbauer cells until birth without destruction of the host cell.

The antigen loss from placental to fetal tissues is striking. This finding suggests that the viral load in the fetus is extremely low. This theory is supported by the observation of Scott (1989) that HIV can be isolated from more than 80 % of seropositive older children and adults but only in 20 % of infants younger than 6 weeks who later proved to be seropositive.

Therefore, our immunohistochemical examination methods are too crude to permit calculation of the frequency rate of intrauterine infection in the abortion material examined by us.

Also correlation of infected birth placentae and infected children remains problematic. The Berlin Perinatal Study includes several children who except for one positive viral culture remain well by clinical and other laboratory standards. These children may be "non-responders" (Grosch-Wörner et al. 1987) i.e. virus transfer occurred before development of fetal immunocompetence and therefore was not recognized as non-self. On the other hand the findings of Imagawa et al. (1989) suggest that silent infections might be more common than suspected up to now. These investigators found that almost 25 % of a high-risk group remained seronegative during the 7 to 36 months follow-up period after the original isolation of the virus.

Hazeltine (1989) offers a possible explanation for this phenomenon. He found that one of the regulatory genes of HIV, namely nef, retards replication of virus. Silent infection therefore may result from high levels of nef, especially since nef is the first gene made in abundance in infected cells.

In addition to the possibly silent infections 5 of 56 children whose placentae we examined suffer from symptomatic HIV infection, 3 of them showed p24 positive Hofbauer cells.

CONCLUSIONS

1. HIV antigens can be demonstrated in formalin-fixed placentae and fetal tissues.
2. They are found in Hofbauer cells as well as lymphocytes of thymus and spleen.
3. HIV antigens in Hofbauer cells show that the fetus is at risk of infection, they do not prove infection of the fetus per se.
4. The viral load in the embryos and young fetuses of therapeutic abortions seems to be extremely low and therefore by our methods hardly detectable.
5. Of the immunohistochemical procedures used by us the detection of p24 antigen is satisfactory, the demonstration of p17, p18 and gp41 has to be further refined.

ACKNOWLEDGEMENTS

We thank Mrs. C. Lausch, Mrs. B. Ossa and Mrs. U. Lüneburg for their technical assistance, Mrs. R. Tiburtius for photographic work, Mrs. R. Franke for typing the manuscript and Mr. W. Trosch for computer layout.

REFERENCES

Artigas J, Freund K, Grosse G and Niedobitek F (1989): Immunohistochemical demonstration of HIV p24 antigen in formalin-fixed paraffin-embedded sections of brain and spinal cord. Pathologe 10: 60 - 63.

Cordell JL, Falini B, Erber WN, Ghosh AK, Abdulaziz Z, MacDonald S, Pulford KAF, Stein H and Mason DY (1984): Immunoenzymatic labelling of monoclonal antibodies using immune complexes of alkaline phosphatase and monoclonal anti-alkaline phosphatase (APAAP complexes). J Histochem Cytochem 32: 219-229.

Grosch-Wörner I, Koch S and Stück B (1987): Berlin prospective study on 35 HIV (human immuno-deficiency virus) antibody-positive newborns. In 3. Int Conf on AIDS, June 1-5, 1987, Washington DC, Abstr. pp 179.

Haseltine WA (1989): Silent HIV infections. N Engl J Med 320: 1487-89.

Hsu SM, Raine L and Fanger H (1981): The use of antiavidin antibody and avidin-biotin-peroxidase complex in immunoperoxidase technics. Am J Clin Pathol 75: 816.

Imagawa DT, Lee MH, Wolinsky SM et al. (1989): Human immunodeficiency virus type 1 infection in homosexual men who remain seronegative for prolonged periods. N Engl J Med 320: 1458-62.

Jimenez E, Unger M, Vogel M, Lobeck H, Wagner G, Schwiermann J, Schäfer A and Grosch-Wörner I (1988): Morphologische Untersuchungen an Plazenten HIV-positiver Mütter. Pathologe 9: 228-234.

Jimenez E, Unger M, Bläss G and Schäfer A (1989): Immunohistologische Untersuchung zur Bedeutung der Hofbauer-Zelle für die transplazentare HIV-Infektionskette. Arch Gynaek Obstet 245: 544-545.

Scott GB, Hutto C and Parks WP (1989): HIV-1 infection in infants: practical laboratory diagnosis. In AIDS and Obstetrics and Gynaecology, eds Hudson and Sharp, pp. 95-100. Berlin: Springer.

Töndury GT and Smith DW (1966): Fetal rubella pathology. J Pediatr 68: 867-879.

Wood GS and Warnke RA (1981): Suppression of endogenous avidin-binding activity in tissues and its relevance to biotin-avidin detection system. J Histochem Cytochem 29: 1196.

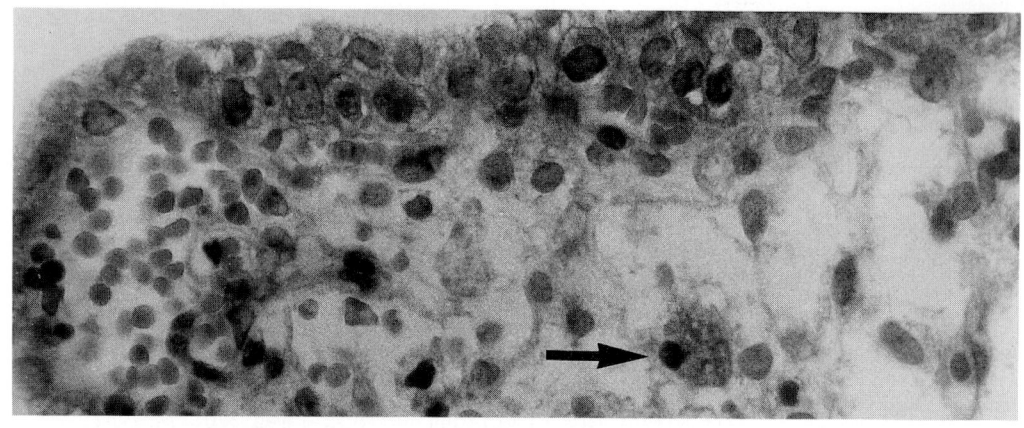
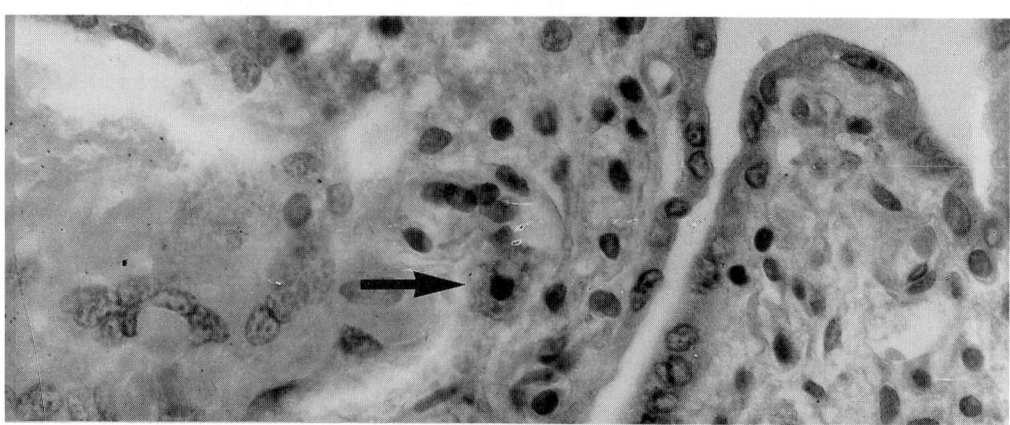
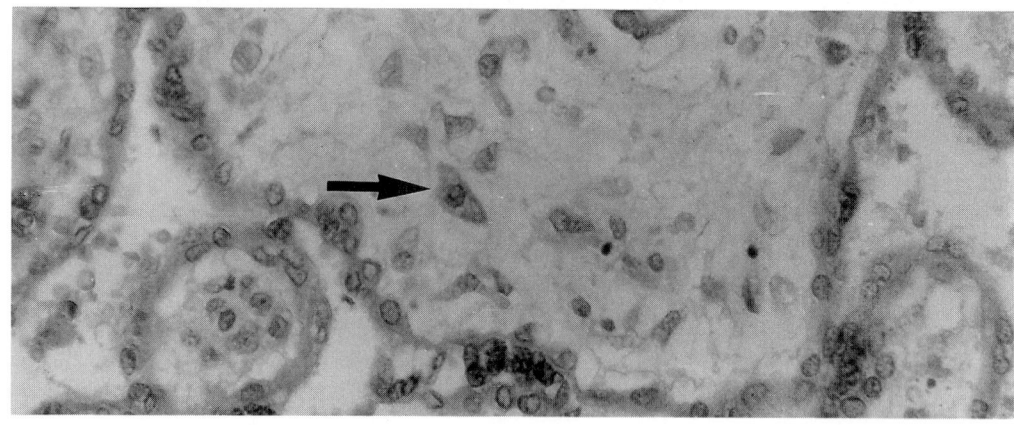

Fig. 1: Intermediate villi with P24 antigen-containing Hofbauer cells
(➤). HE. Original magnification 400 : 1
Fig. 1a: 13th wk of gestation,
1b: 21st wk of gestation
1c: 38th wk of gestation

Résumé

Mise en évidence d'antigènes HIV dans des placentas à terme et dans des avortements thérapeutiques.

Le but du travail est la mise au point d'une technique de "routine" pour la détection d'antigènes HIV dans des placentas et des tissus foetaux fixés au formol et inclus dans la paraffine. Les auteurs ont analysés 51 placentas à terme (17 coupés au cryostat) ainsi que 23 placentas et organes foetaux d'avortements thérapeutiques (11 coupés au cryostat), issus de grossesses exposées au HIV. Les anticorps HIV associés anti-p24, p17 et gp41 ont été utilisés (Du Pont, Niedrig). Une lignée cellulaire H9 infectée par le HIV (Pauli, Robert Koch Institute, Berlin) ainsi que des noeuds lymphatiques comportants une lymphoadenopathie associée au HIV ont été utilisés comme contrôle positifs. Les contrôles négatifs sont d'une part des témoins "salins" et d'autre part des tissus biologiques négatifs. La détection immunoenzymatique s'effectue par la technique de la Péroxydase Anti-Péroxydase ou de l'Avidine Biotine Complexe. Les résultats montrent que sept placentas foetaux fixés au formol, trois foetus et dix placentas à terme ont été détectés positifs par l'anticorps anti-p24, l'un des foetus étant âgé de 19 semaines. Les antigènes étaient localisés dans les cellules de Hofbauer, dans le thymus et dans la rate.

En conclusion l'antigène p24 peut être détecté dans des tissus fixés au formol. La validité clinique de la réaction HIV positive des cellules de Hofbauer est encore discutable. Les réactions peuvent être interprétées comme un risque accru d'infection intra-utérine et non comme une preuve de l'infection du foetus en lui-même. Jusqu'à ce jour, seulement trois enfants de notre étude sont symptomatiques, deux d'entre eux montraient la présence d'antigènes HIV dans leurs placentas.

Placental structure and pathology, including trophoblastic neoplasia : summary of a workshop

H. Fox, G. Roeckelein*, M. Wells**

*Department of Pathology, University of Manchester, Manchester, UK *Pathological Institute, University of Erlangen-Nurnberg, Erlangen, Federal Republic of Germany and **Department of Pathology, University of Leeds, Leeds, UK*

Under the chairmenship of H. Fox, Manchester UK, and organized by G. Roeckelein, Erlangen FRG, and M. Wells, Leeds UK, the workshop "Placental structure and pathology, including trophoblastic neoplasia" tried to be a forum of the morphologists interested in placenta.

RN Laurini, Lausanne Switzerland, showed the significannce of the centrocotyledon haemorrhage as a morphological marker of diabetic placentas and correlated it with the Doppler ultrasound of the umbilical vessels. In study included 20 cases 5 of them showing also a mild hypertonus. In 8 cases with centrocotyledon haemorrhages Laurini found a high pulsatility index in the fetal aorta; all of the cases developed fetal distress in labor. The placental immaturity found in 11 cases was not associated with any flow anomalies. In the discussion the term "centrocotyledon haemorrhage" was questioned because the pictures seemed to show intervillous thromboses.

T Hitschold contributed with a histomorphometric analysis of the placenta in comparison to the Doppler ultrasound of the fetal blood flow. He morphometrically analyzed the fetal vascular space of the placenta and found a decreasing impedance in the fetal placental circulation during normal pregnancy. Since there is a high correlation between the flow velocity waveform and the villous vascularization in cases with a high ratio of placenta weight and birth weight, the Doppler sonography of the umbilical vessels is a non-invasive method to estimate the fetal risk during labor.

BL Sheppard introduced to the changes of the plasminogen activating system in pre-eclampsia and intra-uterine growth retardation. In hypertensive pregnancies and in intrauterine fetal growth retardation the walls of the spirale arteries are containing large amounts. Immunohistochemically plasminogen activator is localized in the syncytiotrophoblast as well as inhibitors of fibrinolysis. In preeclampsia the levels of plasminogen activators in the maternal serum are elevated, while the concentrations in the myometrium are reduced.

Elevated maternal serum levels of alpha-fetoprotein (AFP) are in most cases indicative for fetal abnormalities of the neural tube. E Jauniaux presented a group of 16 cases with elevated AFP associated with umbilical or placental pathology while the fetuses were normal: chorioangioma (3 cases), chorioangiomatosis (2 cases), placental hypertrophy (4 cases) or large subchorial lakes (4 cases), angiomyxomas of the cord (2 cases) and one case with a membranous cyst. The ultrasound examination revealed in most cases jelly-like enlarged placentas with a uniformly decreased echogenity, placental masses with a variable echogenity or giant enlargement of the placentas with multiple sonolucent spaces ("swiss cheese").

Scanning electron microscopy allows extensive areas to be viewed, but in most preparation methods only the surface of the structures can be examined. The interpretation of the surface features in correlation to the internal structures seen in the transmission electron microscope is in doubt. GJ Burton gave a method which allows the correlation of scanning and transmission electron microscopy of individual placental villi. The combination of the three-dimensional view and the analysis of the serial sections of the same placental villus is a very sophisticated method to get a very good impression of the villous structure.

Another scanning electron microscopic study was contributed by G. Roeckelein. 61 placentas of karyotyped spontaneous abortions and 7 of legal abortions were analysed and 5 finding groups established: group A with regularly branching of the villi covered with a velvety surface (4 cases, all legal abortions), group B with large amounts of villous sprouts (12 cases, containing 4 out of 6 of monosomy X), group C with very thick and often bulbous villi (12 cases, containing all triploidies), group D with cylindrical villi covered by a velvety surface (20 cases, 13 out of 25 trisomies) and group E with often slender villi showing a destroyed and fibrinoid covered surface (20 cases, 14 out of 25 euploid cases).

Are pregnancy-associated plasma protein-A (PAPP-A) and CA 125 measurements after IVF-ET possible predictors of early pregnancy wastage ?

P. Bischof, T.M. Mignot*, L. Cedard*

*Department of Obstetrics and Gynaecology, University of Geneva, Switzerland and *INSERM U.166, maternité Baudelocque, Paris, France*

Pregnancy-associated plasma protein-A (PAPP-A), a macromolecular glycoprotein of placental origin, was reported to be depressed in established ectopic pregnancies. CA 125 is a known ovarian cancer marker which has been found elevated during the first trimester of pregnancy and in women with pelvic inflammatory disease. The present study investigated the usefulness of these parameters to predict the outcome of pregnancy in asymptomatic patients with a positive pregnancy test after IVF-ET.

Blood samples (N = 159) were obtained at different periods of time post ET from 39 women, 21 of them experienced a normal pregnancy, 12 had an intrauterine abortion and 6 had an ectopic pregnancy. PAPP-A and CA 125 were measured by radioimmunoassays.

It was observed that from day 30 onwards PAPP-A was significantly increased over non pregnant controls, in normal pregnancies. In the group which experienced a spontaneous abortion, the levels of PAPP-A were significantly lower than in normal pregnancy but higher than in non pregnant controls. In ectopic pregnancy PAPP-A remained at the level of non pregnant controls throughout the entire observation period. CA 125 was significantly increased in all types of pregnancies. However, in 2 cases of hyperstimulation followed by a normal pregnancy and in 4 cases of ectopic pregnancy with signs of peritoneal irritation (hydrosalpinx, ruptured ectopic or salpingitis) the levels of CA 125 were 15 to 50 times higher than in normal pregnancies. PAPP-A levels < 10th percentile when measured after 30 days post ET were an excellent diagnostic parameter for ectopic pregnancy or intrauterine abortion with a sensitivity of 87.5% and a predictive value of disease of 100%. In contrast, CA 125 determinations had no diagnostic values and were only indicative of peritoneal inflammation either in normal or pathological pregnancies. It is concluded that PAPP-A is a good parameter to monitor post-implantation embryo viability in IVF-ET patients.

Placental lipid contents in preterm labor complicated by chorioamnionitis

J. Delmis

Department of Obstetrics and Gynaecology, Medical School of Zagreb, Yugoslavia

Recent studies have suggested that preterm labor and delivery may be associated with intrauterine infection. Microorganisms from vaginal flora might be a source of phospholipase A_2, which can induce the release of arachidonic acid (AA), leading to prostaglandin production. The present study was designed to investigate the effect of chorioamnionitis on the content of AA in decidua free placental tissue. Fifteen decidua free placental samples from nine preterm deliveries without chorioamnionitis were analyzed for free fatty acid content, and that was group 1. Group 2 consisted of 6 preterm deliveries with evidence of chorioamnionitis. In order to be included in this group, membranes had to have been ruptured at least 24 hours prior delivery, and at least 2 of the following clinical signs had to be present : fever more than $38°$ C, uterine tenderness, foul-smelling vaginal discharge and leukocytosis.

Extraction of lipids from placental tissue was performed with thin-layer chromatography on silicic acid-impregnated paper (Gelman Instrument. Co., Ann Arbor, Mich.). This procedure allowed separation of methyl esters of fatty acids from eight to 22 carbons. Peak areas were measured by electronic digital intergrator.

Results were tested for significance by two tailed Student's t-test, with $p<0.05$ providing evidence for significant differences. The correlation between proportion of AA in lipid groups and gestational age were assessed with linear regression and in 2x2 tables, for which significance was determined by two-tailed Fisher's exact test.

There was a highly significant increase in placental AA content (1297 $\pm$181 ug/g placental tissue ; t=5.13 ; p=0.0032), AA in phospholipids (1027$\pm$137.3 ug/g placental tissue ; t=4.48 ; p=0.0065), AA in free fatty acid (168.2$\pm$63.1 ug/g placental tissue ; t=4.24 ; p=0.0082) placentas from pregnancies with chorioamnionitis.

It is clearly showed that AA makes a significantly higher proportion of phospholipids and triglycerides in placental tissue from preterm pregnancies complicated by chorioamnionitis, while the proportion of free fatty acids and cholesterol were not significantly different. The data from this study reveal that AA is present in significantly increased amounts in the placentas of women with chorioamnionitis and preterm delivery, as compared to those women with preterm delivery w/o chorioamnionitis. Our data suggest that AA from amnion and decidua may be taken up by placenta and thus enter the fetal circulation.

Alternatively, the placenta itself represent a possible production and storage depot for prostaglandin precursors, which may be mobilized by various stimuli in cases of intrauterine infection.

Ultrasound/pathologic correlations of placental anomalies associated with elevated maternal serum alpha-fetoprotein and a normal fetus

E. Jauniaux*, G. Moscoso, S. Campbell, D. Gibb, K.H. Nicolaides

*Fetal and Perinatal Pathology Unit, Department of Morbid Anatomy; and Harris Birthright Research Centre for Fetal Medicine, *Department of Obstetrics and Gynaecology; King's College Hospital, King's College School of Medicine and Dentistry, Denmark Hill, London, UK*

Abstract

Prenatal determination of maternal serum alpha-fetoprotein (AFP) levels, aims primarily to identify fetuses with abnormalities affecting the neural tube, the abdominal wall, the kidneys or varying degrees of growth retardation. In 18 cases with elevated maternal serum alpha-fetoprotein referred to our unit for ultrasound examination we identified a variety of unusual or abnormal placental/cord features. These included: A) Gigantic enlargement with multiple sonolucent spaces of different size and shape (n=2; Swiss cheese); B) Placental masses of variable echogenicity (n=4); C) Cord masses with central echodense zone and peripheral echopoor areas (n=2); D) Enlarged placentas with patchy decrease echogenicity (n=6; Jelly-like); E) Large sonolucent spaces with turbulent blood flow surrounded by normal placental tissue (n=4; Placental lakes). After delivery detailed pathologic investigations were performed and the findings were related to the antenatal ultrasound features. Examination of the "Swiss cheese" placentas revealed diffuse mesenchymal hyperplasia of the main villi. The placental masses correspond to chorioangiomas (n=2), infarct (n=1) or subamniotic hematoma (n=1) and the cord masses to angiomyxomas (n=2). The "Jelly-like" placentas were associated with subchorial thrombosis (n=2), massive fibrin deposition (n=1) or generalised enlargement with no obvious abnormalities (n=3). In the group with placental lakes, examination revealed large subchorial thrombosis (n=2), one of which was associated with chronic villitis or no obvious abnormalities (n=2). These findings suggest that, familiarity with the different sonographic appearances of placental anomalies and detailed pathologic correlations, with special attention to the vascular problems, can be helpful in the management of these pregnancies.

Amniotic anomalies and lamellar ichthyosis. A case report

H. Moirot, E. Thomine, M. Martin*, C. Labadie*, C. Fessard, P. Ensel,
A. Pellerin, T. Ducastelle, J. Hemet, M.C. Bourreille, M.C. Boullie

*Laboratoire d'Anatomie pathologique A, C.H.R. Rouen, hôpital Charles-Nicolle, 76031 Rouen Cedex, et *hôpital Le Belvédère, 75130 Mont-Saint-Aignan, France*

The association of polyhydramnios and lamellar ichthyosis is established. A case is presented and polyhydramnios physiopathogeny is discussed.
A 35 year-old woman gravida 2, para 1, was admitted to Belvédère Hospital at 30 weeks' gestation because of onset of labour with polyhydramnios. On admission, pelvic ultrasound demonstrated a flocculent amniotic fluid. At spontaneous rupture of the membranes, amniotic fluid was noticed to be yellowish. The patient delivered a 1700 g. female infant who presented evident lamellar ichthyosis.
The placenta weighed 600 g. and was hydropic. Histologic examination revealed membrane abnormalities : epithelium was pseudo-stratified and eosinophilic granular material was found between some cells. At immunoperoxydase study, this material was decorated by cytokeratine antibody. Ultrastructural changes consisted in sparse microvilli covered by keratinised malpighian skin cells, dilated intercellular canals containing a fibrillar and granular keratin-like material. In this study, it cannot be determined wether these morphologic alterations of amniotic epithelial cells are due to the effects on the amniotic cells of fetal skin disorder and prolonged fluid imbalance or to the cause of abnormal amniotic fluid volume.

The distribution of fibronectin and laminin in the placental bed of patients with different hypertensive disorders of pregnancy

R. Pijnenborg, V. Ballegeer, D. Davey*, M. Hanssens, B. Spitz, A. Tiltman*, L. Vercruysse, A. Van Assche

*Department of Obstetrics/Gynaecology, University of Leuven, Belgium and *University of Cape Town, South Africa*

Endovascular trophoblast invasion into the spiral arteries of the placental bed is restricted to the decidual segments in cases of gestational proteinuric hypertension (preeclampsia). Little is known about mechanisms that control trophoblast invasion. Extracellular matrix proteins such as fibronectin and laminin could be involved, as they are in other invasive processes. In addition fibronectin deposition occurs and can act as a marker substance in different vasculopathies, including atherosclerotic lesions. The observation by Ballegeer et al (1989) of rising blood fibronectin levels in patients that will develop preeclampsia, prompted us to study the distribution of these extracellular matrix proteins in the placental bed of normotensive patients as well as patients suffering from different hypertensive disorders of pregnancy.

Paraffin embedded sections were stained for fibronectin and laminin, using the indirect peroxidase labeled antibody technique of Nakane & Pierce (1966). Spiral arteries undergoing physiological changes were negative for fibronectin. In vessels with subintimal thickening an occasional fibronectin positive lining of endothelium and positive areas in the fibroblastic subintimal cushions could be found. The most intensive staining for fibronectin however was found in the arteries with acute atherosis, i.e. the vascular lesion that is associated with preeclampsia. Preliminary observations suggest a similar distribution pattern for laminin, except for an absence of positive staining in arteries with acute atherosis.

Ballegeer V. et al (1989). Am. J. Obstet. Gynec. 161, 432.
Nakane P.K. & Pierce G.B. Jr. (1966). J. Histochem. Cytochem. 14, 929.

Histological characteristics of placental bed spiral arteries related to the clinical classification of hypertensive disorders in human pregnancy

R. Pijnenborg*, A. Rees, J. Anthony, A. Tiltman, D. Davey, L. Vercruysse, A. Van Assche*

*University of Cape Town, South Africa and *University of Leuven, Belgium*

During normal pregnancy endovascular trophoblast invasion in the spiral arteries up to the myometrial segments is associated with physiological changes in the arteries. Restriction of the normal invasion pattern has been documented in cases of preeclampsia and diverse vasculopathies have been described (Robertson et al, 1986).

Recently a new classification system for hypertensive disorders in human pregnancy has been developed by Davey & MacGillivray (1988). It was the purpose of the present study to relate histological characteristics of spiral arteries to this new clinical classification.

	Number cases	Normal changes	Acute atherosis
Normotensive	6	5	0
Gestational Hypertension	1	0	0
Gestat. HT + Proteinuria	28	0	6
Chronic Hypertension	7	2	0
Chronic HT + Proteinuria	5	0	3
Unclassified Hypertension	1	0	0
Unclass. HT + Proteinuria	2	0	1

Robertson W.B. et al (1986). Am. J. Obstet. Gynec. 155, 401.
Davey D. & MacGillivray I. (1988). Am. J. Obstet. Gynec. 158, 892.

Early spontaneous abortion with different karyotype : microscopic and morphometric investigations

G. Roeckelein, J. Schroeder, R. Ulmer*

*Institute of Pathology, Krankenhausstr. 8-10, 8520 Erlangen and *Human Genetics, University Erlangen-Nürnberg, Federal Republic of Germany*

The most useful classification of spontaneous abortions of early pregnancy depends on cytogenetic chromosomal analysis, but in most cases the only investigation is a pathomorphologic. Most institutes only document intrauterine pregnancy. In order to obtain more information about spontaneous abortion in relation to karyotype, we studied histological sections of aborted placentae.

Material: Placental tissue from 124 spontaneous abortuses which were recovered mainly by curettage were used in the study. Occasionally we used spontaneously expulsed tissue. A part of the tissue was placed as soon as possible in cell culture medium for cytogenetic analysis whilst the remaining tissue was brought to histological investigation. All abortuses date from 6th to 16th week of gestation. 20 placentas dating from the first trimenon which were recoverd from induced abortions (interruptions) performed for social reasons were used for comparison. The induced abortions were not karyotyped.

Methods: The cytogenetic investigation was generally performed as so-called direct preparation of spontaneous metaphases. As a rule, we evaluated mitosis from cell cultures as a cross-check. At least 5 metaphases were analysed photographically.
Placental villi were processed in routine histological procedure. In HE-stained histological slides we semiquantitatively estimated height and condition of the trophoblast layer, stromal density of the villi and the quantity of intervillous fibrinoid. The scale used for trophoblastic height and condition was from 0 (low/bad) to 3 (high/very good), for the fibrinoid according to the quantity from none (0) to 3 (considerable), for the stroma according to the cellular density: hydropic (1), embryonal (2), reticular (3) or condensed (4). The quantity of villous vessels was evaluated according to a scale from 0 (no vessels) to 3 (at least 2 vessels in every villus).
The villi were measured using the morphometry unit TAS plus (LEITZ, Wetzlar FRG). 100 joined and non-selected villi were evaluated per case. The mean values and the median of the areas and the vertical diameters as well as the standard deviation of the means were calculated for further analysis. Statistical procedures were taken from SPSS/PC+ (Statistical Package for Social Sciences).

Results: We found significant differences in all parameters investigated: fibrin, stromal cellularity, height of the trophoblast. Also the means and medians of the areas and the diameters of the placental villi were significant different. To demonstrate the diagnostic value of the parameters multivariate discrimination analysis was used: we found a correct classification up to 73.6 % of the cases. The discrimination of euploidy and monosomy X was difficult, while trisomies were recognized correctly.

Discussion: As a result of direct preparation of the metaphases we were mostly able to avoid contamination with maternal cells which could have lead to erroneous results.

The placenta is a relatively simple, robust and conservative organ which replies to chromosome defects which are lethal for the embryo with little and only quantitative structural deviations.

Our proceeding of measurement is the same as used by Philippe (1973) without considering the orthogonality of the section of the villi. The advantage in our method is that a selection by the investigator does not take place. The results accord to the sizes of the villi reported by Philippe (1973) and Rehder et al. (1989).

The semiquantiative criteria used have already been noted by other investigators; Geisler & Kleinebrecht (1978) reported the surprisingly good trophoblast condition in the trisomies, Vogel (1984) and Rüschoff et al. (1988) an increase of fibrinoid deposits in abortuses with a normal karyotype. Our semiquantitative evaluations of the trophoblastic height and condition, of the fibrinoid deposits and the stromal cellularity showed significant differences. Neither a single criteria nor a combination allowed a completely correct diagnosis. Recent studies of Novak et al. (1989), Minguillon et al. (1989) and Rehder et al. (1989) failed to show any correlation of karyotype with the placental morphology in spontaneous abortuses. Successfull diagnosis rates between 64.6 and 73.6 % do however show that an approximate diagnosis in most cases is possible.

References

1. Geisler M, Kleinebrecht J: Cytogenetic and histologic analysis of spontaneous abortion. Human Genet. 45 (1978) 239 - 51

2. Minguillon C, Eiben B, Bähr-Porsch S, Vogel M, Hansmann I: The predictive value of chorionic vilus histology for identifying chromosomally normal and abnormal spontaneous abortions. Hum. Genet. 82 (1989) 373 - 376

3. Novak R, Agamonalis D, Dasu S, Igel H, Platt M, Robinson H, Shehata B: Histological analysis of placental tissue in first trimester abortions. Ped. Pathol. 8 (1989) 477 - 482

4. Philippe E: Consequences des anomale chromosomiques sur le development. In: Boué A., Thibault C. (ed.): Les accidents chromosomiques de la reproduction. Inserm, Paris 1973.

5. Rehder H, Coerdt W, Eggers R, Klink F, Schwinger E: Is there a correlation between morphological and cytogenetic findings in placental tissue from early missed abortions? Hum. Genet. 82 (1989) 377 - 385

6. Rüschoff J, Chudoba I, Köhler A: Zur Histopathologie von Chorionzotten im ersten Trimenon der Schwangerschaft. Gynäkologe 21 (1988) 104 - 106

7. Vogel M: Pathologie der Schwangerschaft, der Plazenta und des Neugeborenen. In: Remmele W (Hrsg): Pathologie 3. Springer, Berlin 1984.

The correlation of placental and decidual histology with karyotype

C.M. Salafia, J.P. Burns, A.M. Vintzileos, L. Silberman

Departments of Laboratory Medicine and Obstetrics and Gynaecology, Danbury Hospital, Danbury, Connecticut, USA and Department of Obstetrics and Gynaecology, Division of Maternal-Fetal Medicine, University of Connecticut Health Center, Farmington, Connecticut, USA.

The purpose of this study was to correlate decidual and villous histologic characteristics with karyotype classified as euploid, viable aneuploid and nonviable aneuploid. Since October 1983, 384 pregnancy losses of less than 20 weeks gestation had chorionic villi set up for cytogenetic study. Tissues were examined under a dissecting microscope, and the presence of villi was confirmed histologically. Cytogenetic analysis was performed using standard banding techniques 5,6. Of these 55 failed to grow, 163 had normal karyotypes, and 166 were chromosomally abnormal. The chromosomal abnormalities are presented in Table 1. Analysis of the results was confined only to 143 cases which fulfilled the following inclusion criteria: (1) a date of last menstrual period was available to allow calculation of gestational age. In all cases the calculated gestational age was less than 20 weeks, (2) hematoxylin and eosin slides were available and contained a representative sample of chorionic villi, and (3) all cases were clinically symptomatic, and were not electively induced for fetal anomaly. The 143 pregnancy losses included 71 with euploid karyotype (46, XX or 46, XY), 14 with "viable aneuploid" karyotype (such as trisomy 18 or 21) and 58 with nonviable aneuploid karyotypes (such as trisomy 5, 7 or 16) 5. Histologic examination was performed without knowledge of the karyotype category. Characteristics of the villous circulation, placental erythrocytes, villous morphology, trophoblast regressive changes, and decidual vascular histopathology were compared among the 3 karyotype classifications. Decidual vasculitis was observed in 20 of 59 cases with euploid karyotype (33.9%) and 5 of 14 cases with viable aneuploid karyotype (35.7%) as compared to 6 of 49 cases of nonviable aneuploidy (12.2%, $p<0.05$). Increased degrees of hydropic villous degeneration was more frequent in cases of nonviable aneuploid karyotypes (24 of 56, 43%), as compared to euploid (17/71, 24%, $p<0.05$) or viable aneuploid pregnancy losses (2/14, 14%, $p<0.05$). Decidual vasculitis and hydropic changes were not related to each other. Decidual vasculitis may indicate the presence of subclinical maternal-fetal immunopathology, and this histologic feature may be useful in targeting patients at risk for recurrent pregnancy loss on an immunologic basis. This study suggests that histologic examination of products of spontaneous abortion may indicate the likely karyotype of the miscarriage, in terms of euploidy, viable aneuploidy, and nonviable aneuploidy.

Placental pathology of idiopathic intrauterine growth retardation (IUGR) at term

C.M. Salafia, A.M. Vintzileos, L. Silberman

Departments of Laboratory Medicine and Obstetrics and Gynaecology, Danbury Hospital, and Division of Maternal-Fetal Medicine, Department of Obstetrics and Gynaecology, University of Connecticut, Farmington, Connecticut, USA

We examined the placentas of 130 cases of idiopathic IUGR at term (defined by birthweight <10th percentile in our hospital population) and compared them to 177 gestational age matched cases with birthweight >10th percentile (non-IUGR). Placentas were examined after removal of adherent blood, and the following tissue samples examined microscopically: two sections from umbilical cord and extraplacental membranes, four samples from grossly normal villous parenchyma, and one sample of each grossly evident lesion. All gross diagnoses were confirmed microscopically, and diagnoses of chronic villitis, hemorrhagic endovasculitis, and thrombi within placental vessels were made according to standard criteria. Maternal and neonatal data were extracted from medical records. Mean prepregnancy weight and mean maternal weight gain during pregnancy of IUGR cases were both significantly lower than for non-IUGR cases. While lack of prenatal care was associated with IUGR, only 10/130 (7%) cases demonstrated this feature. No other maternal characteristics were associated with IUGR. The placental lesions studied included: placental infarction, chronic villitis, hemorrhagic endovasculitis, and placental vascular thromboses. 1 or more of these lesions were present in 77/130 (60%) of IUGR cases, and 66/177 (37%) of non-IUGR cases ($p<0.05$). 38 of 72 (53%) cases with chronic villitis were IUGR (29% of IUGR cases). 32 of 49 cases (65%) with placental infarction were IUGR cases (25% of IUGR cases). 19 of 32 cases (59%) with hemorrhagic endovasculitis were IUGR cases (15% of IUGR cases). 12 of 17 cases with placental vascular thromboses (71%) were IUGR (9% of IUGR cases). Relationships of all placental lesions to IUGR were independent of each other, although chronic villitis and hemorrhagic endovasculitis were associated with each other. The significant maternal characteristics were independent of placental lesions, except for associations of low pregravid weight and lack of prenatal care with an increased incidence of placental infarction. Cases with decreased birthlength were significantly associated with presence of placental infarction. Cases with chronic villitis had significantly lower ponderal indices than cases without chronic villitis, or with other placental lesions. Our data suggest the following: (1) gross and microscopic placental lesions are significantly associated with growth impairment in a low-risk population (2) different placental lesions show different patterns of related growth failure, suggesting different times of onset of intrauterine stress. Recognition of placental lesions may identify infants at risk for longterm neurodevelopmental complications.

The placenta in human spontaneous abortion

A.E. Szulman

Departments of Pathology, University of Pittsburgh and Magee-Womens Hospital, Pittsburgh, PA, USA

Embryonic demise is the prime-mover in most abortions, to be followed by spontaneous uterine emptying − usually at 7-14 wks menstrual (MA) − after a variable retention period of several weeks. The embryo is usually poorly preserved and may be fully absorbed ("blighted ovum"; empty chorionic sac) during the retention period, while the placenta (maintained by maternal circulation) undergoes secondary changes that reflect embryonic death and allow for its approximate timing. Such assessments rely on microscopic features seen by the pathologist: a) collapse of the small vessels in the chorion and villi with retention of embryonic erythrocytes therein, producing a "peas-in-the-pod" appearance, b) slow fibrosis and/or edema of villous stroma, the latter of rather fast onset, c) the presence and nucleation status of embryo/fetal erythrocytes. RBCs appear in full force in the villi during the 7th week MA; nucleated at first and coming from the yolk sac they give way to the non-nucleated population, derived from the liver − during the 9th and 10th week MA. Attention to the above items allows for a fairly accurate timing of the embryo/fetal exit since upon its occurrence the normal erythrocytic developments come to complete arrest, to the accompaniment of a slow autolysis of the vascular endothelium.

The above method allows for correlation with and supplementation of data from sonography and from cytogenetics (since early embryonic demise correlates closely with aneuploidy); it lends insight into abortions following IVF and CVS, without reliance on state of the embryo proper, the latter most often unavailable for examination.

Dynamics and synthesis of new placental tissue protein PP19

M. Takayama, K. Isaka, Y. Suzuki, H. Funayama, K. Akiya, H. Bohn*

*Department of Obstetrics and Gynaecology, Tokyo Medical College Hospital, Tokyo, Japan and *Behringwerke AG, Marburg, Federal Republic of Germany*

PP19 isolated from the human placenta by H. Bohn (1985), was immunohistochemically localized in nucleus and cytoplasm of villous syncytium, extravillous interstitial trophoblast and maternal leukocyte in human placenta (Takayama et al. 1988) and choriocarcinoma cell (Takayama et al.,1989). In this study, we studied the dynamics and synthesis of PP19 in pregnancy and other diseases. Methods: PP19 concentration was measured by RIA, where purified PP19 (225/245) and anti-PP19 antibody (632ZA) were used; the minimum detectable dose of standard was 1.5 ng/ml. Results: Serum PP19 concentration was increased by hemolysis. In blood cell fractions separated by the Ficoll-Paque/Macrodex method, polymorphonuclear leukocyte fraction contained the highest PP19 concentration. The circulating serum PP19 concentration was detectable but less than 10 ng/ml in nonpregnant women (n=15) and men (n=12). Seminal plasma (n=8) contained 212±99 ng/ml. The mean PP19 concentration was higher in amniotic fluid and retroplacental blood, but lower in umbilical cord blood than that in circulating maternal serum. Serum PP19 concentration in hydatidiform mole, where vesicular fluid contained high PP19 concentration (1154±659 ng/ml), was not statistically higher than normal pregnancy range. The chorionic villous trophoblast contained more PP19 (1643±859 ng/ml in late pregnancy) than decidua, chorion, and amnion. The sera from malignant neoplasma and DIC showed elevated PP19. Ascites (2/5) and fluid content of ovarian tumor (6/8) contained high PP19, ranged from 28 to 185 ng/ml. Placental explants in tissue culture secreted PP19 into medium and was suppressed by 0.1 mM cycloheximide. Our results suggested that PP19 has an extraplacental source, even though the chorionic villous trophoblast may be the main source throughout pregnancy.

Closing lecture

Allocution de clôture

Non-invasive methods in placental research

M. Panigel

Biologie de la reproduction, Université Pierre et Marie Curie, 4 place Jussieu, 75252 Paris cedex 05, France
(Invited closing lecture of the European Placental Group meeting in Dourdan, France)

When it comes to the right to survive of the human conceptus, the first commandment "Thou shalt not kill" brings violent philosophical and political dispute. World War II was however responsible for the holocaust of millions human beings. A stricter second commandment is now intimated to medical research workers and clinicians: "Thou shalt not harm" (the old "Primum non nocere"). Before starting any research project, one must indeed consider the prenatal risks incurred, liable to jeopardize fetal development in utero. The placenta at the fetal maternal interface is at an especially dangerous location to experiment on, either for diagnostic or therapeutic purpose, and even more so, for basic research.

Forty years ago, the difficult problem to solve, was to obtain legal approval to sample early human placentae for investigative purpose. Only in very few countries like Switzerland was abortion authorized by law and provided valuable material for trophoblast research. Nowadays, in many countries, placental specimens can be obtained post abortum, legally but a new legal problem arises: to avoid malpractice accusation for experimental intervention during pregnancy. The peaceful minded, scrupulous medical scientist has to beware of any perinatal risk incurred following the clinical application of any new investigative tool or active substance to a pregnant woman. Administration to animal models is advisable before any clinical trial in the course of human pregnancy.

Here again, another interdiction, a third commandment sometimes difficult to obey, condemns the protocol of many a research project: "Thou shalt inflict no pain" especially to domestic animal friends cherished by "humane" societies who would favor the use of consenting human volunteers rather than accept pain or sacrifice in any protected animal. And when it comes to a human pregnant volunteer prepared to suffer some pain for the sake of Science, (usually a medical worker herself), the question remains "Will mother's and father's O.K. replace the unobtainable consent of the defenceless human fetus?" The answer is "No" of course. Any invasion of the human uterine "black box" for experimental purpose should be refused except when it is needed for the sake of both maternal and fetal organisms linked by the placental area of interchange.

There are only two ways to solve this difficult problem: to perform the research in vitro on tissue of placentae obtained post cesarean or post partum, or to make the choice of an optimal animal model on condition the results of the experiment can be extrapolated to the human species. We are going to consider these two possibilities together with a third choice: the use on both human volunteer and animal models, of non invasive methods harmless and painless.

The Three R's alternatives for animal experiments remain:
1) Reduce the number of animals by choosing the best diagnostic method for the appropriate species
2) Refine the experimental procedure by perfecting painless non invasive methods
3) Replace, when possible, animals by in vitro studies on human tissues.

The pros and cons in the choice of a relevant animal model have been considered by many authors and referred to in two previous papers (Panigel, 1980; 1983). Pigs, sheep, dogs, cats belong to another placental type than the human hemochorial one. Even for animal species with hemo-chorial placentation, electron microscopy of the placental "barrier" as well as metabolic and endocrinologic study of the trophoblast have pointed out many differences of the placental zone of maternal fetal exchange. For placental studies, only a few non human Primate species do approximate the human condition. But here once again, in addition to painful surgery or inhumane captivity, another commandment (the fourth) discourage scientists and stop scientific progress: "Thou shalt not spend too much"(!) The upkeep of Primates in "humane" conditions is indeed most expensive. Only indomitable research workers will ignore these "Dont's" and faithfully pursue their pilgimage towards the promised field of Placentology.

In vitro human material is now freely available in obstetrical labour rooms, and whenever placental material can be used without danger and discomfort to the patient and in sufficiently satisfactory conditions for the research, it should be preferred to any living animal model. When one looks at the list of this pre-congress workshops, at the papers submitted and discussed at the plenary Symposium on Placental Signals as well as to the posters exhibited during the meeting, it becomes obvious that the present trend of research relies mostly on in vitro experimentation using human and animal placental tissues: pre-implantation embryos, trophoblast cell cultures, organotypic survival of placental villi and slices, microvilli preparations and placental homogenates, dually perfused human placental lobules.

   In vivo, animal surgery in acute experiments or in chronic preparations (the old jewels of the crown of prenatal life studies of Barcroft (1946), Barron (1970), etc.) has become the goal of research of a minority group of research workers in peri-natology and of very few placentologists sensu stricto.

   The advent of molecular biology probes (c-DNA and m-RNA probes) for *in situ* hybridization techniques, the discovery of trophoblast cell markers (enzymes and antibodies) adapted to the study of trophoblast proliferation normal or neoplasic, the use of radionuclides (Tritium) labelling chromatin, ribosomes, nucleoproteins, to be localized by high resolution autoradiography at the ultrastructural level, all these methods among others, provide new opportunities for *in vitro* investigations on human placental material surviving outside the uterus. These investigations can be considered as "non invasive" as long as they are applied to a tissue expelled from the uterus under natural conditions, especially term placentae collected immediately after spontaneous delivery.

   We shall rather insist in to-day lecture, on the application *in vivo* of recent intra-uterine technology which has made available new non invasive methods to study pregnant women and animals (mainly Macaque monkeys and Rodents) without any invasive intra-uterine surgery.

   All of these investigations should not bear any risk or cause any pain. The two commandments to the prenatologist have to be kept: "Thou shalt not harm", "Thou shalt not bring any suffering" together of course with the first commandment "Thou shalt not kill". This is why should be discarded the conventional radiology of the pregnant human uterus which has first unveiled the mechanism of utero-placental hemodynamics (Borell et al., 1958) due to the risk of irradiating with ionizing radiations the fetal gonads as well as the blood stem cells. Of course these methods can still be applied to analyze the heterogeneity of both maternal and fetal blood circulatory patterns in the isolated dually perfused human placenta during its *in vitro* survival.

   "Scanning" the placenta with X-Rays densitometry can still be performed using computed tomography (CT-scanning) in pregnant animals and human placentae surviving *in vitro* to evaluate circulatory changes as well as placental tomo-chemistry (calcium or fibrin deposits). The interdiction of X-Rays also applies to conventional radiological exploration of pregnant women, as well as to X-Ray scanning during pregnancy.

   Radionuclide studies of placentae using aggregates or microspheres labelled with short-lived radioactive isotopes such as Indium-113 or Technetium-99m have in the past been applied in routine monitoring of placental hemodynamics but this method is also now widely discarded in obstetrical practice.

The success of ultrasonography for monitoring the fetal placental development in utero, has completely modified non invasive clinical and scientific research. This method is no longer applied for placental localization only but for many fruitful investigations on the morphologic evolution of the fetus and of its placenta and also to monitor the changes of fetal and uteroplacental hemodynamics (velocimetry and blood flow). Nowadays, the new equipment called "Color Doppler" permits very fine resolution studies and will in the near future for placental blood circulation studies replace all previous invasive surgical procedures ("acute" or "chronic" preparations). The possibility to obtain information directly on the human species, will render obsolete intra-uterine surgery animal experimentation.

The application of Magnetic Resonance imaging, Magnetic Resonance spectroscopy, Microscopic magnetic resonance, has been considered in details during a workshop of the last Rochester Trophoblast Conference. The reader can refer to the general report (Panigel & Mattison, 1988) as well as to the different papers which have been published in the last volume of Trophoblast Research.

Two pathways for further research have to be specially mentionned here:

1) the use of non toxic para and superparamagnetic contrast media which not only improve resolution for MRI but also permit placental volume and placental blood flow measurements.

2) the study of ^{31}P MRS spectra which opens a new field of research on ATP and phosphate metabolism linking histological, pathological and biochemical studies by in vivo non invasive screening of energy levels in the placental tissues (Malek et al., 1989)

3) the development of MMR methods which allows the study in situ of the first stages of placentation and embryonic development (Panigel & al., 1989).

The general conclusion about the application of all of these non invasive methods to placental research is indeed full of optimism and hope. New progress seems at reach on condition one keeps a fifth commandment "Thou shalt for better or for worse continue to work on the placenta".

References

Barcroft, J. (1946). Researches on prenatal life. Blackwell Sci. Publications, Oxford.

Barron, D.H. (1970). The environment in which the fetus lives: lessons learnt since Barcroft. In "Prenatal life", Mack, H.C. ed., Wayne State University Press, Detroit, p. 109-127.

Borell, U., Fernström, I. & Westman, A. (1958). Eine arteriographische Studie des Plazentartereislaufs. Geburshilfe und Frauenheilkunde 18, p. 1-9.

Malek, A., Miller, R.K., Mattison, D.R., Bryant, R., Panigel, M. & Neth, L. (1989). Energy production by the dually perfused term human placenta in vitro: (^{31}P) nuclear magnetic resonance spectroscopy AT 2.0 and 4.7 Tesla. Placenta 10, p. 515-516.

Panigel, M. (1980). Le choix d'un modèle pour l'expérimentation animale en toxicologie pendant la gestation. Sciences et Techniques Anim. Laboratoires 5, p. 435-439; (1983). Monitoring human and animal intrauterine development with non invasive methods and perfecting in vitro placental perfusion techniques for toxicologic and teratologic experiments. p. 147-160, in Animals in Scientific Research: an effective substitute for man? Turner, P. ed., Macmillan, London; (1986). Anatomy and Morphology in Chard, T. ed. The human placenta, Clinics in Obstetrics and Gynecology, 13, p. 421-445, Saunders, London.

Panigel, M., Cho, Z.H. & Lee, S.C. (1989). Nuclear magnetic resonance microscopic observation of rodents implantation sites and early embryonic stages. Trophoblast Res. 5 (in press).

Panigel, M. & Mattison, D. (1989). Magnetic resonance investigations in the maternal-fetal-placental complex. A workshop report. Trophoblast Res. 5 (in press).

Colloques INSERM
ISSN 0768-3154

Other *Colloques* published as co-editions by John Libbey Eurotext and INSERM

153 Hormones and Cell Regulation (11th European Symposium). *Hormones et Régulation Cellulaire (11e Symposium Européen).*
Edited by J. Nunez and J.E. Dumont.
ISBN : John Libbey Eurotext 0 86196 104 8
INSERM 2 85598 324 X

158 Biochemistry and Physiopathology of Platelet Membrane. *Biochimie et Physiopathologie de la Membrane Plaquettaire.*
Edited by G. Marguerie and R.F.A. Zwaal.
ISBN : John Libbey Eurotext 0 86196 114 5
INSERM 2 85598 345 2

162 The Inhibitors of Hematopoiesis. *Les Inhibiteurs de l'Hématopoïèse.*
Edited by A. Najman, M. Guignon, N.C. Gorin and J.Y. Mary.
ISBN : John Libbey Eurotext 0 86196 125 0
INSERM 2 85598 340 1

164 Liver Cells and Drugs. *Cellules Hépatiques et Médicaments.*
Edited by A. Guillouzo.
ISBN : John Libbey Eurotext 0 86196 128 5
INSERM 2 85598 341 X

165 Hormones and Cell Regulation (12th European Symposium). *Hormones et Régulation Cellulaire (12e Symposium Européen).*
Edited by J. Nunez, J.E. Dumont and E. Carafoli.
ISBN : John Libbey Eurotext 0 86196 133 1
INSERM 2 85598 347 9

167 Sleep Disorders and Respiration. *Les Evénements Respiratoires du Sommeil.*
Edited by P. Lévi-Valensi and D. Duron.
ISBN : John Libbey Eurotext 0 86196 127 7
INSERM 2 85598 344 4

169 Neo-Adjuvant Chemotherapy. *Chimiothérapie Néo-Adjuvante.*
Edited by C. Jacquillat, M. Weil, D. Khayat.
ISBN : John Libbey Eurotext 0 86196 150 1
INSERM 2 85598 349 5

171 Structure and Functions of the Cytoskeleton. *La Structure et les Fonctions du Cytosquelette.*
Edited by B.A.F. Rousset.
ISBN : John Libbey Eurotext 0 86196 149 8
INSERM 2 85598 351 7

Colloques INSERM
ISSN 0768-3154

172 The Langerhans Cell. *La Cellule de Langerhans.*
Edited by J. Thivolet, D. Schmitt.
ISBN : John Libbey Eurotext 0 86196 181 1
INSERM 2 85598 352 5

173 Cellular and Molecular Aspects of Glucuronidation. *Aspects Cellulaires et Moléculaires de la Glucuronoconjugaison.*
Edited by G. Siest, J. Magdalou, B. Burchell
ISBN : John Libbey Eurotext 0 86196 182 X
INSERM 2 85598 353 3

174 Second Forum on Peptides. *Deuxième Forum Peptides.*
Edited by A. Aubry, M. Marraud, B. Vitoux
ISBN : John Libbey Eurotext 0 86196 151 X
INSERM 2 85598 354 1

176 Hormones and Cell Regulation (13th European Symposium). *Hormones et Régulation Cellulaire (13ᵉ Symposium Européen).*
Edited by J. Nunez, J.E. Dumont, R. Denton
ISBN : John Libbey Eurotext 0 86196 183 8
INSERM 2 85598 356 8

179 Lymphokine Receptors Interactions. *Interactions Lymphokines-récepteurs.*
Edited by D. Fradelizi, J. Bertoglio
ISBN : John Libbey Eurotext 0 86196 148 X
INSERM 2 85598 359 2

191 Anticancer Drugs (1st International Interface of Clinical and Laboratory responses to anticancer drugs). *Médicaments anticancéreux (1ʳᵉ Confrontation internationale des réponses cliniques et expérimentales aux médicaments anticancéreux).*
Edited by H. Tapiero, J. Robert, T.J. Lampidis
ISBN : John Libbey Eurotext 0 86196 223 0
INSERM 2 85598 393 2

193 Living in the Cold (2nd International Symposium). *La Vie au Froid (2ᵉ Symposium International).*
Edited by A. Malan, B. Canguilhem
ISBN : John Libbey Eurotext 0 86196 234 9
INSERM 2 85598 395 9

Colloques INSERM
ISSN 0768-3154

194 Progress in Hepatitis B Immunization. *La vaccination contre l'hépatite B.*
Edited by P. Coursaget, M.J. Tong
ISBN : John Libbey Eurotext 0 86196 249 4
INSERM 2 85598 396 7

198 Hormones and Cell Regulation (14th European Symposium). *Hormones et Régulation Cellulaire (14e Symposium Européen).*
Edited by J. Nunez, J.E. Dumont
ISBN : John Libbey Eurotext 0 86196 229 X
INSERM 2 85598 400 9

Reproduction photomécanique
IMPRIMERIE LOUIS-JEAN
BP 87 — 05003 GAP Cedex
Tél. : 92.51.35.23
Dépôt légal : 286 — Avril 1990
Imprimé en France

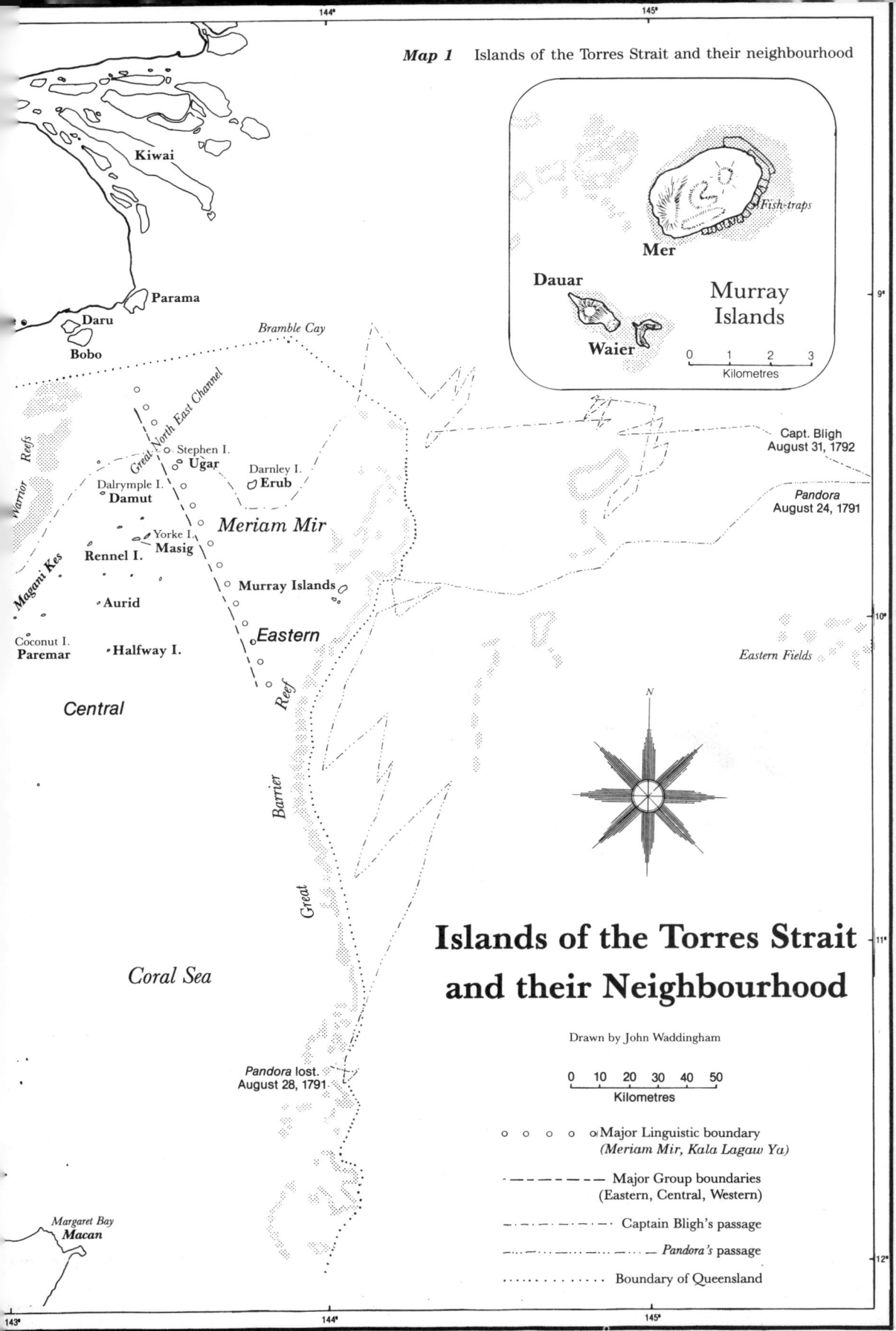

NERUT ORWAIR ORWAIR AMIRARET

Mabo, A Mer Lera Ged Asared